AF506981

HELLO PLANET EARTH

By

Richard Dean Neff

In Memory of:

THESE WRITINGS ARE DEDICATED TO SANDRA L. ALLEN - With all my heart filled thanks to one who is no longer with us in this world, a former school teacher who worked for me as my secretary. A woman who refused to stop her efforts daily to see my success. And when I wanted to stop so many times, too many to count them. She would say: "Richard the world needs you" And I would go back to work. She began as my co- worker, and then she became my friend, then my sister. She was like a mother to me. I grew up on 9 Maple Ave Ripley New York, and Miss Sandra L. Allen of Ripley, New York, who grew up around the corner from me at 53 Burton Avenue.

I was born in 1953, and we became like a family for 30 years. It was exactly 9:53 at night when I got the phone call that Sandra had died. 9:53, - my former home address number 9, and her former home address in Ripley number 53.

Sandy allowed me to tap into my life's thoughts as to be able to focus without the pressures we all face, Sandra you gave me time to think about our world, our lives, and what is possible. She inspired me to write this book and much more.

Wherever your spirit is- "Thank you Sandy". (With Love.)

Copyright © 2024 by Richard Dean Neff

All rights reserved. No part of this publication may be reproduced, distributed, or transmitted in any form or by any means, including photocopying, recording, or other electronic or mechanical methods, without the prior written permission of the publisher, except in the case of brief quotations embodied in critical reviews and certain other noncommercial uses permitted by copyright law.

INTRODUCTION

Welcome, and allow me to thank you for your support & curiosity and what I hope will be considered one of your better discoveries within these pages; also, there are many issues that may be of great concern to you. My thoughts are on many subjects in which I am making my best effort to fill the need to repair our planet's issues and even some people's issues.

This book was started in 1992, and ideas of invention would appear to me from time to time, and I would log away the idea and then add it to the book. So, the book has been in the making for over 32 years.

But, before I begin, allow me to explain that some designs presented are in concept. Whereas the technology is already established, it's just not used in the manner that I designed.

For example, I may say in the reference that refrigeration is used to do a certain application, being that refrigeration is already in use, as the technology to build it is "established."

There is no need to elaborate on the application of the equipment required. So, there is going to be some information withheld from the content of some of the subjects, but the majority of the content will be what we can call an "eye- opener."

The reason some information is being left out is there are some ideas formed that I am seeking funding to build, and they are extremely close to my heart and, yes, have a potential for my personal profits, as any real entrepreneur of any standing will agree that to give away the farm is not wise. So my ideas, as they are seemly complete, are with added features to insure the possible success of the Inventions.

In bringing that concept to creation, it is difficult to see all new ideas presented to the public. Be it known that my content is without the information from a third party, to my knowledge, any similar application that has already been formed is not my responsibility.

Be it known that all the content referenced in my applications to the best of my knowledge are my creations placed into this content book as dates of art designs are logged and these dates go back from over 37 plus years as conceived by many ideas.

This book is a way to introduce the issues or problems facing our world and the solution to repair the problem by either new ideas of how or by inventions presented.

But within these pages are changes that can help our world not only to understand it better but could change the ideas formed by acceptance to give a person greater understanding.

Some content is similar in development from other inventors, but my concept designs depicted herein are what I consider more appropriate to keep the costs down or making a more effective application to increase productivity. I'd like to begin by saying "Hello, Planet Earth".

Here is a collection of my observations, theories, and considerations of behavior. Ways to improve our world is the main focus, and my many possible invention fixes, and remember many of the issues are what is called theory, "like Einstein's-Theory of relativity."

A theory is not always something you want to test but is a prediction of what "COULD" occur on the basis of facts already presented or connected by the association of the content to form a conclusion of that theory as close as possible.

This said: "Hold on, get ready to enter into the future".

TABLE OF CONTENTS

1. SUBJECT: GROUND EROSION CAPTURE & FUTURE FARMING

Ground erosion prevails in many parts of the world and along our oceans, lakes, rivers, and creeks. From a satellite view of large river outlets, we can see thousands upon thousands of tons of top soil that is displaced, which has been removed and sent to the mouths and bottoms of many rivers and even into the oceans; this also carries a great amount of pesticides from the farms that infect our food directly when consumed. In my designs, I have addressed a way that we can keep the majority of sediment that can be very beneficial for farming, and I have also invented a new way of reducing the space of farm lands as they are being taken by new housing as populations increase so goes good lands that could be making food. My idea is that we build what I call "STACK FARMING "for short, "SF." The [SF system] is using methods to remove pesticides and bug damage to farm crops; with stack farming, we can remove the need for pesticides completely. So let me first start with the new form of farming, which does not use oil based tractors that also can leak oil into the land. Instead, all applications are electric and automated, so farms are now 90% removed from old methods, which will make farming not only safer, but faster, and economically greater profits because these farms, when developed, are not with just one crop per year. So let's visit a stack farm. To enter the building there are three stage entry rooms. All items entering must be inspected and cannot enter until the product coming into the foundation has been bombarded with bug sonic sound and bug herbs. There are many that repel bugs, and also kill bugs once again without the use of pesticides. Sage, Rosemary, Catnip, Mint, and Basil are just a few that will be in the air release systems that will be loaded with fans that will release these odors whenever there are any items being moved into or out of the food production floor. These odors will be just what the doctor ordered to maintain clean food production. So the presents of Marigolds, Chrysanthemums, Geranium, and Garlic (some of which are harsh odors) will be used a further distances away as the front line odor. Other flowers will be in the landscape surrounding the buildings to gain a lovely presence will be welcomed, but more importantly, their ability to deter bugs will be a benefit, so these plants will make these odors into separate rooms to deter the entrance of any insect. The next room surrounds items in black light to kill fungus and, germs, and even viruses. The last room is called the white room and is the holding area warehouse for items that will be entering the grow level, such as dried fertilizer with other ground nutrients that will be required. There are sound devices to remove mice and insects and repel them from the building whenever there are no people inside. We have a system in place which can be with a silent alarm that shows a location to any intruder, no matter the size, as motion sensors are on the farm to locate any movements and can be seen by AI detection of unwanted elements, like rodents or insects, 24/7. Sensor drones are activated to remove the threat and are controlled by AI. To capture and destroy, with a bug vacuum powered by a solar-powered system, with a hose connection and camera to seek the intruder. These multiple bug killing drones are able to detect and are backed up by camera monitors that can adjust to zoom in on areas to seek the intruder. The fact is this system could be used for regular farms, but the system would require many drones to be able to maintain full control. The topsoil being used is from the process that is depicted in the illustration below. The stack farming growing process will attain quality dirt as topsoil enriched and has other benefits. A method of removing sediment is a movable track system that is able to capture plastics from entering our oceans from rivers and can even bring valuable nutritional sediment, such as topsoil, back to standard farms and to the new futuristic stack farming complexes. As many of you may know, the American Indians used an actual small fish to be placed into the ground with a seed of corn in a similar way, using sediment attained by Creeks Rivers and depth dredging of rivers. The topsoil achieved as it is spread on land will actually contain many decomposed dead fish and other elements of a compost nature, a natural fertilizer that will be included in the sediment mix. A few advantages of "SF" are 1st. Dirt placements on growing levels allow the land mass to be less in area consumption. 2nd Temperature and lighting control.

The third is nutrient-based, these farms don't just use any common dirt, but the dirt with the highest nutrients added, and our crops rotate as never seen in history all in one place. The temperature control is in massive rooms with 3 feet deep massive planting trays, the length and the size of football stadiums with reflective natural lighting inside a structure that is from mirrors used on the outside to reflect light to disbursement lighting fixtures that spread light evenly. Ten even twenty-story level structures with steel beam reinforced floors that can hold dirt and crop growth, these tiers are the harvest areas of planting in a way that green housing is like a comparison to the horse and buggy, which was replaced by the introduction of the automobile. To include again the advantages that the harvesting is within weeks of each other, to have one crop yielded. In addition, as used in today is farming. Instead, the yields of one crop per year are replaced with many crops being harvested simply because the crops are planted per level at different times to allow the seed planting two weeks from each other, and there are yields from items grown weekly, all because there is no actual growing season. The light and temperature is always at a steady exactness for the plants to be able to maintain proper growth, to include that each of the same crop is separated into extremely large harvest areas that could also be designed as in rows. Therefore, as we walk through, we will see Row 1. A crop, say of tomatoes, and then after walking about the four lengths of a football field, then you have Row 2. Planted, say, a week later of tomatoes, And so on because this is not only a massive growing structure but non-stop harvesting. A structure that uses two types of lighting: 1. sunlight, and if these structures are in desert areas, there is an abundance of direct lighting, which, with reflective panels, bounces the light from one end of the structure floor to the other side of the growing floor level. And a most abundant use is to have all the power from the sun to produce solar lighting for grow lights to operate at night. To include keeping air quality control with air-conditioning and upper sun shields, I call- "SUN SAILS" because they look similar to massive sun blocking but are actually the solar panels creating shade, which lowers temperatures in the desert on the buildings, and these are adjustable even to allow sunlight to be adjusted to more or less as needed for each level of the plants being grown. To give more detail of the use of natural sunlight, the light once again is directed over all crops evenly as the use of mirror reflection, the reflections spread from the sides of the structure which face the sun rise to sun set the most, and to be bounced from reflector to reflector and eventually to the center of each crop row, which each outside reflector has a programmed time movement that is all connected and actually follows the path of the sun to provide maximum light, so unlike normal sunlight that may be touching one side of a field, yet some plants are not getting as much light because another group of plants is blocking light. This system makes all lighting to be an upper 100% evenly lighted to all plants. It can also be controlled by movable shade controllers, similar to a window shade used in the home but on a much larger scale. So our light is natural during the day, but at night, the solar-powered batteries are now turned on. The night grow lights continue on timers to allow our crops to grow around the clock for the fastest yields possible. So we can say that these structures are all like giant potting plants, which are controlled by A.I. to many of the functions, yet have people who manage the yields to check plants in inspections. So one crop is being harvested, and the item removed, another floor is now harvesting the next week, which is controlled by warehousing personnel and possible pickers from time to time. Prior to replanting, the soil is always enriched with either a full removal of old dirt or a new topsoil is applied and entered into the mix by an automated system that only needs to have the dirt dumped into conveyors that spread the new dirt from overhead moving conveyors dumping dirt into the lower trays evenly with a shaking motion. While this is happening, we now have the next harvesting in motion on the next level. Each harvest crop building has the ability to handle up to ten different crops, all planted at the same time if different crops, but the structure has the flexibility of row-timed planting or entire floor weekly planting of a single crop. Either way, the crops are yielding massive amounts of food for a starving world. Some of the equipment for some floors may vary in design, for items such as wheat, we may have a massive ten floors, but each is grown once again weeks apart so that harvesting is done year-round. The landscape potential of 50 buildings that cover, say, 100 acres if this dimension of harvest area layered out, it would take thousands of acres more to harvest, and with buildings with a floor space the size of four football fields, each with ten or more floors would provide, possibly, more food than a hundred large farms could provide.

This is why stacking farming is so important, and it gives us the needed land space while increasing crop yields and quality of food without pesticides. Even more important, water distribution has no way to evaporate fast from climate conditions that may harm growth; instead, it allows a direct application of water distribution so that plants maintain their growth perfectly as an indoor greenhouse with temperature control.

Description of STACK FARMING in more detail: Farmlands we have at present and in the past are being taken by the extending housing projects being built on what was once prime land for food crops, as farmers lose their business to large corporations, and the farmer is forced to sell. The types of equipment used in "Stack Farming" are structures with a remote control ground grader that breaks up the dirt; this is a massive long grader that, with the single turn of a security key code and release of the lock system and an activation with a push button, and the ground is turned to be ready for planting for each individual row, or for an entire crop being planted on the entire floor, with a automated process now started, and A.I. is completing control regarding the proper grading level needed for the item entered into the computer system. There are no tractors, the ground is broken, and there is also a mix of vitamin liquid poured as the ground is lifted, and the seeds are dropped into the area directly after the ground is turned. All the elements are on one sliding track grader to turn the ground and drop the seeds from the seed batch dispenser, one of the few manual applications to dump seed from 100 pound bags into the vats. Then, with an automated flat paddle we can cover the ground by pushing the ground over the seed. All this is connected to the automated system. The operational equipment looks like massive, long electric, controlled slide tracks from one side of the building to the other side, with graders attached for digging seed drops, and seed input covering paddle. The seed planting into ground with dirt covering travels all at the same time down the entire length of the building to the other side of the building. So we have done the entire ground process in one application, but before all this happens, we, of course, need a very important element, "the ground," which is brought in only after it has been treated for bug removal by the use of high voltage being delivered into the ground with electric prods that transfer electric through the dirt as to kill any bug larva, any bugs that may be dormant are killed. Low voltage would be used before high voltage to extract worms.

The worms will be added back into the dirt later to be placed inside to keep the ground aerated. As you will see later in this book, water is now not a problem even in the desert. The water used for these stacked farms will be from massive underground water tanks that supply all water needs because the water being underground and in tanks stays cooler and does not evaporate, and the water is pre-set by timed settings that use above sprinkler system to give a mist of water that is automatically done by computer settings, as to when and how long the water is used. The water applications are injected with nutrients from time to time, so we no longer have the worry of crop droughts; GONE! (You can applaud later, chuckle- chuckle!) So let's go fast forward for just a moment- we have this one structure that is making ten items: corn, wheat, potatoes, beans, cabbage, carrots, lima beans, lettuce, tomatoes, and celery, each of these crops is on different levels of this ten-story building.

The structure is designed over time, and can be built even higher as it always has the ability to be with top supports already for additions when needed. So, the dirt needed for indoor crops will become richer as time goes on. At this time, we have many elements in our food that we are not even aware of, items too many to list of chemicals and hazardous waste, much worse than we even realize. So don't frown on the use of sludge attained from river run-offs because it will have more nutrients than even regular dark soil being used as fertilizers. Let's face it folks, they use cow manure today. And it's mixed with a high concentration of acid from cow urine! Not to mention that this urine is absorbed into the food we consume, which, once again, is not a pleasant thought. (Oh yummy), but we don't seem to mind that now, do we? As I spoke earlier of the use of a fish planted with seed by the Native American Indians, I propose a similar but on a much bigger scale could be achieved. It goes like this: Our world has many rivers and lakes that need dredging. The layers of sediment are "filled" with wonderful proteins and nutrients that can be attained and placed back in our food farming lands even at present. Stack farming is the future because there is no need for pesticides. Entry into the farm levels is made to repel all types of insects, and these farms are like clean rooms where everything is inspected prior to entry. The three entry documents have some serious problems for insects and even insects' eggs. The technology used to repel includes the inside use of infrared high-intense light and UV lighting to kill bacteria and molds. The use of night time grow lights will be a big plus for the crops to grow even faster. This idea isn't new, it's a method which has been used by drug dealers for years who had basement pot plants growing without direct sunlight, so now we can adapt this for the use of quality fruits and vegetables and make food produce higher yields in faster time frames. We have many such rivers that could be dredged, and only every twenty years will the dirt inside these stack farms need the soil to be cultured outside once again.

The containment holders will have fresh dirt placed inside to change the dirt so it is filled with proper growing elements and added nutrients. Some of the smaller plants can have a rotational system, and under the growing areas is a conveyor. The growing trays are 110' x 20' x 4' and have a connection that when the power is turned on, the trays can empty by hydraulic slanting similar to a dump truck; it empties the entire tray for new dirt to be entered. This would be done every ten years. This action causes the entire dirt to fall to the conveyor, which will take the dirt outside to be re-processed by adding new dirt from dredged rivers, all plants such as that. This rotation is done after the harvest of that particular growing floor to remove the old ground layer and topsoil and replace it with newer ground that has been given higher amounts of natural fertilizer. Items such as dead fish, dead vegetation, and nature's chemicals were added in larger volumes to allow the mix to be measured for acid levels and to place top-quality enriched soil back to be planted with a new crop. People need proper nutrients in their food to be strong, and at the present time, farmers use methods that do not give the soil enough nutrients. So, for smaller crops like berries, beans, broccoli, lima beans, and such, these containers are like massive flowerpots with just enough dirt fill to sustain the growth.

But for items such as corn and potatoes, where the roots need to be deeper, added dirt is placed on a massive conveyor with a four-foot depth, as the rotation of dirt is with a hinged 100' wide conveyor with rollers under the bottom that support the platform cad n rotate like a massive loop when the dirt is removed to be renewed. When the dirt needs to be spread over the platform, another conveyor which is above the platforms now makes a slow fill of the dirt, which is being dumped onto the platform floor trays and is transported by dump trucks outside, which dump from the lower level onto a conveyor that now carries the new dirt to the desired floor where it is spread across the platforms evenly by a moving back and forth dirt conveyor spreader. Even the picking of the plants is done by robotics that is programmed to remove the proper food element.

This process could also be done by humans. So the entire Stack farming is almost automatic, the difference is the ability to maintain a land base in an ever-growing world in which people have turned good farmland into homes and business locations, and in their desire to grow, never considered what they had taken from our need of food areas to grow. For enriched soil to be made, dirt is dumped into a massive rotational barrel to add nutrients with extremely bright UV lighting applied at the entrance after the dirt has been placed in. Dirt is being dumped into the barrel by a Mini Skid loader with a dump bucket.

The soil needs to be turned to allow the UV lighting and airflow to reach different areas. So the mix is made complete to have all the soil enriched. The dirt is moving in the same manner as a cement mixer, rotate of the soil to aired it and keep the fertilizer from sitting and causing what is known as burning, this can happen when we allow things such as mulch to sit for long periods of time, and the gases now produce methane that can explode, and cause fires.

Any methane will still be slightly outside the building but not allowed to build up because of the soil rotation being used. The image depicted removing sediment, which should have been done years ago, is with a possible attachment to be used in front of a bridge to allow pressure of water passing through the filters to be supported already by constructed bridges which will lower the cost of development as to have a way to attach the old construction with the new filtration system and use actual strong bridges to have the stainless plates move and continue to take sediment from the water- debris like logs and tree limbs, will be detected removed before they reach the smaller filtration moving trays, by the use of larger sifters that are located in front of the waters flows coming toward the filters.

Any living fish has the ability to navigate through side openings that do not interfere with wild life but allow fish to swim under the moving filtration trays, and these filtration trays are not moving so fast that it captures fish or animals; it captures items that harm our waterways, or capture quality sediment from leaving our farm lands and are able to be used. But to use bridges as to connect the systems, the need for boats to travel over is so simple concept of the system is shut down and when it is shut down the ramps loosen and sink to bottom completely as to allow a boat to travel over the top, after the boat has passed the system turns back on the ramps raise upward from the back side as to lean and now the trays move slowly to capture sediment again.

Items such as plastic have a top area that closes like a clam shell to keep any plastics from escaping before it lowers into the water way, this done by a plastic-coated metal cable that pulls the top over, and closes it when a boat needs to pass, when it is reactivated the cable is now pulled from the other side as to open the enclosed trap and now the large trays will begin moving again toward the dump areas on land.

So items can be sent to the other side of the bridge at to waiting dump trucks sitting on weight scales to determine a full load which is then notified at the distribution contact office by filled alarm, which could be road crews as to have someone deliver the load filled to its predetermined farm or dump location. In the event that sediment trucks are not being filled, the sediment can be with a conveyor system that drops the sediment into massive dump piles away from the waterfront.

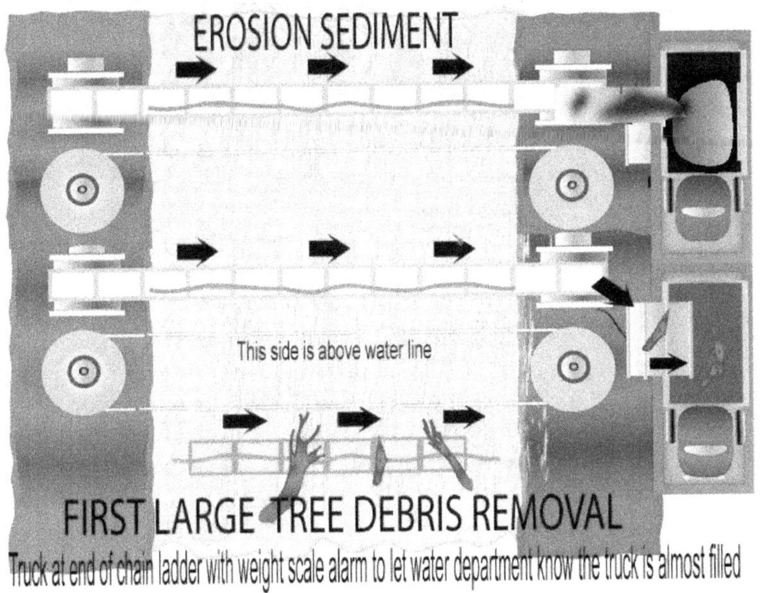

These soil drainage devices will help restore old farms to better crops with a topsoil distribution and help many industries, such as forestry, agriculture, and even home and garden supplies. During the winter months, the sediment ramps are closed temporary. Some of the time, the sediment shall be allowed to settle not at the mouth of rivers but in the rivers, and dredging should be done to retrieve this sediment, but let's consider the water life below, so the dredging of any areas should allow an escape method for living organisms to have a way to avoid the dredging. So dredging should be to only one side of a river or creek during spring and summer, and then two years later, allowing the population of whatever grows to come back, the other side is now dredged, as two years later, the center of the waterways is done in sections also, this way we are not disrupting the natural order of life below, and the equipment should move extremely slow so not to stir the water as less as possible. We need to consider that disrupting the water means life is also being killed, which may be essential for the river, and allow the life below as much opportunity to get away as possible. So actually, for a river the size of the Mississippi for example, that river should take about ten years just to dredge the sediment and nutrients in that sediment. These types of rivers need time for sediment to rest in between. Now, to farmers who say that "stack farming" is just too much money to establish, and then what I propose is the larger farmers invest together, and consider that this system is less than your massive equipment you may already use and you have this equipment exposed to the weather conditions that harm the equipment right now. The difference, it does the majority of the farm work and equipment and the food is not exposed to any weather conditions.

As stated before, the loss of topsoil from natural farmland has become, over some time, an extensive problem. I addressed this issue earlier, and know that the impact of erosion is greater than many people realize. Just along the shores of Lake Erie, I have seen land mass depleted in just a few years, and this will not stop unless real actions are taken. I also stated that a filtration systems placed in waterways is a good start, but is not the entire solution. A sifting sediment device is effective, a metal sheet that allows water to flow through it but catches the dirt and sediment.

As continued effort should be made to have large rocks and boulders on our creeks, rivers, and streams, for erosion to decrease. We should also realize that efforts to regain the erosion materials can be made as they are removed by nature and can be retrieved by man. So if we created sifter systems that are like small dams that allow water to settle, then is also filtered before the water exits these small dams, the dredging could be sent to enter the base of these dams, so to pull out sediment like a conveyor that is down at the bottom. The conveyor now drains the sediment coming upward slowly and the composite of sediment is dumped automatically into a dump bin that will be picked up weekly from these creeks and streams, we could capture the majority of sediment. Sediment that can be reintroduced back into the environment. Now, we can increase this concept to even our largest rivers, with water being dammed this action slows water down to allow water sediment to settle. This large sediment container is placed on the bottom of a river bed at the bottom areas of the dam wall where water has been slowed down most. With large chains that pull this container again to shore. Still, the difference is sedimentation that has settled for the bottom and will include living and dead organisms that will now be in rich soil and make it more fertile for land; it would be called "plus segmentation." It would be worth considerable profit from this soil, because it would be the richest soil available. A real no-no is concrete sidewalk blocks, once used for sidewalks tossed into creeks to stop erosion located in many areas throughout the U.S. and are also one artificial element that is used very freely to stop erosion, but when just placed in any manner, leave the waterfront looking like a garbage dump, and placed without any consideration of its appearance, but instead what does look good is when concrete is formed with a stone look and placed in slants under the water deeper than the bottom of the water and the water is re-directed away so the concrete can dry, then once the concrete has dried a small layer of rock is placed above, this stops water from traveling under the concrete, and the edges could even have areas that allow people to sit, so this will not look ugly. Instead, concrete should be broken down into large blocks and re-broken into small chunks for reuse.

This would be better for this situation than whole pieces of concrete that are sidewalks, mostly broken into large pieces and still looked like sidewalks. Instead, formed concrete complements the water area and gives the desired function, stopping erosion and creating an appearance pleasing to the eye. It could be functional as walking alongside a creek or river and providing easy access to the waterway and a place to sit down. If my ideas offend, I'm sorry, but I will tell it like it is. But people who have been given a great amount of responsibility for the appearance of our country should consider every aspect, not only its function but how they present our country to people who come here daily. Please, to those in charge, take more pride in how we take care of our creeks and rivers and the time to remove tires and garbage from the creeks and rivers is undoubtedly needed. Lastly, fines should be a minimum of $2,000. And 60 days of community cleanup for dumping in any waterways. Just a suggestion, but one worth noting.

3. SUBJECT: FOOD INCREASE – MORE FOOD FOR THOUGHT?

We all have seen the massive indoor arenas of our desire to watch sports and these structures can contain thousands of people. So, is watching a sport more important than eating food? That is an obvious mistake to build a structure that can house a few thousand people on a weekend for the gain of money yet has no real gain for humanity, and please understand I am very much someone who likes to watch a sport from time to time. But we have placed more importance on the ability to house people to watch something, rather than paying attention to the need to make food for the future that cannot which be affected by weather and insect damage or even birds eating the seeds or the crop, not to mention flood, and excess rain, droughts, all of these actions could be reduced 100% if our structures that we use for games could be developed to house levels of field inside for growths to be able to yield food not every year, but a weekly basis.

So to these owners of sport teams, maybe you could see the potential of having money made daily in hundreds of thousands of dollars from weekly growth cycles and I am pushing this, and I admit it, because what I call "stack farming" is the saving element of a bounty of food harvesting, and extremely profitable, but helps people save food costs, at the same time investors make a bundle of returns. That is correct. Imagine a food production system that plants food on a weekly foundation and yields crops every week in rotation.

Now that food supplies are abundant, growers maintain good profits because they sell not yearly but weekly, and when I say weekly, I mean massive amounts of food as in a yearly harvest. The indoor farm is the answer to remove many problems that affect our world; right now, we have prices shooting upward in the cost of foods; the reasons are many, but mainly the weather changes cause considerable harm to farmers' yields and the cost of transportation.

This brings us to the next input of "localized "productions, being these strictures are a massive indoor greenhouse but like on steroids, as the size and the ability to be year-round is made as even temperature control, and the use of solar panels to accomplish the winter months, with other energy electric designed by myself as XCESS ENERGY which is being presented. However, food delivery is no longer being shipped hundreds of miles away, and it is within 30 miles of cities. They are to deliver to a center hub that is now distributed to other retailers. This means the cost of delivery is reduced greatly! And the cost of food is now decreased.

Quality control is also increased because the food is not in transit for long distances, which can harm food quality. All in all, the processing is also close, so we are now increasing local jobs. To conclude, the sooner we as humans understand our needs should outweigh our desires for fun, and we should have our government be the leader in the development of our taxes to bring costs down and make quality our goal, I hope you can agree. For your future, and the future of your children. So call your congress man and tell them, "Richard sent you."

4. SUBJECT: ASTEROIDS IN SPACE

The following is called theory as to an observation of our universe, and I call: "HOW ASTEROIDS ARE MADE!" These flying groups of devastation are from a process that took many millions of years, when a planet has reached it's potentially largest size from the lack of vegetation on the surface which causes planets to grow, and this planet's growth dies and enters the ground which then is covered over time from layers of dust settling this vegetation is now under pressure and over time the layers increase and the pressure increases to become the nutrient of the planets core which uses the development item that mankind calls oil.

This substance is used inside the planets core to be burned, and this also causes the inner core to expand and volcanic acts are the action of the outer surface increasing in large diameter, and with a disruption such as sun going nova, or say, an asteroid caused a massive cloud of dust that blocked all sunlight for many years, these actions can stop planet growth that a planet needs to sustain its growth and its inside core.

The fact is we have many planets that are different sizes, but if our planet is allowed to grow to its potential of the outer core could become a greater size than even Jupiter. Once a planet is depleted of all vegetation, the entire oil base from the vegetation is also depleted. This process now causes a massive cavity to start to form inside the planet, and the core crystallizes. As this happens, the inner core can have a massive cavity form. The planet may have a great mass from many years of growth; this can sustain the planet as not to collapse, provided it has a thicker layer formed on the planet's surface.

However, many planets, when the inner core begins to cool, can have a total destruction, you see it is in a planets destruction that asteroids are sent soaring outward through space. A relatively small planet may have a thinner surface mass which increases the chances of the planet not staying intact. Simply, when the inner core is without any vegetation from the actions stated of a natural disaster, but any event that may cause the vegetation to die, over time the planet's food source diminishes. And in this case when this type of planet with a thinner surface, when it begins to cool from no oil to continue the cores growth has been disrupted to stop the burning process.

The core now becomes brittle and the planet will "implode" from the pressure of gravity. This implosion is why these asteroids are in many different size, and this action can be like dominoes as asteroids fly through our universe, they also can cause planets to be destroyed and, so we have constant movement happening throughout the universe with great potential for destruction.

As I described earlier, this is how asteroids are made, but there are also other ways, which include massive collisions in the space of an asteroid hitting another asteroid. So let's take a closer and longer look at our planet, and we have been depleting oil from our planet, and if our planet could talk, it would say: "Hey I'm hungry! Why are you taking my food?

Yes, oil is it's "food" and since times when pool oil was used which was found on the surface to light oil lamps, and even earlier provided torches as lighting at night, but in the early 1900's the invention of the automobile caused the demand for oil to rise as population of vehicles increased globally, and mankind made many elements from oil to include some of the obvious such as plastics. As this demand has increased, we now have many areas that used to have oil wells, but today, that oil is no longer available in large quantities. So, as we use this substance, we are also placing our planet at risk of no longer being able to sustain its life span. In short, without the oil, our planet dies, so goes mankind. It would be wise, if not essential, that humankind stops this process and allows our planet to continue living.

But my fear is that men of great greed will never allow change as long as they are making money. Now, this action of depleting oil also needs to be understood. Our planet is alive, yes, our planet grows and science has proven this fact by the expansion of the continents and observations from space.

The planet has been recorded to be in the process of growth, but as all things that grow, one factor is always present.

The item of growth must have a form of nutrition to sustain the growth; simply, a living thing of any type requires this. Some examples: A bee gets its nutrition from honey made, all plants gain nutrition from the grounds, minerals, and water, and our planet gets it's nutrition from the millions of years of plant life that died and is pressed into a thick substance once again , we call oil. So the fact is, this substance is our planet's lifeblood. And here is a strange fact: when an oil well is drilled, the drillers of the well call the oil connection a "vein. And how true this actually is, they are draining the life's blood from that vein, and the patient is dying from bleeding to death, so an oil vein is what our planet feeds from, and when it has no food, it has a thing called a rumbling stomach, that we call "earthquakes" and so we are seeing the results of our taking food from our planets tummy, that can cause some serious disruptions, and if we can use a comparison to when someone is hungry, their stomach will be making some very disturbing sounds and eruptions inside.

Our planet has that same need for its food. The planet also has a waste system similar to our own, we call it planet waste. "Black tar," and it comes from tar pits. This development of tar is actually from oil that has been processed by our planet and is now a waste result pushed upward away from the digestive areas from the inner core exits. We use this waste to make our roads, yes folks, we cover our highways with "earth poop". The planet has another interesting feature that we can consider like our own body, it's body is covered with small hair growth that we call trees, and yes, this growth keeps the planet cool just like the hairs on our body, and these hairs (Trees) also keep the planet warmer in winter, and so the planets body temperature can be in a somewhat control as it burns oil, or should I say it's (food).

Earthquakes are caused by the need of its food required, and if the need is great, it makes a greater rumbling, and this action can displace plates that are movable and all the life on the surface to be in the ocean. Mankind is very similar to our friends, the ants, as we have mounds of buildings that can come crashing down from earth quakes which once again, I believe could be directly connected to the oil being depleted from our world. So those of you profiting off the oil consumption to gain money know that future depletion of the planet's food will have a very deadly result for your grandchildren.

Possibly even you and I could be caught in a stomach rubble almost anywhere. As events as of April 5th 2024, a never seen prior. Earthquake was felt in New York City. So the next time you use a can of oil, just know your taking candy away from a very big baby called Earth. If we actually saw our world from space, humankind is no more than a parasite in size comparison, and it feeds off of all life, kills and grows to the point that if not stopped in growth will even do what other parasites do, consume itself when there is nothing else to consume. I can remember looking down many years ago from riding in a jet or looking down from the Empire State Building in New York City.

Looking down at all the people who looked like just like ants that I had watched travel as a child, ants that were busy navigating to go out to get what they wanted and return home with what they attained- just like people, which is really what we are, but mankind up close considers it's self as "BIG !"- And "IMPORTANT"- be bold, be proud, and yet, is small, and minor in the scope of the universe.

But we have greater regard for ourselves than others do, and that may be in our universe. Maybe what we need is to re-think the way we treat each other, and see our existence as what it really is- a group of parasites that have need of better self-control, our behavior on this living rock traveling at 67,000 miles per hour. Yet mankind thinks they deserve the new dawn each day, and does not see this miracle, or how important life of each of us is, and why our planet should not be stripped of its life. We need to be aware of our true nature and seek to improve ourselves, and find purpose for this mere short time we are here.

5. SUBJECT: ASTEROID SPACE TUG

Moving the monsters. I call these asteroid movers "space tugs" and these crafts built-in space because they are massive in size similar to the Titanic of spaceships. These massive rocket ships have massive extension push rods that can be extended when needed by hydraulics. They are docked at outer areas around our planet hundreds of thousands of miles away. Space stations that can detect far away objects and because they are far from each other, these crafts docked, can reach an object faster.

Others can meet them later to help if needed. We will extend our views of our universe in the future to what I call "visual space drones " that are sent outward and also are able to look and detect objects even further, these "visual drones" can first detect any object, but more importantly, warn mankind far ahead as they extend outward into our universe in every possible direction. When they arrive at an area of predetermined stationary distance can now even halt their travel to be a watch area to any activity. The onboard system sends images back to other drones that receive the signal and now resend that signal as to leapfrog the information that also carries a signature of the area drone's location, so they are able to report any and all large objects detected by the onboard systems.

These devices should have been considered long ago, but only now do we have the technology to build monitor stations. This ability could give us long distant views and an early warning signal of possible threats to our planet, which can include but not limited to asteroids , and asteroid groups, and alien encounters, and we can even monitor activity we are doing and send information faster because each solar operated system has a power transmitter system that can move signals by increasing the signal strength from the furthest monitor drone to the next monitor drone, and can also retain information as by recording any activity sent and can be resent if needed.

The docked SPACE TUGS, are also equipped with battle equipment to protect our world in any and all incoming threats. They are equipped with extra fuel, but another feature of the monitor drones could be access to stored electricity that may be needed by a space traveler, and have storage compartments for dried foods, and water tanks, possibly greenhouse hydroponics for air development, so these floating devices could not only be a detection system, but a life preserver for anyone who may need the resources. When these resources are used, a signal is sent to inform that these items need to be replaced.

Now, back to the docked space tugs, they will have plenty of time to prepare for any event that may be traveling in Earth's direction, with multitudes of months or even years to push the oncoming asteroid or asteroids away from our planets gravity and the course needs to be calculated as the direction that cannot be in the path of another planet in our solar system that may be in harm's way.

We need to consider the nearest solar system, also after it has been pushed off the course. If possible, a survey team may land on the asteroid. This team can investigate any minerals and any forms of metals that may be helpful for development after they have passed Earth's gravity and are on their way away from Earth. The team on the asteroid can mine it and break it into smaller sections.

The asteroid could even be slowed by the space tugs with cable attachments that can drag it very slow, reducing the speed. Testing the asteroid, a science team could investigate the age of the asteroid and its composition. To include the possible origin of the travel. But if the asteroid causes a threat to any other areas of our universe or to our planet, it will be far too large to stop progress.

Move the danger away from Earth gravity

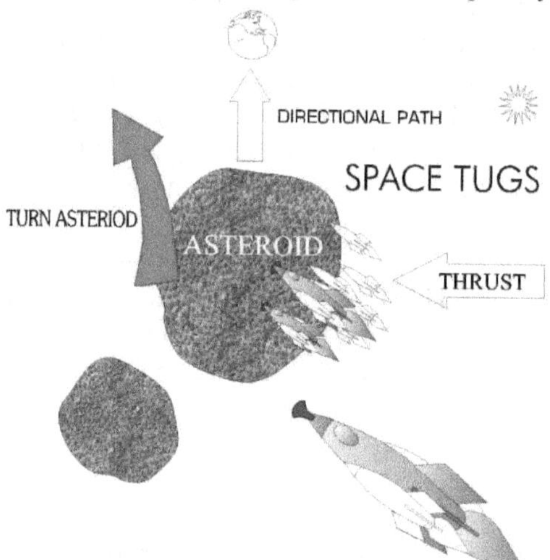

The end of the nose can extend to reach a distance that may be needed to make contact.

The asteroid could be broken with multiples of explosives that do not break the asteroid into small pieces, but the shock waves could push the asteroid away to change its direction. An asteroid of great mass that does not even move from the impacts would then need to be considered for a detonation or group of detonations, non-nuclear but very large explosive devices, to break the content used on Earth in mining. The explosive material would require air, so the explosive would be drilled into the asteroid, similar to we drill on Earth. Once the hole is drilled, an elongated tube is filled with high explosives, and air inside the tube is inserted into the drilled hole and detonated by remote control, which is far from the asteroid. Now, some science fiction has been seen of this event happening where an asteroid is met from a rocket ship launched from Earth to meet an incoming asteroid, and then an explosive device of a nuclear bomb is detonated. Okay, reality check! First, the time it may take to dismantle an asteroid could take months and, yes even years, so we cannot have a close window of entry. So forget the concept of launching from Earth. Second- an asteroid that is traveling at massive speed suddenly exploded in space by a nuclear device will still be moving in many directions-to include now in smaller but still very deadly-sized meteors that are still asteroids size, but now you could have hundreds of meteorites. This is called a moving disaster because since we have just exploded a nuclear bomb, there is now a collected thing called radiation!

Now, that could be attached to this oncoming mass. So this is why we use conventional explosives, and we use them not to explode the center but to break the item into large pieces that can be maneuvered. Eventually, dismantle to a size that is easier to maneuver be considered.

Important-Why an asteroid should never be blown up closely prior to Earth's gravity. The answer is simply that the size, being extreme, can also be broken, but now the problem becomes worse because there are now possibly multiple massive asteroids heading in our gravitational direction. On the plus side, very good use of a massive asteroid moving through space could be our means to attach surveillance equipment to see where this asteroid heads out into space.

We could call it a scout observation that can send images and graphic videos of places it is travels by, so we are now using natural movement in space to achieve information that could be beneficial as the discovery of a new planet or any anomaly; the device placed on the asteroid could be a very powerful telescope with a powered solar transmitter.

And to ensure that we can continue to receive images from this viewing asteroid, we can launch a power signal transfer station that allows the signal transfer to be increased and dropped out years later like a pod opening like a flower as it drops its contents "a space signal booster boey" at a precise time, and becomes a signal increaser to transfer clearer information from even further distances in space back to Earth in a manner much faster because of the signal booster abilities.

6. SUBJECT: GOOD NEWS FOR PLANET EARTH'S CORE

I have some good news for our oil friends, being that I have invented new energy devices that can make it possible for these people to produce " profit "and replace oil as their means of making money. And I suggest that those reading this who know people who invest their money into oil please pay close attention. What I have, yes, is a replacement for oil, which could save our future. I can give you a better way to make money and still perform solid energy, but in a way that is more profitable, safer, and 100% clean. I have included some devices that do not require sunlight or wind, and what all people need is a means to travel. I am aware that there are other processes that can produce plastics through soybean production, so the need for oil can become a thing of the past. I'm not saying it won't be used at all by some people, but the reduction could help many major problems that oil has caused. To mention just a few, how about the BP oil spill and the oil-rigs that have destroyed millions of ocean life? The process of oil used to make gasoline makes CO_2, and for those of you who feel cars making CO_2 is okay, if you decided to lock yourself in a garage and run your car and leave a window down, you may not be making any phone calls in the future, and your subscription to "Good House Keeping" could expire. So, we can conclude that oil is not a fix all. The process of making plastic from soybeans only takes financially serious-minded people to make those changes. So, to all of you who have the big oil money, why not understand the days of the "oil milk cow" being put out to pasture and electric methods that preserve the planet are the new direction? So invest in plastics made from soybeans in the future and leave the center core alone to continue it's growth. My energy devices I call XCESS which is coming up next.

7. SUBJECT: HEY ENERGY DEVELOPERS!

I have a great suggestion for anyone who is seeking a cheaper and better way to make electric with solid profits,

For decades I have seen the need for business of producing energy to make a new direction, and I can say to those who seek the change, well you have found it. so why don't we move those from working in oil and coal to jobs, to new energy sales, and marketing, the need for energy is important, but your old methods are causing a world of problems, so why not move the process, and start training people to build electric production that can achieve greater and cleaner results, without the dangers included. The so called electrical Hydrogen methods and items which is: hydrogen gas (H2). Splitting water molecules into **oxygen gas** and hydrogen gas using electricity is called "electrolysis".

Through electrolysis, we convert electrical energy into a store-able fuel namely "hydrogen gas," which burns, this process is not stable because it has the end result of produce heat and there are people playing with forms of energy developments , this one looks easy , you just need to collect some items and place them in the proper environment , sounds fun- It's called "nanoscale energy-harvesting device" from a piezoelectric material called zinc oxide, which generates an electrical current when under pressure or mechanical stress.

The zinc oxide was arranged into arrays of microscopic wires akin to a field of "seaweed", capable of swaying within the surrounding liquid, I just love the seaweed part. But if you will see from what I present is something that stays stable doesn't use off the wall materials, and uses a thing called – Common sense. Which is the direction this planet needs, simply, we don't need an added heat, and we don't need dangerous items in the middle of deserts, which I explain later, it really is a wild ride this world of "wha-- the heck was that?"

I am about to introduce you a series of inventions, not one, no a group that if used will be around long after I and you are gone. The reason is they do not cause but one factor- clean non-stop energy. So if you want to survive in this developments of energy, you may want to consider that there will be companies contacting me about the full details of these inventions. So why not be ahead of the curve? I know these sound impossible, remember the horse and buggy? Well your energy methods are about to go that direction in history. But I have seen some wild inventions with new energy that is being developed which has some off beat methods and some radical which plays with atoms and potential explosive abilities, so the good news is we can replace them with energy devices that I call: "Continuum Energy" which is titled as: XCESS energy as it supports the use of energy developed with more output than what is used to run the process.

These designs are exact and will always distribute more energy than it takes to run. It has multiple applications, so even in the most hostile weather the plants are still producing energy. All are down to 100% green without any waste being produced. With XCESS "Continuum Energy," we can produce electricity for the history of the universe and clean, non-polluting energy.

"XCESS SYSTEMS" are designed to make an output greater than the amount of energy it takes to perform the needed movement of generators to develop the needed energy. This structure depicted is the control center and solar power tower, so good news for solar panel developers: they can still be used in this design.

This structure includes all buildings designed to face the rising and setting of the sun at peak time. It absorbs the most sunlight, but is not depicted in art design. The solar panels on the rooftops actually follow downward directly with a flood of light from the sun as it travels east to west with controlled mechanics at the base of the panels.

Similar to the controls of a satellite dish, this structure does not have LED lighting surrounding the top, which laminates downward over the structure, and because of the angle, the lights can be in contact with the entire four sides of solar panels all the way to the bottom, so solar panels which during the day do not have great input from sunlight, at night the lights are surrounding the structure and will produce energy because solar panels can absorb natural light and absorb man-made light also to function, which was discovered by observation when I used solar-powered calculators with just an electric light bulb on. So solar panels can be used as non-stop power with overhead night lighting.

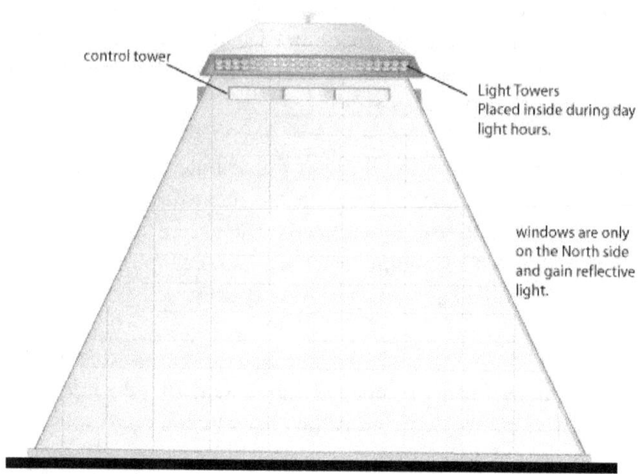

EAST SIDE, SOUTH SIDE , and WEST SIDE OF STRUCTURE IS ALMOST 90% SOLAR PANELS
All buildings even for storage and distribution have solar panels applied to these latitudes.

8. SUBJECT: XCESS ENERGY

With the understanding that many massive water dams have been made to move the turbines by gravity and water pouring over massive generator wheels to turn the generators to produce energy and these structures have actually harmed the ecosystem of many areas, caused server change in water consumption and the distribution of water. So the XCESS water power, Solar power, and wind power design are a triad of energy development that can now release water use from massive dams that have the potential to break one day, as many have done from natural disasters from time to time, like the dams that caused a catastrophic event in Derna, Libya in 10 & 11th of September 2023, and took the lives of an exact 11,300 people, to include missing another 10,100 people. So dams are not a good idea over time; simply, they become damaged without being seen, and the cracks can enter far below the massive structures. So this event of disaster took exactly 21,400 people gone! My designs reflect on this to eliminate the need for water to be collected. An example is Hover dam has a water support of 9.2 trillion gallons and a very close distance of only 290 miles from the world's largest earthquake fault line. The last recorded earthquake where two earthquakes back to back, one in Ridgecrest on July 4th, 2019, a 6.4 was recorded, and also in Ridgecrest, estimated in a straight line distance of only 120 miles from Hover Dam, but on July 5th, 2019 the very next day! A recorded 7.1 Mw was recorded.

Now the report is that in the world an earthquake has been felt some 240 miles away from it's starting source. So these earthquakes could soon be the end of the Hoover Dam. So who is closest to Hover Dam? How about, Boulder City, and Las Vegas, there is about 15 different townships that could be in the path of this dam breaking. And many more that are on the riverbanks of where the water could travel down the Colorado River. This is a warning, one that shouldn't be ignored. But XCESS energy could be established in these areas to develop electric energy safely, and non-stop. Simply, dams and nuclear energy are just far too dangerous to be supported, and mirror refection that has been used as in the Mojave desert is only a development of more heat, which our planet doesn't need, but what we need is to avoid devices that increase the world temperature, this system of mirror reflection that collects sunlight and develops a temperature of 500- to 1000 Celsius, but for our U.S. Fahrenheit users that is just 1,832 degrees. No big deal, right? And who supports these massive heat devices? How about Google's $168 million, an NRG 300 million, 1.6 billion United States Department of Energy?

These and others supported these devices to be built and warm the planet further. A massive boiler, and anyone who understands the word boil, knows that it has to do with increasing temperature. So these people tossed away the facts of the after- effects and only care about one element, production, and the profit it makes by supplying more customers. In the meantime, planet Earth continues to drown in attacks of heat increase, increase that causes weather changes, and massive damage to our ocean life by red algae called Karenia brevis, also called Red Tide, which has killed millions of ocean life because of the toxins and these toxins effect people, wildlife, marine life, and even birds, and the lack of O2 in the sea and fresh lake water. So why is this algae growing so fast? Well, they do not like to report the direct cause, but the growth is evident from increase in water temperatures, which have made them grow.

So let us consider the decisions made to increase temperature in a very big way is what the builders of these device inventions called safe? I think they should just go and cut down a rain forest and heat the water will be just as effective. So why am I saying all this you ask. Because my inventions are practical and if used, eliminate the crazy inventions that cause an after disaster potential. So, it has not been well known that these dams can break over time, break and kill thousands of people who have made their homes and businesses on the waterfronts. Let's not forget to include the potential of billions in property damage.

So, once again, my system does not cause these events but maintains a steady flow of electric development. They do not require nuclear fusion, which in any nuclear reactor, with the byproduct of nuclear waste, a waste that can kill, and has underground storage that, if ever leaked, could destroy underground water rivers that have fed hundreds of thousands of water wells, water if contaminated with radiation will kill millions of people, a radiation in water unseen, no taste, just radioactive. So, my applications are designed to use zero waste in the final applications and 100% fossil-free energy. Now to those of you who would like to see these inventions, a mere online application will be required to see "How these devices mentioned within this book are invented as how they work is also explained More in detail, but here are the outward concept examples as forthcoming and illustrated on next pages. The means of XCESS energy is only one of many types that I have designed, this one shown above I call: "TRI 3 SYSTEM, Gravity Water flow to turn generators, Wind energy with electric fan ability on no wind days, and Solar power with night solar capacity for continued energy production. With this system able to be placed anywhere, it will provide electricity for homes and businesses in a way that is non-stop no matter the time of year, as to be able to rely on more than one method of production. Another form of "energy" development presented is "AIR PRESSURE GENERATORS, as shown in the illustration on the next pages, with a group of condensed air fan pressure similar to a jet engine as the fans increase the air pressure as it is collected.

I'll talk more on this device later. This air pressure device is another one of my inventions. An air pressure fan could be used with boats of any size to place the three air pressure designs drafted in art form. When placed directly into the water will provide the means for the boat to travel faster than a spinning rudder and will not harm wildlife at all. We could even use air travel fans that do not require fuel but are run by electric power, a system of energy development for cars and trucks that could be powered by this system I call: "Continuum Energy," a form of energy not presented within this book. But I do speak of it from time to time. The following image uses solar, wind, and water power.

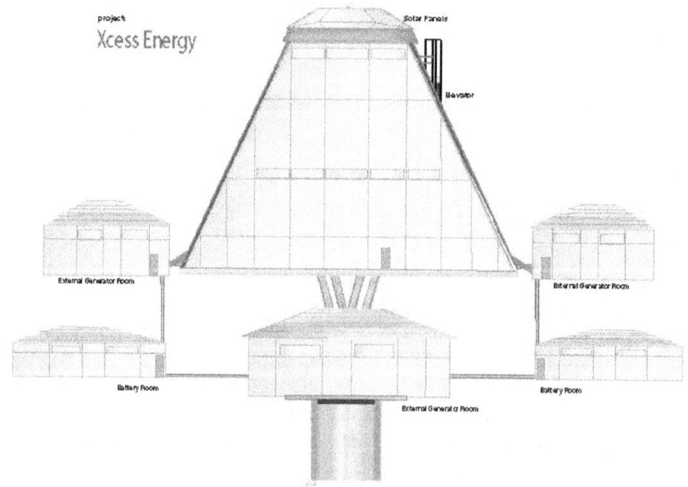

Illustration breakdown of actual components to be used to accomplish the water cycle power function:

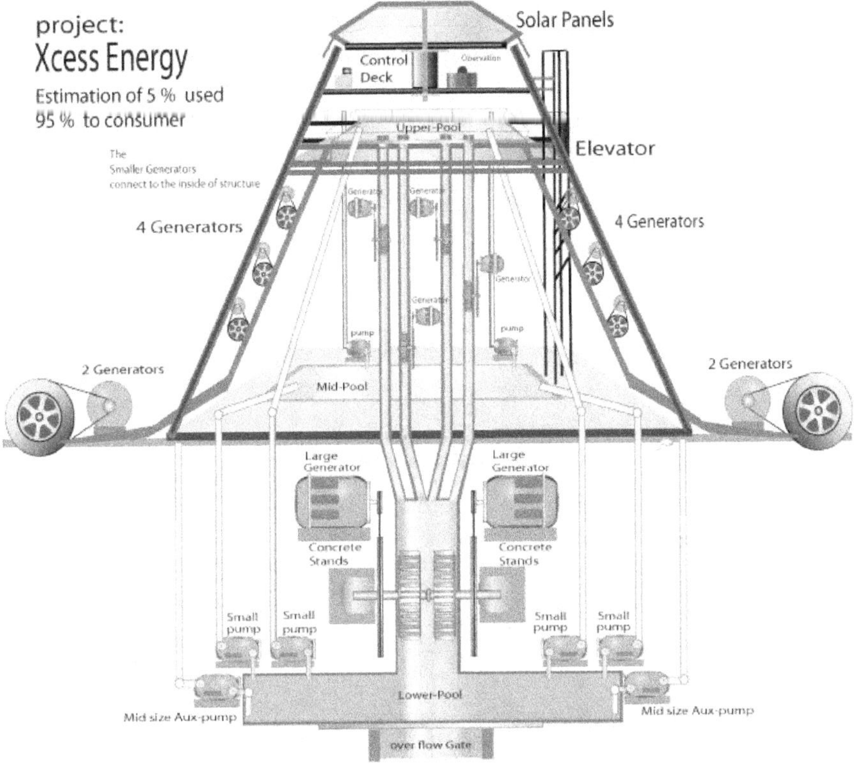

9. SUBJECT: XCESS AIR PRESSURE POWER GENERATORS

The next level of wind energy to be in the XCESS family is " wind energy" made with the power to move the turbine with air pressure is far less than the output of energy being developed, and the energy stored and sent to the consumer. This wind energy device is placed on hills that have been noted as high wind areas and is located toward the top to collect the wind before it reaches the wind generators. This wind design does not require massive ugly towers over our landscape because they are only 12 ' (feet), and the ability to guide the wind into the device with an electronic wind weather-vane that guides a directional outside metal wind sail.

This sail allows the wind to be captured and increases the speed of the wind as concentrated into the fan blades that move the turbine generator, and develop energy even when there is no wind with the illustrations of the concentrated air pressure fans, which can be attached outside or even applied inside. So wind anytime is now possible with the powered air pressure fans that concentrate air pressure to cause an increase in the rotation speed of the fan blades, and air pressure fans power the energy generator when there is no wind present.

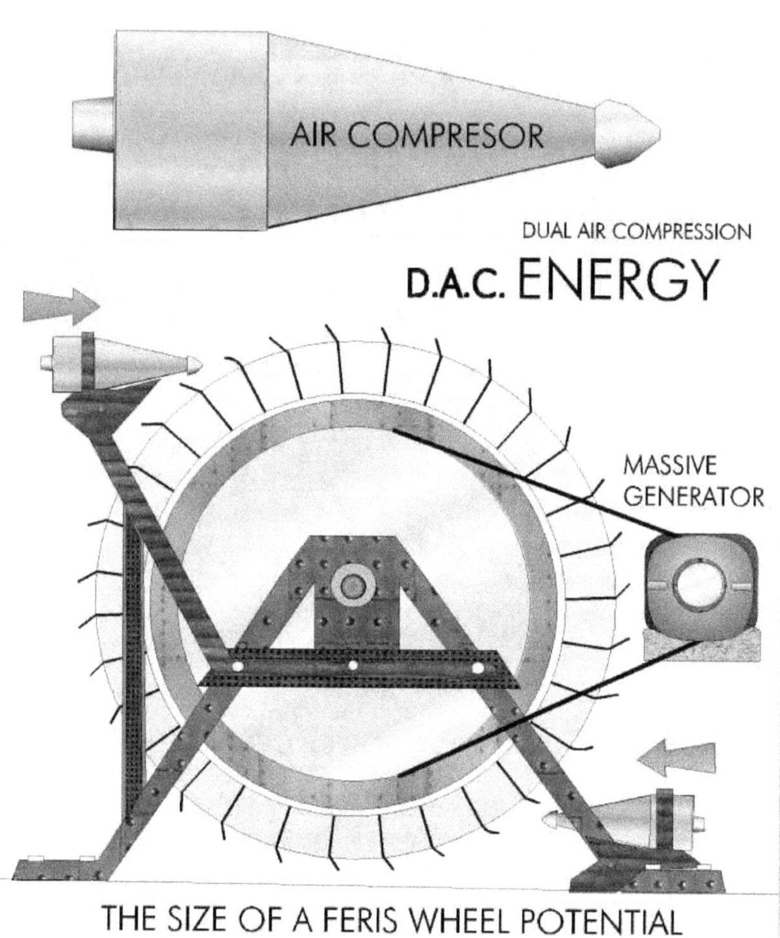

THE SIZE OF A FERIS WHEEL POTENTIAL
WITH MULTIPLE AIR COMPLESORS TOP AND BOTTOM SIDE VIEW

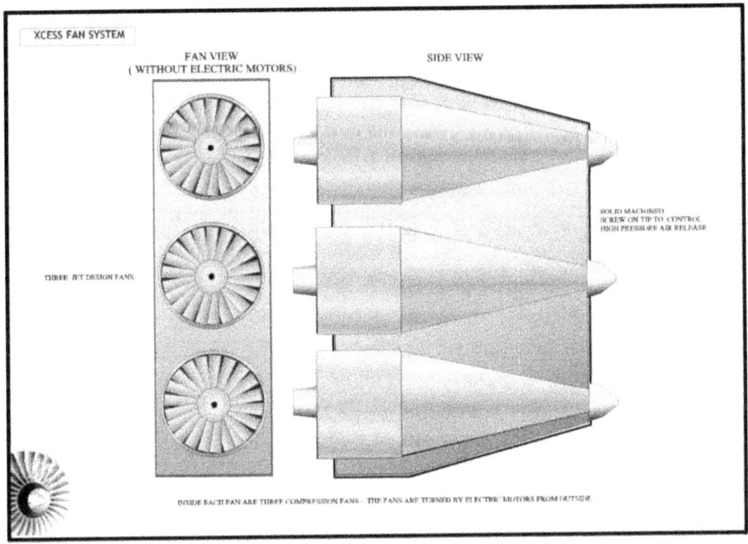

That power is less than the generator output, which is now directing energy to storage batteries. So, we can now have energy by wind with these devices without the limitations of other devices.

We now have new, improved wind development energy. The main difference is these generators are small but develop higher rotation to be a dual action, and the ability of direct connection to the wind direction with the ability of the weather vane to keep the greatest amount of wind direction at its peak level and produce more energy, and secondly when there is no wind outside the electric fans now push the generator production from air pressure created, and because the output of the electric generator has a higher energy output that is needed to turn the air pressure fans turning the generator we now have EXCESS energy, which can be used to bring electricity to all users.

The next SYSTEM-"XCESS MULTI CONNECT SYSTEM" Or an "X.M.C.S." It is the use of a massive electric motor at one end with connections from one long bar and pulley-style belt connections that turn the generators all at the same time. It has few breakdowns and massive output potential being that one electric motor controls many. The use of these XCESS systems- In short, our entire planet can have clean effective energy use without damaging this world or the people. The potential to increase the power output can be accomplished with rows of these systems depicted, and can even have a large spool for a belt to be attached with a very large metal spools with rubber contacts attached to the spools on the shaft so rather than use belts to turn the generators the spools are large enough to make direct contact as rubber to rubber and this would maintain the proper contact to turn the equipment. To make the contact more effective, an outside belt could be attached to both spools as with a belt drive but with a tension bar to pull the belts tighter and allow the second contact as a safety measure to keep spools at proper contact and best performance.

The larger the belt spools or larger spools are on the main shaft, this extends to turn all generators. To increase the output of the generators the long shaft spools are much larger and attach to the much smaller generator spools, which, being smaller, will increase the RPMs and give the electric generator maximum delivery of electric power. As all inventors do, we have moments where suddenly we see that a new addition or improvement can be applied to the idea, and friends this has just happened. Let's both look at the design of the many generators; if we add an additional line of generators that are reversed as to be aligned next to the depicted "generators", we can now add another "Generator " to each of the already existing generator's which can turn the new generator.

If its side is next to the prior generator, the contact places the new added generator spool is reversed. This, my dear friend is stack-able generators. We need to have an added spool as connections on the first generator there are two spools not just one. One is the connection to the shaft as a belt drive, and the other is a large rubber wheel that is now connected to the reversed generator facing the other direction. And the connection to the generator is a small spool that causes the generator to increase the RPM's. This system increases the output of power as doubled by adding the extra generators to be applied. So, instead of having 10 outputs of generating power as depicted in the art, we now have 20 generators. There you have it- a combined Solar, Water, and Wind- and multi-generator system all XCESS ENERGY!

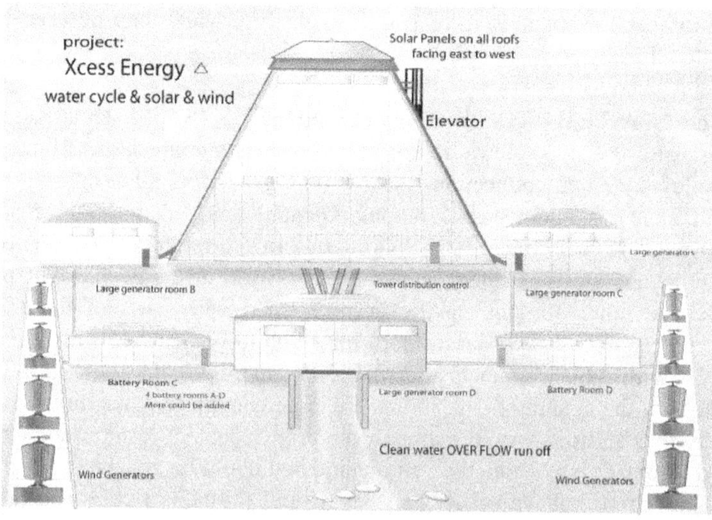

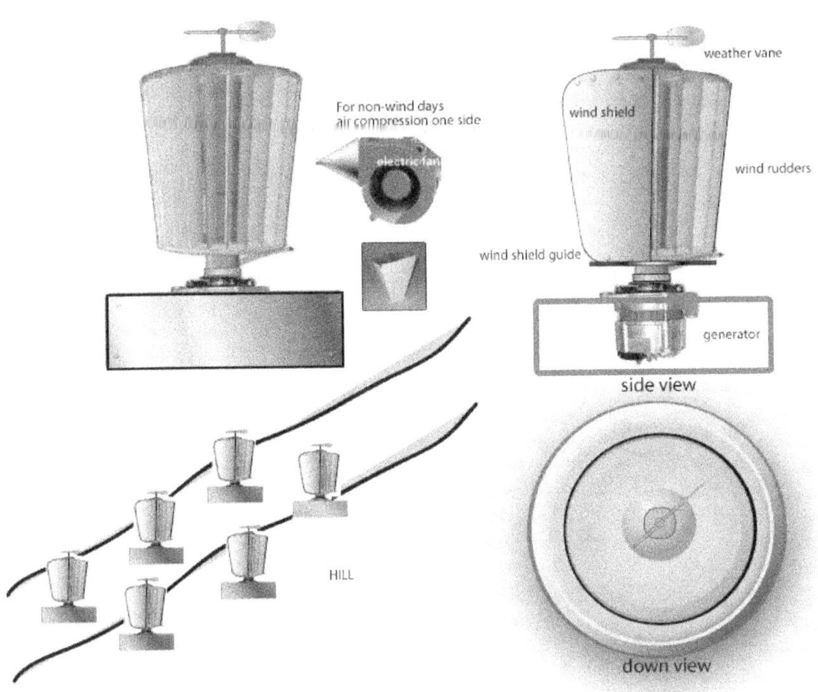

For non-wind days
air compression one side
electric fan
weather vane
wind shield
wind rudders
wind shield guide
generator
side view
HILL
down view

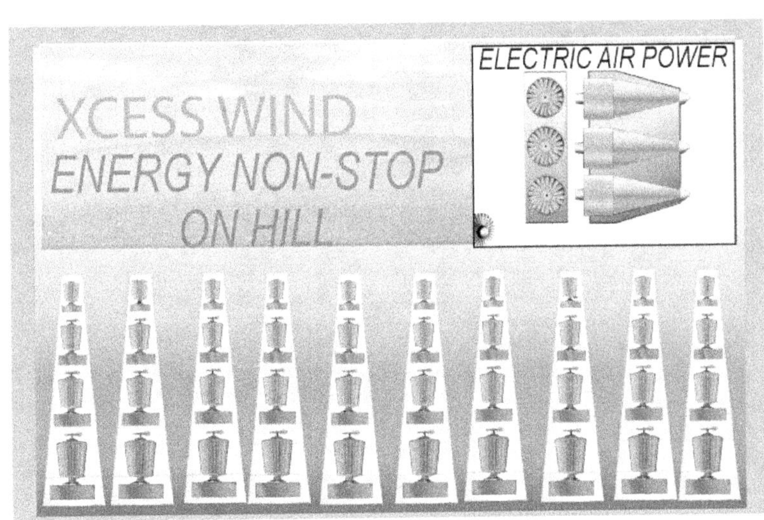

ELECTRIC AIR POWER
XCESS WIND
ENERGY NON-STOP
ON HILL

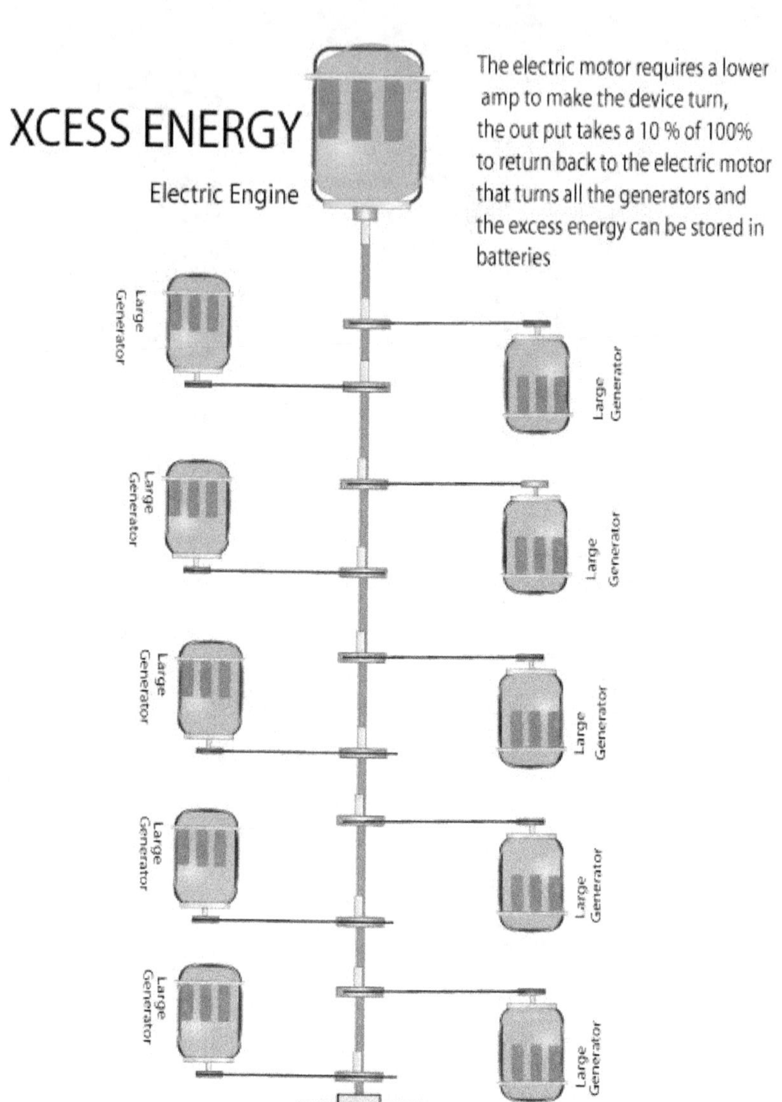

MULTI-GENERATOR PRODUCTION

CONCLUSION of COMBINED TOTAL XCESS ENERGY- To have 95% output of energy none stop, with production goals from 90% to 95% output potential by a 5% to 10% power consumption needed to produce items power output of the equipment.

10. SUBJECT: DROUGHT

I have recently become aware of a desert renewal project being held in UAE, and they have had some relatively slight success. And some other attempts to make trees grow, a massive tree project which, for the most part failed. The problem that planting in the deserts is like placing ice cream in a hot oven. The deserts are growing at a rapid progress from elements of pollution and the causes are in many forms of destruction, a major contributor is CO2, which released into the atmosphere in many ways, but to just say, " Oh look what I found" The main issue is how to prevent the growth of desert land and reverse the lands back to a normal state. The process of cloud seeding has been happening for many years, and this process has been successful for some regions of the world, but it is like putting a Band-Aid on a major wound. And chemical use can only mean contamination is somewhere in the world's water. This next presentation, as boring as it may sound, is essential to fight back against global warming. I call this method the proper "MIX" applied to develop land that has lost its ability to sustain life. Unlike the project I will introduce later in this book, which is related to this first application, has more ability for success as to begin the plant life to be inserted. This first method is only to get the sand in many deserts the preparation for the secondary greenhouse system to be applied. The second greenhouse system is to be a method that once the first method of dead plant life being applied to the sand and the mix has been archived even slightly. The next application inside the greenhouse is the dumping of a mix of mulch and waste material. This now prepares the sand to be a firm foundation for the trees to be planted. And these green houses are each of the size of a football field with slanted walls as to prevent wind damage with a ceiling that is 60 feet high. Trees are grown 30 feet apart to allow them to have wide, perfect growth when they mature. Inside we have a hose system with a water pump that is attached to our underground water supplies, and the plants inside are on an automatic sprinkler system high above the ceiling of our greenhouse. And yes, our greenhouse has electric temperature-controlled shading and indoor lighting. So, once again, the installation of these massive greenhouses is a secondary application and is for TREE harvest development only, but it is a system that has a rotational yield that allows trees to be maintained in the deserts at all times. This system is also depicted later as a method that will allow mankind to never be without wood or the development of O2 and removal of Co2. This method is extremely needed because the deserts in some parts of the world have no firm land mass, and tree development reduces wind activity, which can be very damaging. So, the creation of the first layers of the mix is essential to give the sand resistance first to help keep the sand from blowing around as much in high wind conditions. In a way, the desert mix application to develop the ground to be stabilized requires that the applications are ready and are done as fast as possible. I can compare this process to a war. It would be almost like a war against the bad elements of nature, and I can imagine planes loaded with dead wood flying over, bombing the ground with wood chips and mulch, and even planes with low-flying waste being spelled like a massive flying spreader. All these applications are coordinated to launch within the right conditions.

Let's return to what is a first need, and without this, we would have no growth. The displacement of water only to fix a problem temporally is not the solution. We need a long-term method of water that can be used weekly to give our desert some serious help in growing back to the days when it was a thriving land. So what we need to compare this action of desert renewal and forest development is to consider it like the building of a high rise. We first start with the foundation. Without the foundation, nothing can be achieved, and this foundation requires some added applications that come even later. But we first need to establish our water source.

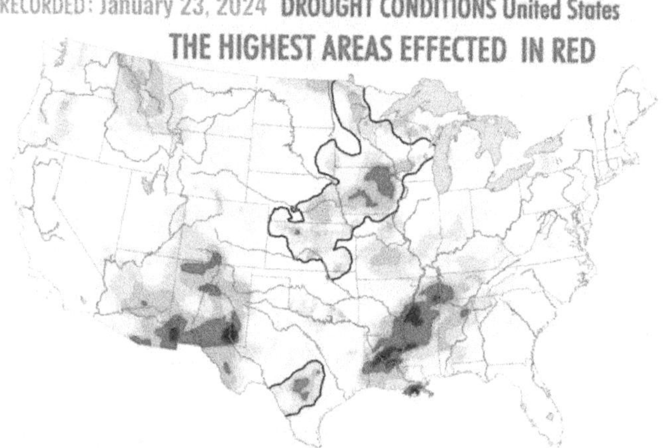

This is our main element for this foundation, so we developed a placement of manmade underground lakes that are established as drop points from water delivery. Water in abundance will provide a long-term future for plants to be maintained and grow. To deliver water from areas that have a great demand is not the water we use; we use water that is transported from flood seasons, an overflow system which I have placed in this book for your review. And we transport water from areas of abundance by water pipelines to areas that need water. In short, the world is blessed with many rivers that simply take water downstream and never is used properly to achieve growth but is only flowing into our seas and oceans. And we need to respect these flows, but we can use the overflow systems that capture water and still give the rivers their needed flow, but it would be decreased at the exits slightly. When we remove floodwater or level water, which is also used further down rivers or streams, unlike many water storage lakes that could dry up in a desert very easily, the water placed underground is inside tankers. These tankers can even be shipped by rail and truck to their needed destination to be stored once again underground. From the tankers, we attach pump stations, which work on electric solar- powered batteries to draw the water to the needed areas.

We don't need anyone there to turn on the system, and it's set with an automatic timer to displace water from hoses that are out inside the greenhouse, which has been built to retain the moisture. You may ask, "Why build greenhouses if you've placed down a layer of wood, which is the application before we build green houses, but hoses systems are placed in later so as not to have them crushed by trucks moving in a mix to the land. So here is the real next step of soil preparation before greenhouses are built over the mix. It will be a big help to allow the structures some means of staying stable. The greenhouses will require a certain amount of ground strengthening from time to time due to wind storms. Even when we have the greenhouses built, the applications of further desert mix is always applied yearly to restore the sand base back into a solid base outside the greenhouses, and yes, also inside will be spread and will have the ability to maintain soil to be hydrated whereas the moisture allows root formation to stay intact. These soil developments, as to the processes of applications of the mix being done at the best time of any year to reduce evaporation as much as possible, should be done at the coolest time of the year, where temperatures are at the lowest in deserts. The second and even more important aspect of desert renewal is the placement of mulch and wood chips into the sand, as the process must consider in all these applications the factors of wind direction, and the items placed must be heavy enough not to be blown around in high wind storms. If it is blown away, which will happen, the weather conditions should be considered prior to dumping to allow the material to be blown inward toward the desert core not away from the desert.

The fact is sand which is a fine stone that has been eroded, and some have even become so small in size as to develop a dust-like component that has very little substance of which plant life can't make a connection because the water can travel through sand with very little absorbing properties in the sand material, and does not allow any adhering or absorption to the roots of plants. So actions such as rainfall or water placement on sand without a strong adhering property or material that has an absorbing ability simply evaporates water much faster, which does not allow roots to absorb enough water for plant life to be sustained. We are not finished with our process of building the sand base. The next step is after larger materials are prepared for placement in our desert. Materials similar to mulch that is used in home applications, but slightly bigger. We must always consider that some of these deserts have sand that is like massive hills. And we will be using heavy equipment to level these dunes as much as possible so that the mix, when applied, gives the land mass an even maintained layer. So the mulch has to be large enough not to be just absorbed without the purpose of making a firm top layer that, with more and more applications of layering, makes a topsoil layer over a period of years. This desert sand will become saturated with firm compositions of decomposed plant life and waste that will then support plant life that can be planted. As this happens, soil testing will be done on the next greenhouses to expand the growth of trees. Every two years, this should be accomplished and done exactly like prior years as each step is essential to maintain the proper advancement of the land mass of sand is to be leveled with compression added when the land mass has been saturated and reformed to withstand weather sand storms rather than graders used, a roller system could be applied by massive dump trucks pulling massive rollers to push the soil downward and flat again. This would be done just before any mix application.

Another support material that could be used is trees that have fallen in lakes, streams, and areas where wood has absorbed a great deal of moisture is preferred, or fallen trees in the woods could also be used. These are not only good to pull out of our lakes, which in a lake a tree roughing can cause the water to be polluted and discolored, so the fallen trees that are removed have a great deal of water absorbed into the wood, and when dumped will break into many pieces that can be a great help to our foundation. To change the subject slightly, the need for our forests and creeks to be cleaned is essential for trees to develop without damaging smaller trees coming up and cleaning our dead trees could give the forests better access and improve the conditions of trees growing. I have seen many times where trees have fallen from a lighting strike and now damage other trees if we took the time to have people designated to go into our wooded areas and remove these items, which could include removing of grape vines or large vines that kill trees, but those vines could be used in the deserts as a quality retainer of moisture. The quality of our forests would be improved, and our deserts would have better sources of energy to gain direction and be renewed. And let's not forget our creeks need proper water flow for fish , and when trees have fallen some fish they make their journeys to survive and find the water flow is blocked, so to have these trees removed is important for natures wildlife, and let's not forget garbage!. Please let me thank any stupid people who consider dumping tires and junk into our waterways. Would you poop in the middle of your living room? It's no different. (not sorry) but we as people need to understand that what we consider as a small thing, when you have thousands of others doing the same bad idea, the outcome is pure damage to the world. You're drinking bottled water to prove that statement. So here is a good idea which we could improve, water quality. To you farmers, be aware, why not move your waste from animals to these future desert renewal projects, and in the meanwhile, how about pumping water up to your cows and keep their poo further away from water creeks or lakes, be mindful that your so called minor adding of urine in the water only helps pollution, give kids who swim down creek a serious illness, or can kill wild life, yeah it does, fish are found all the time around lakes dead from pollution. So move your animals away further, fence an area at minimum 80 feet or 50 yards away from the fresh water so the animal urine and poo has a chance to deplete into the ground, and not back to these creeks. This should be maintained as the responsibility of our towns and cities to oversee these ideas, and maybe some laws need to be enforced for people who don't respect our fresh water. If we clean out or creeks (HINT-HINT!) and then transport these materials by dump trucks to the proper dump stations for desert use would be the ideal condition. If each town did this yearly even for one time a month, we could revitalize thousands of acres of desert land in just one year. And you may be saying- "Who is paying for all this extra applications to build up deserts"? Well, we can have people who help our forest departments and our roadside departments be the supervisors, and those individuals who have just be fined for some illegal action, say a kid made graffiti or damages property, someone causes a fire, or has an auto fine, or a property damage or illegal dumping, and now they get a lesson of proper behavior for a week or two weeks of "community duty" And, the fine is reduced if they agree to half of what they have to pay. And later you will hear of how prisoners can be a help to our world.

Now, we can also consider presenting this idea to a major group or single corporation to do some serious investing in sections of deserts for the building and transportation of the desert renewal materials and the structures needed to keep them intact. To be exact, there are billions of dollars to be made from these applications within a relatively short time for some structures, such as "stack farming," which I address within the book.

These desert territories are considered worthless as landmass, so the cost to attain desert areas are minimal compared to established land that does not need any development. The cost of regular land would be over the top in price to attain. However, desert renewal with the projects I have stated within these lands as developments would yield everything from tree harvesting in 20 years to food production within two years, and even possibly a one-year period on a scale never imagined. Food alone would be so abundant in an area that has the highest amount of sunlight it would become a major export for investors globally, making profits.

Let's face it: this subject is a real bear. As we have seen, the devastation caused by global warming, and the question of it is happening is long gone, so we need real answers. The answers are not with a single application; what we need is to go to the root of each cause of the increase of CO2, and I am very sorry to report that there are many different things happening that have been the helpers of this increase over many years.

Here is a small list of the bigger contributors, but one at the end is a serious one. They are our burning of forests and fuel burning from cars. There is a massive amount of heat released from each engine; just touch your engine sometime after you have just traveled a few miles. I wouldn't recommend it unless you like hospitals. Animals, yes even pigs, and chickens, all put off harmful gases, forest fires, black roads, and buildings without shade, and could grow plant life on top to reduce the inside temperature of the building. It wouldn't look so bad, either! I have just had that thought: why not place masses of shrubs on top of buildings to shade the buildings and even shade the buildings below. We should also make trees along our city streets that shade the sidewalks and reduce sunlight heat. Imagine trees that have massive pots, which can allow a planted tree to grow onto the street. Placed in movable plant holders when they begin to get too big, we bring in the new smaller tree to replace it, take larger trees away on felt beds, and plant them for tree harvesting. The city collects the tree profit from the harvested tree.

To prevent the trees in the city or township from over watering or under 7watering, a simple device is entered beneath the soil that shows the city road crew- "Hey, this tree needs watering "! There is a water-filled window on the roadside showing the amount of water inside the pot. So a water tanker is set out. Every day, the water tanker drives use a water hose to water the tree. We can plant flowers inside the tree pot also to make the water stay inside longer or a grass that will grow long and look super cool at the same time will keep moisture from evaporation, and the tree will be able to have water longer. When the trees become too big as to be close to power lines or buildings, we could place them in the center of our highways and, inside inner state highways and even along regular highways on the side of the sunrise and setting to allow shade to be dominated over dark heat absorbing highways.

Inside the cities where the trees separate the traffic flow, we could place bright neon reflective chains with headlight reflectors in each chain with an LED light for a solar-powered chain link to stop cars from turning around, but low enough for a pedestrian to walk over the chain.

After years of growth, these almost fully grown trees could then be loaded onto trucks and sent out to be planted in the rural areas people seeking trees could be on a list, and they pay a small fee for the tree. Possibly with an added delivery fee, so a tree, which helped cool the city, could be outdoors in an area that needs shade and some added beauty or even given to a home near the city to be planted to grow to its best potential.

This image shows why all stated is so essential as the increase in desert land is due to many factors that some of us are aware of, which are helping global warming and the destruction of rain forests and O_2 depletion from fire, just to mention a few of the effected conditions to our planet.

The increase in desert land is due to many other factors as stated, but the rain forests are dwarfed in comparison to the countries raping the wooded areas for resources. It is evident to this writer that the rate of destruction far outweighs the rate of growth.

If we go fast forward into the future where plant life no longer is able to support life because of the extent of a global warming we have now crossed into what I call "The Easter Island effect". Now to be exact, this is a theory which means a possible, but has a strong basis. But my thoughts from what has been shown is this place is where eventually the inhabitants no longer were able to provide food or shelter and revert to cannibalism.

Because the food had been depleted and eventually all life ends because it's kind of like the person who eats others then soon has nothing left and he also dies.

Not a pretty picture for our youth coming into this world. But this is what happened at Easter Island as it used all its trees for transportation of their idols and eventually had no more trees which then once the tree population was reduced to zero, and fishing boats broke over time, and this prevented the needed food source needed for the inhabitants, so on a larger basis the world would do the same to mankind once food production is in temperatures that exceed 140 degrees in the shade. I know these sounds like gloom and doom, but by my calculations, the process is actually in start mode, and we will have what is called the boil point not very long from now.

Before I tell the reasons of why things will be so horrid, also know that I have invented ways to make these events go away, but as we all know the applications are not a single fix, but many items that do a special application to gain the proper results, at present our planet is a patent in a hospital who is bleeding to death, and the doctors are using Band-Aids to stop the cut that requires 130 stitches to stop the bleeding. But in the very near future, the use of air-conditioners will be futile as the outdoors now enters the indoors, so in short the worst is about to come. Imagine being on this planet so hot that you cannot even go outside and any water you have is not fit to drink.

We already live with bottle water, but now the water that used to be in reservoirs is dried up. So how do we avoid this, you now know the only person who has the answer to the correct way is I. So do nothing, and "nothing" will be your world, and your loved ones world sooner than you will realize. Much sooner, to be 100% truthful, I could be too late already, but it may not be if people actually wake up and do something rather than talk.

So, if you're able to imagine these conditions, which are not very far from happening, recently, as in 2023, the temperature of 127.7 for 53.3 Celsius was recorded in Death Valley. In that same desert in the early 1970s, I was there in the middle of 114 degrees, and I barely made it out alive as I was hitchhiking through that desert from California.

I was on my last leg when an Air Force retired soldier saw my condition and decided to give me a ride. He saved my life because I was heading back to my home in New York. To that person, whom I do not even know his name- thank you. But I know that 127.7 f is unlivable, but wait!

The Sahara desert has been recorded with 136 degrees Fahrenheit and 58 Celsius, so as we can see the temperatures of increase if we subtract years of highest recorded in 1970's and temperatures recorded today, we have 22 degrees increase! Another 22 degrees means we are no longer able to live outdoors or indoors, but all remaining wealthy individuals will go underground. However, far sooner, there will be world destruction from the migration of staving peoples seeking refuge and finding none.

Another issue will be clean water, which was once in our freshwater lakes, will

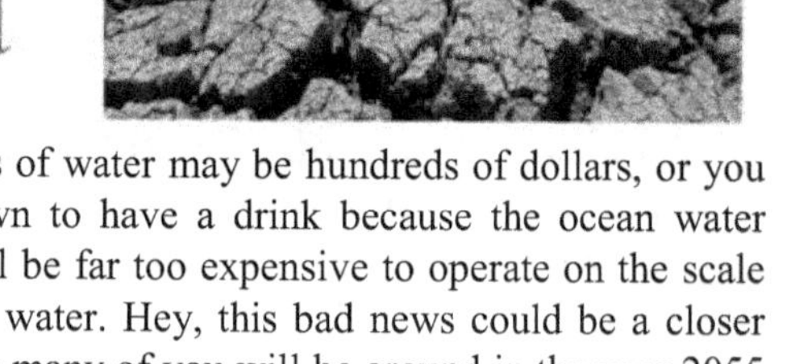

become so rare that the cost for a glass of water may be hundreds of dollars, or you are willing to give everything you own to have a drink because the ocean water processing for a global community will be far too expensive to operate on the scale required of billions of people needing water. Hey, this bad news could be a closer event than you may be considering, but many of you will be around in the year 2055 or around that period.

Will it affect us now? It already is; the costs of food increases, fuel costs, travel costs, and the rise of housing and our basic needs are all being affected on every level, but what is coming very soon is unimaginable for our children, and their children will be in a living hell. So why is this a fact that is not some made-up hype to get attention?

Simply, our land mass for food production is decreasing at an alarming rate, and each year, fossil fuels are being used and increasing the air temperatures, which now increase the ground temperatures, which in turn, the ground plants in our deserts that did have a tolerance to high temperatures, no longer can withstand the higher temperatures. As of now, their roots die because the temperatures are too high that the ground is not allowed to absorb the water, but as it rains, the water evaporates at an exceeding normal speed.

And get this news: our forest preservation, with the help of the existing Government, has approved the burning of underbrush so that they are able to move equipment easily into forest fire areas, never considering that burning the underbrush now increases the ground temperatures in forest areas and increases the air temperatures. That affects the already high air temperatures in desert areas, which are now increasing further. As we have witnessed recently as 2023 from Canada, they also had massive fires in 1989, 1995, and 2014; this, my friend, is the tip of the iceberg that is coming. As the temperatures rise, so do the dry grass and trees, which now become an easy fire hazard.

With your support, we could make these events reverse, but maybe your too busy getting season tickets to your favorite football team, or a donation to your political party. Unless your political party has an inventor with top ideas, they can't help. And can your sports team can't help you survive, sorry no, too busy trying to impress your life of entertain you into believing that they have the better life, which is not improving global life at all. Plus,-we all know there are scientists who have degrees from some college, that say they are qualified, but let us be really understanding about a piece of paper that says you have the understanding of a subject. When in real life the subjects of life can change in a blink of an eye, we saw what happened with so called experts of the virus that was made, how many people died believing that they had the protection? Not to sound like a complainer, but I made many attempts to contact government programs and States with problems of drought, yet they only wanted to hear from those who had a so called expert ability.

Instead, all they could do is state the problem, but have not fixed it at present. Well until now a real answer is found, they could do something positive but they didn't as I contacted the Governors of California, since around 2009, and the White House was warned of magnetic train travel, and told about new energy to reduce global warming, but what have we witnessed from our so-called caring government of the United States?

Our Department of Energy invested billions to promote a device that exceeds 1,835 degrees to be added to our deserts daily. A project called "The Ivanpah Solar Project" which spoke of earlier a massive mistake of increasing our planet's atmosphere like a blowtorch from hell. Does not seem they have any real idea of what is right. I have known of this issue for many years as the science community has bounced or should I say dropped the ball on making firm commitments of global warming and the future simply shouldn't be oil based.

Big oil corporations refuse to change gears, and they could, they could make the change so they still could maintain profits. They don't want us to change their filled pockets, instead they want to keep flowing toward our destruction. The money is more important to a bunch of greedy people. The older global maps prove my thoughts as they show the great decline of ice in the poles and compared to the reduction of size shown today.

More than this is the increased storms from higher than normal temperatures in the central areas of the planet. I have had ideas since 1992 that would reverse global warming. The designs I have already been confirmed as workable, I can't say who this was, because I had them sign a NDA which prevents them from stating anything, but the project ideas have increased over time. But what we need is a start to slow down this global warming disaster, and with my other inventions to come we could actually turn the planet around to be stable.

So as you will see I may have ideas that can reverse it to be back to what it once was as livable. Not to beat a dead horse, this lack of tree growth also increases the air temperatures that effect the already high air temperatures in desert areas, which are now increased further. I would like you to know that at this present time as I live near Buffalo- New York, The date is January 20th, 2024. The entire month of January has traditionally been a time of extreme snow in our area, so extreme that it is considered one of the coldest and most increased snowfalls of all the winter months. In 1977 which was recorded to have 100 inches on average snowfall, and snow drifts up to 40' feet high. Today, it has been now weeks of zero snow, and today is the same; in fact, I just went outside of my home to feed the birds some bird feed, wearing my T-Shirt, cool weather, and no snow. This is proof that global warming is happening and will destroy our world if the plans made within this book are not implemented. If people ignore these facts, please allow me to give you fast forward, here goes, I don't want to scare people but here is what you need to know.

As I stated before there have been attempts to fix our deserts, but they continue to fail, and require more needed attention than what they (the supporters) considered. The desert heat still growing also causes storms to increase and become more frequent and larger in scale. The ocean waters are hotter, and our waters in some areas are growing red algae that kills fish because when they enter this area inside the ocean where there is no O2 in the water they simply die. The growth of these algae is increasing yearly and will eventually control the sea as a dead zone for living creatures in the near future.

We need to understand that even the small considered area at present effects humankind. Over 12 percent of the world's population depends on ocean fish as a source of protein. To change the subject slightly, we need to understand that it is essential for our planet to attain global harmony, and it is also necessary for mankind. For our world to survive, ocean life must be maintained. This means stronger regulations against the dumping of chemicals and pollutants into lakes and rivers. We need a global committee to enforce these laws that can be made by all countries. To mention just a few names of rivers that are in need of repair from massive amounts of waste being dumped into the rivers are known as the Mississippi located in central United States, and another example is the Nile River these are just two examples of rivers that have been destroyed by pollution, and let's not forget the Amazon.

Sadly the Mississippi alone dumps tons of nitrogen pollution into the Gulf of Mexico yearly, creating an area bigger than Massachusetts as a dead area for any wildlife. We need a global alliance globally to structure laws to maintain this fragile ocean food yield. A global court made of all nations to have representatives to decide the fate of nations who infringe. At present, China has become a major infringement of fishing territorial rights; this action, if not stopped, will cause a major conflict that could destroy the world. If we consider the expense it took to build the Alaskan pipe line, then we can ask ourselves which is more important an oil that delivers us to our work and build plastics,- or,- A way to grow food, and be able to have clean water to stay alive. The answer is clear, we need to stay alive. So this is what could be made to bring water from flooded areas to areas needing water and be able to store it under ground.

DESERT RENEWAL- New Water Development

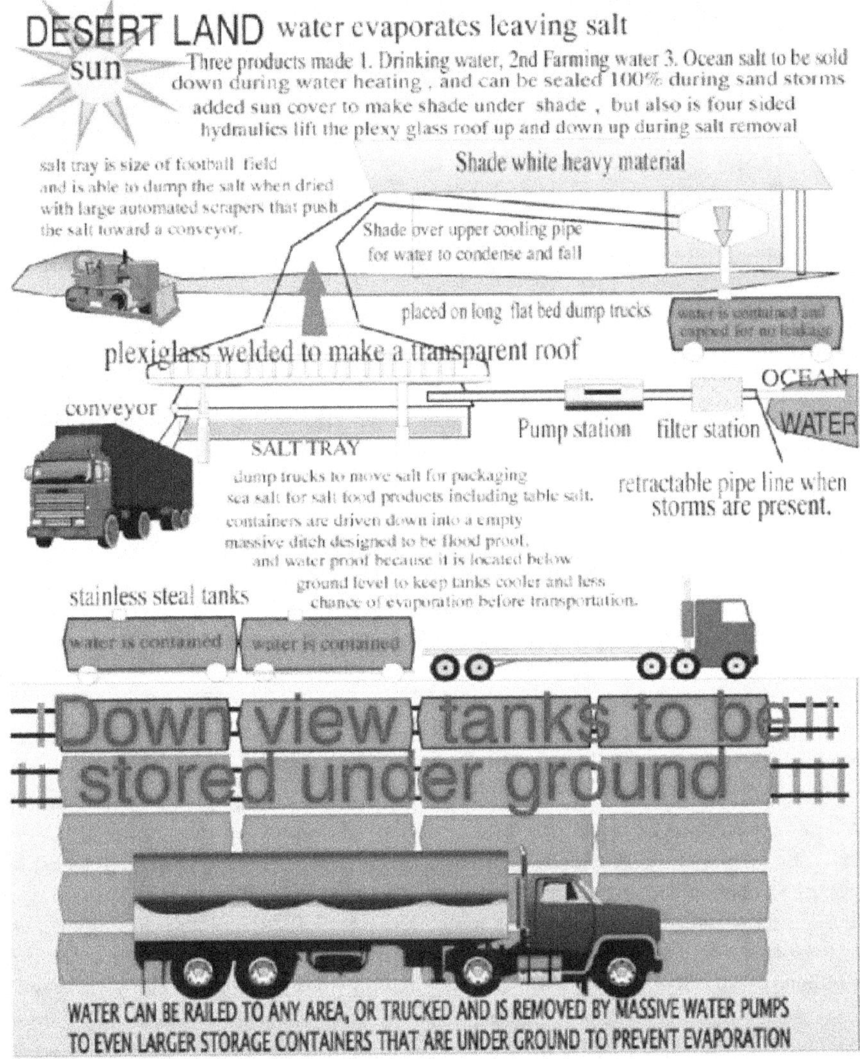

[WATER TRANSPORT TO DRY LANDS]

11. SUBJECT: PEEK INTO TOMORROW

TOMORROW the need for electric cars is essential to remove the massive amount of Co2 being processed by the fuel being burnt, but these car batteries are not the savior we thought.

Electric cars with lithium batteries have been reported to actually have a worse carbon foot print as emitting 75% more Co2 in production; this is totally unacceptable as the car manufacturers have boasted of climate change ability, which is a farce; so, we need an electric vehicle that has a lower Co2 rate a conventional car battery. Electric travel is our best solution, but I have been seeing different batteries from time to time, which make claims of better performance. The one battery that is in development is called a solid- state battery. This battery has the potential to be the golden ring of progress. This battery has an ingredient called the halide electrolyte, which this battery has a much lower Co2 footprint, yet it still has a high heat index, and the battery development has been proven to be more harmful to the atmosphere than the actual Co2 released in fuel- generated engines, and have not yet been perfected as to not overheat and explode. If you're wondering why I am addressing these battery issues, the answer is simple. The battery is the most important function of a clean traveling ability, one that was completely ignored by inventor Mr. Ford, and he never considered that billions of fuel- burning vehicles would be a problem. So, I am promoting the use of a battery that has been determined as the future battery that can be used without heat buildup and is able to recharge in a very fast way. This battery is called RT the Helena project. A Halide Solid-State Battery with a very low carbon footprint, it is an ALL-solid-state lithium- Metal Battery. So, to date, the

39

winner is this battery, which will go very nicely in the new vehicle I am designing. So, my hope is to introduce a vehicle of wonderful applications that have never been conceived; you could say this is one of many goals, but unfortunately, the design for this car is not to be included just yet; I need a patent application.

The last subject of CO2 is fires, is more complicated; simply, we know that trees absorb CO2, but when burnt, they release the stored CO2 back into the atmosphere. So large forest fire are making not only massive heat from burning, but are also injecting CO2 at the same time. But in the burning process, that fire puts off a gas that increases the greenhouse effect heat by absorbing infrared radiation, e.g., Carbon dioxide and chlorofluorocarbons, which are emitted from the forest fire. But the destruction of trees by fire is not alone. An estimated 1.83 billion trees were planted just in 2023 alone, but wait! The estimated number of trees cut down that same year- 15.3 Billion! A number far higher being destroyed than allowed to grow. The fact is mankind lives in a growth mentality , this growth is only a focus of the immediate and the inward demand, which seems to have a never ending increase, This can be even seen in the new types of business's that emerge. And as business after business emerges many are willing to take forests, and land that could be for food, and they put up a warehouse and op offices, and shipping and receiving docks; the point is mankind has no care about the space they take for the world as a direct effect on the condition being harmed in our world. Instead they raped the once growth areas with parking lots and massive stores traveling along our highways. So if we are to survive, yes survive! We need to "consider" what we are doing that may have a future impact. So business needs high rises for offices only, and stores need to return to malls, but high-rise malls with elevators, this will shrink the land mass being taken by business and home developers. Today everyone builds homes that take away former farmland. Farmland that could save trees and yes save the people of this planet, because the future of homes is living in a building that when you need anything there is great low cost goods directly in shopping areas located in the lower level. Imagine top quality stores with all the things that you desire are there, to mention just a few of what you could expect- top grocery stores, mechanics for cars , and doctors, legal help, even police security are in the same building, exercise companies, there is a bike park that goes from inside to outside on good days. The needs are with everything imagined all in one location, but! let's say you're seeking an adventure, so you take the sky way that is a combo walk and run path, and an electric seated transport which will stop where ever you say because it only seats two people, but they are controlled to be released by a sensor that detects how many potential passengers are going, and you simply tell the system your destination, and off you go. When you get there, you can walk around or take the moving chairs, which move slightly faster than a person walking have a stool that can be sat on, the journey is a simple walk off the ride onto a slide rail, which slows your exit, and now you can walk into any store. If you remember, I stated that even the security police live in each complex, when there is an issue- bada-bing! the police are there, not one police officer, but four, when a dispatcher is called, the police actually answer, there is no delay, the dispatcher is only making sure all is needed and that other back up is not required, which is only about five minutes away from another building in the event of a serious problem, so no longer are the police far away and 90% of the time never get to a scene of a crime until it's over and the art designers are making chock marks. But in the future, the police have a backup called a robot cop, no it's not the guy in the movie, but are AI robotics able to be utilized if a situation has greater danger, but all commands are made in real time. Now these massive hotel style structures are round because of less wind damage, and better reflection of high winds, these structures are with over 100 floors each, and they are in a very wide circle. Each floor has 40 families, or 40 residents per floor if a single person complex, that's 4,000 families living in one building, and if each family has one child, or one child on the way, which is required to live in family housing. Every floor has its own elevator which is from basement car garage to floor A- The food court and skate park, to floor B , with human services, as doctor, dentist, lawyer, security, banking, electric, are a few to mention , and every floor has a park in the center, now each park has different applications. Some examples, - Top floor can be the indoor beach and pool with lounge chairs and mini-bar for adults, and food vendors for all. Plenty of sun time and life guard. Let me not forget to say that all the people who are mentioned live in this place. Just imagine this, you go to work, and you're not doing a three hour commute. I once did that from Toms River New Jersey to New York City. Crazy, so yes, this make life a bit easier. Another park can be basketball and tennis court combo, and has water fountains, and bench areas, Another park can be swings and small child park with benches for parents to sit and observe children, and this is combined with a dog walk, and "dog wash" spray pool, that can wash an animal of any size in seconds to include soap or any choice added extras into the water such as flea soap. So the vision is to build upward our homes and communities, yet provide a wilderness to be maintained and areas to farm can be directly over underground roads. Okay I'll be more direct, we must use this system or solution so our home developments are concentrated, not spreading out, not killing trees, not making more heat, not making people travel far to get their needs, yet have the ability to travel for pleasure destinations, and those areas are built in different styles, yet are exactly the same applications as to be easy to navigate, and there is a map on each floor which shows exactly where you are in 3D, all exits, and all named items as to locate, you can even ask the computer at the "find it station" where is the child nursery? It will tell you and show you in real time. But these structures are made to last also, made to be in the world long after we are gone, stainless walls in kitchens, sound proof walls so absorbent your music would never bother anyone, in fact the AI controls volume whenever the door knob is turned it actually turns down the volume to your stereo instantly by Wi-Fi sender. But close the door, the sensor connects the Wi-Fi back to its original sound volume setting. These structures are bigger than regular apartments made today, actually much bigger, because we are making room that gives people that feeling of space and not being congested, instead your life is full, expanded, and relaxed to have every luxury imagined. And how is this possible? It'll get to that, but first let's see what else we have in our new future home a baloney with a slide open sides so if you want privacy the neighbor can hear or see you, the panels are designed as sound absorbents. But let's say you like your neighbor, well, then it's an open sliding door, and you can even invite them over through the balcony, so these balconies are not small by no means, you could hold a party of 20 people easy on one of these balconies. Do cook outs, and every balcony can have a full screen television for those sports fans. And this one is in the slide away wall! Totally wild, and you can customize these homes any way you desire, The packages are full of ideas, you only check mark what you desire, and it is now added to your home. You pick your furniture, you choose from a list extremely long with pictures and when you sleep, your bed has built in foot warmer, body vibration massage, and your selection is a simple touch screen to where it hurts. Yes, this will be the future, so these are actually homes you can purchase and all are welcome with different pay methods, as to go by income, and be stretched out for persons with low income, but are sold as homes are sold. You can be in a family style home or a single person style home with one bed room ,but affordable and modern, suited to your design, you pick the style of furniture and the style of all appliances, including dish washer, bathroom design, living room design, kitchen design, window design, overhead lighting style, all as to have choices of size, comforts added, and even a rotational view that allows an over look of a fantastic view, not the big city apartment with the view of a fire escape, or a brick wall ! What you see below is further away around 70 yards away is the other structures that look like your structure, and a sky way floor connector for travel to and from each unit. The best part is not all units have the exact same thing, so people will travel from one to another, but your travel time is a few minutes, not I have to drive clear across town. Now to enter another structure you need a pass key that you swipe to enter another structure, but that pass key is your entrance to your car port in the basement, and you swipe that card when you get in your elevator, remember, elevators go directly into your home and AI will be watching if it is not someone who you know enters the elevator the elevator won't budge until security has arrived and you verify that this person is your guest, and they do a quick search of the person to see if they are forcing you to comply.

Now let's look outside from any window you have, flowers can be planted outside windows, fields of flowers bloom at different times of the year, and are planted everywhere, and there are forests trees of every type everywhere, these structures have a roof electric travel from one complex to another, and an underground electric car storage for each home up to four cars per car port The family can be with storage compartments which have a card entrance I spoke of. and the card can be opened as if a lock down happened and your locked out or even locked in, there is a two way contact to security which shows them automatically where your calling from with a LED light over the car port indicator and number. But let's say you want to take a drive, cars exit through the underground road to then enter the outdoor areas, which are for industry, and large offices, food production plants, and business areas. This is pretty much what I call the perfection of living with views of forests, and farm lands far away, a perfect life , where even everything you need is shipped in daily from an underground highway that enters the structures, and our children attend schooling at home, and the structures have parks on each floor , swimming pools, tennis courts, basketball courts, and even 3D events , music room, art room, exercise room, sauna, massage parlor, let's not forget learning tools, such as in-house library, and technical developments as teaching platforms. The lower floors with stores, and restaurants, and each hotel structure is connected to a skyway travel that comes from the rooftop. So this form of living helps maintain more forest areas and help reduce the ground temperature, so the air temperature is cooler, this cooled temperatures helps maintain the soil wetness, which is essential for trees to maintain a good moisture to grow, and we do use many of these forests for harvesting trees, which is presented, but the trees around living areas are in cycle formation to match the circle structures and from above the roads that are in the woods are not easy to see by most residents because the trees block the lower road view when people look out their home from above. What many will see is simply wonderful.

12. SUBJECT: PLANTING TREES FOR HARVESTING BY MASS PLANTING

This subject is not easy to get super excited over, but it is extremely important. So immediately. I spoke of greenhouse planting but this process is for open planting in deserts with thousands of trees being planted at the same time. The planting of trees is needed by our planet, but we need access to trees, unlike many forests that are a massive bundle of trees close together with no entry point until trees are cut down, and then when they do this, many younger trees are damaged. So, with my tree planting, we will use dirt roads that can reach inside each row with a cutting tractor. Also, trees must be aligned in the same direction so that wind and erosion cannot affect the trees, and the rows must be wide enough for heavy equipment to enter and exit when trees are cut and moved to near roads. So that loading does not disrupt or trees move and does not damage other trees. Trees that fall on other trees damage or destroy the bark to kill a tree not ready to be harvested. These trees are being planted in tree farms that could be used to restore deserts. However, the design presented as illustrated is a depiction of the outdoor planting of rows of trees. So this system of logging can also be used in regular non-desert areas, including underbrush areas, which are grown at the outside edges of forest development to deplete the wind and rain effects and help retain moisture. Mature trees must be removed in their 20 to 25-year development stage to attain the maximum growth potential, and the distance of each tree planted should be a minimum of 30 ft. from each planted tree so that the branches can extend to their potential distance and allow sunlight to be most utilized in photosynthesis. Vegetation of tree farms that are not in deserts can have bushes grow as long as they do not obstruct the tree growth, and any of nature's devices, such as a vine of any type, should be cut down in regular forestation. In this system of harvesting one group of trees are processed that were planted every six months to one year, and then the next six months to one year is planted. There are younger trees already in progress as seedlings in greenhouse nurseries. These trees will be inserted in the new rows designated after trees that have reached their potential are removed from the row. The remaining tree crop that was not harvested remains intact and maintained. So, where the tree was cut, it was also cut at the base to remove the root base and plant the new tree seedling. So, the rows cleared are already being re-planted.

To establish desert growth for trees without greenhouses is a harder process but this would be how it is accomplished. We must consider sunlight in deserts can be extreme, so we need to shade the trees with large overhead wind sails at the same time; so we plant a tree called "Acacia Raddiana" just in front of our new trees; this tree will become a massive shade to allow ground temperatures to decrease, and also help retain water into the ground. The wind sails used first will allow light to be filtered, and these wind sails drop down during a wind storm, but during the high sun, light can be automatically lifted into place by cables with electric powered motors at the end of the rows, and rows are extremely long, so to lift the sail will require a strong cable only when the wind is not present. We will use solar electric machinery to accomplish any motor required, such as a water pump to access water. This is essential so as not to pollute the ground or the air. This can be accomplished as long as the water hose sprinkle systems are placed down first and marked as the locations in the event they are covered in sand. So, to start a tree growth, we have a similar need for water retainment, but the mix to make this absorption will be a much larger amount, as the water retainment inside a greenhouse can be regulated. But to keep costs at minimum the mix material can't be shipped in the same as greenhouse delivery because it has no way to be inside to keep it from being disbursed by high winds. But we bring the mix in after the hoses are in place, and now drop the mixed material made of dead wood and wood chips from helicopters with large basket dumpers; the mix is dropped from a height that does not disrupt the mix below, and the pilots drop on flagged areas only, this could be done by trucking also. So, this material mix to be dumped could be larger, and then we could add smaller materials later. Then we place cypress trees that are already pre-grown in a plant nursery that are adapted to high temperatures, and undergrowth plants that are already adapted to high temperatures also, and as time goes forward, a new series of other undergrowth such as moss or weed cover, or grass is added to the undergrowth so that the strength of water retainment begins to increase, and they are planted as to cover the entire crop of the trees. The desert is a harsh place for plant life, with sand storms and high winds being destroyed in many ways; the erosive nature of the sand can kill trees planted. So the main thing is we make rows of hard brush, and with the regular cypress tree, which has a fine wood for use, there will be another tree growing right beside the cypress, the Doom palm, which has a very good application; they are high shade ability, once again if planted properly could lower the ground temperature greatly, and they have used like apple size fruit that tastes like gingerbread. The fruits inside the nut are used for making buttons, which I am sure, will be in small demand, but the rind from the nuts is used to make molasses, a real moneymaker, and the groundnut can even be used in medical wounds. More money making. The desert tree, adapted to high temperatures, will be planted previously. As said, this would be a thick mastic tree. Tree, once planted, has a strong ability to not need the water demand as other trees, so the tree will be a good ground preservation root system. All nursery trees are shipped in on flatbed trucks, which are the beginning of windbreakers to keep the soil of trees intact. Cypress trees are planted last, so after the heavy brush rows are in place, a grass called love grass would be the growth; the Shepherd's Tree is a very solid tree that has a deep root system and to be watered daily from the sprinkler system which was first laid down. These plants will keep the plants alive, as the heat is a ravage even for plants used to high temperatures.

Now, the strange fact is there are over 1.8 billion types of tree growth in the world's largest desert. The problem is that the trees are widespread, and the sands dominate the ground temperature. So, the need to place the proper tree growth with the proper undergrowth could bring this land back from the dead. So, the combined smaller plants mixed in with the shade, and the larger tree growth can be accomplished as bushes behind the rows are now planted. These are to be thicker bushes, and this will give the trees even more protection.

One point to consider is that the rows should work like a line staggering as to keep winds bouncing and allowing the line to face the wind direction most actively so as to block the winds as much as possible, but are close to the trees as to entwine the root systems, which will make the rows stronger. Last, we place poles as deep as possible to allow trees to grow straight. It would be wise to apply smaller fence poles to even bushes planted to secure twine loosely and keep them from having the root system pulled up. If we are planting trees in regular temperatures non-desert, the woods will not need to be cleared of underbrush when the trees are planted, this will help the trees stay intact, and should have already been started at a tree farm prior to planting. One of the main problems of tree loggers is they have not considered that trees should be like any other crop. They go into forests that are not organized and have to be actually destroyed to even enter; the trees cut crash down many times upon other trees, so instead, let's consider making trees like any other crop; the difference is we plant monthly new rows, and each month the trees are always allowed to grow to 20 years before harvest begins. Trees should be spaced enough to allow growth to be wider, and the rows should be made so that equipment for harvesting can be accessed easily; even a section should be removed without disruption to any other tree not being cut. (See Illustration)

The vehicles must be used to not dig up the ground around trees to not damage roots, and even in wet weather, the use of wide and thick rubber tires with deep tread on the vehicles will be able to have them move freely, and they will be equipped with metal clamps that can drag a dead tree out or even carry it out with a cutting device to make it smaller if needed. A heavy cage protects the cutting driver, and the cutting tractor is electric and solar-powered. And if we are planting in a desert without a greenhouse this should also be applied in All trees will be carried out with the tree being cut on sight and always cut in the direction of the exit of the vehicle path so that trees do not fall on good trees that are not being harvested. Each harvest is a row removed one tree at a time in the same manner that a vineyard is cleaned, but only trees are designated at the time they were planted so as to keep a full rotation of harvesting. Our harvesting will create an abundance of small limbs that are chopped and made into mulch, which is carried to new-planted trees just replanted.

The mulch can be sent to desert areas that have vegetation by truckload and distributed as evenly as possible. The best result for desert mulch also is when there has been a rain fall on the mulch; the ideal mix of mulch would be branches and leaf material shredded, and the areas far from the desert could have a leaf and branch pick-up once a year from towns and cities which have leaf fall and dead limbs or trees for the needed compost to be delivered to the outskirts in edges of all our deserts. People have not realized that these materials like banana peels, leftover lettuce rotted, and anything else we have tossed out in the past were actually something that could save our planet. Because if we take these compost items into our deserts, they can be a good source of ground restoration.

So, what we can accomplish is turn garbage into something positive. The process is not just planting a tree and leaving it to nature to grow. The deserts are harsh areas and will require intensive care, but it is essential we stop desert growth because it is one of the top reasons our planet is failing. So this is why I am presenting the elements of 1. Water development, 2. Water Transfer 3.Proper growth devices.

4. Temperature control and proper lighting. The use of ocean water desalination as removal of salt from water by thermal heating, does not need this heating if used by sunlight. My purifier system is designed to not use one system to use water for food and drinking, but the systems placed in desert conditions are perfect for the water to evaporate in multiple units that are aligned to provide water for massive greenhouses and tree harvesting. This would include purified flood river water pipelines to dry desert areas to be stored underground for farmers in large truck-style tankers.

Earlier, I addressed the three major contributors to global warming, and so population control is one; Forest Fires and fires of any type are second, including the billions of fuel-burning devices called cars and trucks, and they also help in increasing heat in our world. Just touch your car's engine the next time you travel over ten miles. I wouldn't recommend it because it would burn your hand. But Co2 also needs to be removed, and if you're not aware of the gas being released, you've been living under a rock.

Now, an interesting statistic that may change your mind about your gas engine: the actual percentage determined by Co2 studies states that our automobiles and trucks are responsible for 30% of the Co2 levels. I am deeply sorry to hear this because in 1980, I went to visit a relative in California, and I was there for a year. Every day, I watch the sunrise like a big ball as the smog (Co2) covers the skyline. The smell of car fumes was everywhere in Orange County.

Since then, I have been to almost every major city in the United States, and all had air quality reports on their news. Air quality, which actually kills people each year. So 30%, that is a joke, try 70% worldwide. So let's face it, you may be a car enthusiast and like the roar of four four-barrel 350 engine, the fact is, I did also many years ago. But kids, it is time to change our toys; fuel-burning engines need to be replaced 100%. ASAP as of yesterday.)

13. SUBJECT: TREE HARVESTING

The trees are harvested every 25 years and each tree has a year mark that shows by color when it was planted, or should I say, "born". Any tree removed is replaced with anew nursery tree which has been re-grown inside a tree green house, and is started there and delivered to the tree farm. In the event that a tree has died in that planting time, the tree with illness, or has not done well, will be removed down to the roots will be cut so the tree will not grow further.

The purpose is so the ground will accept the new tree without too much disruption of roots from old plant being a drain on the water. All trees are harvested only after they have been alive 25 year to mature. The tree that is not the best will be removed premature so that the other may grow outward and the root system will not be hindered. Trees removed with a good size can be used for small wood items, or chopped for mulch.

The rows of trees are by year and removed by the oldest being used first; again,-no tree is allowed to be cut before it is time. Wood that is used must be cut and stored as soon as they have been cut to not waste the tree-to-tree rot. The land is used not all at one time to grow a tree like the one it is done now. But first one row in one month, then another row another month, and then another row another month, so on and so on, until a desired full year of rows have been established, but this can even be weekly if the farm has enough land mass, and enough nursery starter trees. The rows should be as minimum length of 2,000 yards per row so that the harvest crop is very large per forest area, not saying it couldn't be larger, but the process to cut these are in one month of each other, so to keep the process in a timed method as to not get behind in harvesting and planting, a month should be adequate pre crop being harvested.

The depiction of the amount of trees is not showing the hundreds of rows duplicated in another, which would best be shown with an aerial view when planted, but there could be potentially hundreds of these row groups, so what I am showing is the correct way forest can be accomplished without destruction of other plant life and a way for mankind to have forests and proper O2 being developed Co2 being removed, trees to give mankind needed products, yet also to maintain the planet's atmosphere.

Tree harvesting that does not take massive trees all down at the same time, which cause floods, and massive mudslides, and increase the ground temperature because trees are not there to shade the ground. The possibilities to build these trees in my method will bring millions of once thought dead areas of our planet, back to life,

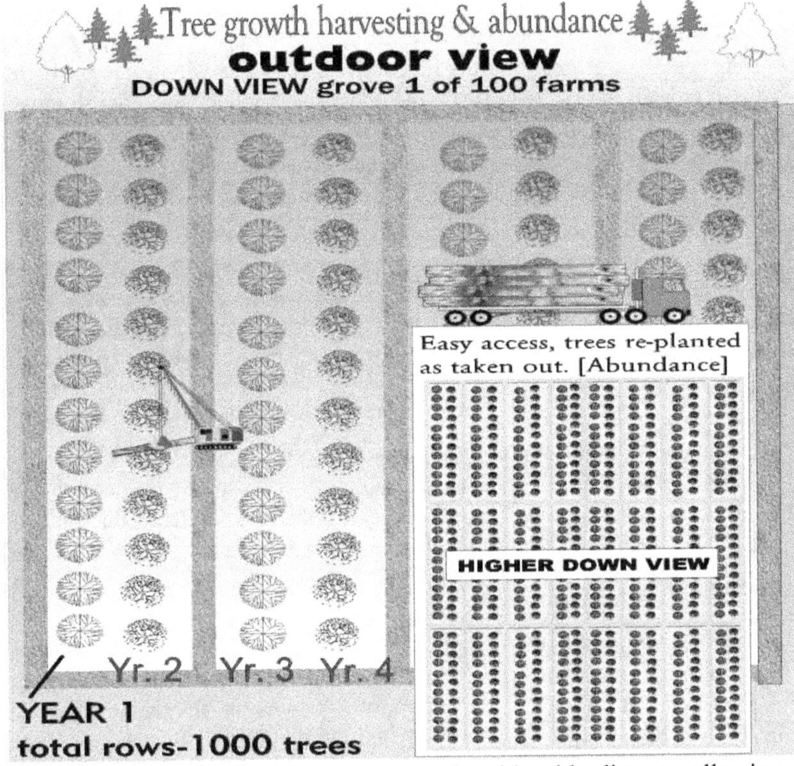

acres strictly for trees, all with easy access roads with wide distance allowing the trees to expand to their potential.

Also these trees can be grown inside nursery conditions that remain inside our deserts, Just think trees growing in what once was only sand, now in massive green houses and these greenhouses can grow other items right near the trees. Such as berry bushes could be planted as a yearly crop, straw berries, blue berries, many types, and now the areas are producing food that has no harm to the tree and is once again easy to be reached, and when the entire desert has been filled with these green houses, as the trees then must attain the increased temperatures where changed ever so slightly, so the trees could maintain their ability to grow in warmer climates.

Certain trees are planted after tests have been made to see which trees are able to be used for the populations needs which have the greatest resistance to heat, now this test is actually easy, the growers only need to find tree's that have survived best in the climate already, and see which does benefit mankind most. The next important factor to have these trees to stay alive is that plants that need to establish as "underbrush" is required to maintain the moisture. The desert greenhouses also collect Co2 to be transported to outside our planet, and we can do something to make these green houses have more water than they will ever need without drilling for wells.

We will transport the water as depicted with trucking to drought areas, but we can even go further as to make areas where flood water happen , these areas can now have redirected water entry points that now send water which has been filtered to area thousands of miles away by pipe lines that connect to underground massive tankers, and the method is to use major rivers as to send the water in the season of most rain, and we must learn to build water structures that are along the rivers to keep water even in dry seasons, massive tanks that are underground so the water cannot evaporate.

So the main application to attain water in dry climate areas is to use transportation of water no differently than what was established called the "Alaskan pipe line" which was used for oil delivery, so why not use this method to move water?

The cost to build these long distant pipelines will pay for themselves in relatively short time from the fact that unlike oil that can deplete and dry up, these pipelines are to sustain food production, tree production, cattle production, and life in general. If we can do this for oil developers, we can certainly do this for water, which is our life's need.

The starting point is much higher for water to travel downward so there is need of water purification before transport, and the water is pumped upward to extreme height to travel. This means we are now using many floodwaters that are now controlled on their banks as to collect that extra water to be moved thousands of miles in many directions.

The creeks and rivers will all have this ability to make a fruitful world and a better life for our entire world. No longer will water be a problem, instead we control the distribution when it is at the peak rainy seasons in one area, and yet it is dry in another, the excess of water is captured and stored and is ready for use.

The more we grow, the temperatures of our planet will decrease. And the planets ability to stay in its regular weather conditions will return. This means water temperatures of the ocean will be depleted and the strength of storms will decrease along with the size of those storms, because the high evaporation along with the higher temperature change has been the causing many of these massive storms.

In short- we will be on the right path.

EXAMPLE OF PROPER DISTANCE OF TREE ENTRY
FOR CUTTING

14. SUBJECT: BAMBOO & SWITCH WEED

These are two elements I researched as the best remover of CO2 from our planet-but they have a negative point when CO2 is released when they die. All wood, when burnt, does this, but the amount of containment is the gem in these elements. These plants have the ability to absorb CO2 in a higher capacity and store the CO2 like no other. Bamboo and Switch weed are two very attractive forms of plants, but I have another which is just as pleasing to the eye; it is called "The Empress Tree and is a tree touched by God. My research for these items is a double-edged sword. First, we can consider growing these plants as a standard use at our homes as the areas of the world could use them; not only do they do great good for our planet, but they are very attractive outdoor plants to have, so they are a luxury potential standard for any home. They have been tested to produce 35% more O2 back into our world, with a recorded amount of C02 in one bamboo plantation that can capture over "60 tons of Co2 yearly." With this news- the greenhouses depicted in this book illustrate a Co2 greenhouse capture system;- These plants would be most effective if bamboo and switch grass were combined in the greenhouse to gather the Co2, The bamboo could be used for the following items to store the Co2 as carbon as actual products if not already used in this way. Bamboo has a very attractive decor, with a bright green, ribbed style, almost brick-like, and extremely strong. Some of the uses could be, but not limited to, could be 'In-home construction wall braces with a simple end cap to be added to be placed in supports and even support beams, lighter yet stronger, and being they are already with an air gap if they are placed together to make inside walls, they have insulation built in already. May I mention again that bamboo is extremely strong, so they would only be cut to the length needed for wall brackets or wall insulation? Another use- outdoor yard furniture, chairs and tables, and window shades, to mention uses used today but may not be in your home, but if purchased, you help the environment greatly, but the next use of bamboo is something that would be done global in massive amounts, and capture millions of tons of Co2 by using these very high trees already rounded unlike trees used today, and are just as hard if not harder, so bamboo which has a potential to grow up to 60 feet in the areas of the Pacific North West, these bamboo trees if they have a growing guide pole to keep them straight in the future could be our new inexpressive telephone poles, and get this- to keep these poles strong we dip them into either clear plastic for communities who like the natural tree look, or we can dip them in any plastic color to protect the pole from wet weather. These poles could replace regular trees that can be used for other uses, and we could have lighter transport when putting bamboo to use, and the only difference is the connectors would be needed to be pre-drilled to allow screws to be used, which are stronger than nails. The bamboo tree actually would help linemen who need to go up the pole as screwed-in rims could be placed where the normal rims already are, so the access to an upper line is now faster and easier to climb the top could even have a bracket that allows the worker to stand on it with a round attachable platform. And here is a thought: the bamboo could be made even stronger against storms that have, from time to time, caused poles to break. The bamboo, when in place, can be with a portable support that is removable when the pole is set upright, and then a concrete mix is poured at the base. The pole being set requires that an excavator digs a single hole at four feet deep, and the top dirt is made smooth by the excavator. Then, the bracket is leveled over the hole, and the pole is lifted by a bucket vehicle. The pole, which is lighter so it can be fairly easy to maneuver into place, is now dropped into the bracket holder and lowered until only one foot of the pole is still open at the bottom. Then the pole is locked and the bracket has two levels four for up and down as to be straight, and one for the actual support being level. Once this is placed, the concrete mixer that was standing by now pours a small stone concrete mix into the hole three feet deep. Since the bottom of the pole has an opening naturally, the mix is filled to the top and allowed to settle, and then the mix is poured again to allow the concrete poured to enter the underside and bottom of the pole. The portable brackets remain in place for 48 hours, then removed. We now have a pole that is extremely strong against winds, and the top of the poured concrete is covered with dirt. Now, if we make an opening in the side, we can even run coated wires up the pole, and our future could be that only wires that are crossing a section are where wires go above; all other wires could be in underground walk through tunnels, and the wires are clipped in on the sides of walls, the floor has water drainage and plastic grids to never be standing on metal, in short, 90% of the wires today would be no longer an eye sour and a logistical mess, because all wires below are color-coded, and numbered straps tell the line, and get this, if a line is out, every thousand yards is a light indicator that if out, means the line has been broken, and the area is detected instantly. Entrances to all tunnels are with a master key to a plastic cover door that is with a dumb waiter for tools to be brought in and out, but we even have tool rooms to store wire, and connections, to include tools. These are small storage areas, and the storage areas are next to the entry port so when extra material is needed the placement is easy to get at from above. So one day, we will have attractive streets without massive numbers of telephone poles and an efficient underground electric wire and cable system. But one more thing, keep all items separated as category, such as cable in one area on the wall, which could even be on the cellIng connected, and high power wires could be located on a separate wall. The need for electric lines not to be exposed to storms is very important, and that is the reason today. Computers that have had a line hit by lightning can destroy a very important development; placing lines underground will reduce that danger. Many homeowners will love the fact that no tree has just wiped out a transmitter, which may take some serious long time, and some people just can't afford to have the freezer down for 24 hours or more, and yes, I have seen this happen for a two-day period. And food had to be tossed, which is like tossing money away. Here is another use for bamboo; with these tunnels made to work on electricity and cable, the work areas could be as stated, but in some areas where making a ditch just wouldn't work from an obstruction to maybe a gas line, this could be applied as to get around the problem, so we use the line to travel in the ground, yet the bamboo is like a protective hard cover, and the bamboo is coated or dipped into a waterproofing plastic inside and outside, a PVC pipe is inserted bamboo is extremely hard which means this will protect the line from breakage from pressure above possibly from a large truck running over the line. The poles could be used for electrical wires as they are natural pipes already and can be cut at different sizes. They are marked in growth as to when to be cut as to match the width and the open areas needed for protecting wire inside, and we could even use larger bamboo for water drainage lines also as covers, wherever any line is, above is a line marker, this marker is shown on concrete sidewalks, and yes the sidewalk areas is actually where the wires all travel, and were the entrances are located in all cities, and rural areas. Areas without sidewalks are not safe, and many people have been killed on bikes being forced to travel on roads. So get busy; we need sidewalks that can be used to be our tunnels for electricity, and let's look at PVC pipe for just a moment; why not make a 96-inch pipe that could be square and used to make instant tunnels both for underground electric and separate water lines. Today, we see a 60" round, so increase even the round to 96", and we can now lay down electrical lines much faster. It would be better for people to be plastic, not metal, and we could make on the other side of streets this underground system for water lines for drinkable water, and being the lines are in an open space, they could be heated in winter, and the lines traveling to the home could have an air gap from the tunnel heated by solar power above the tunnel and possibly attached to a bamboo pole. Also, water lines don't freeze and are easy to get at. The street drainage could also be inside the tunnel as the workers walk on a plastic grid attached to the walls. Now, the other fact is that bamboo will break down over time like any tree, but the material has a higher resistance to fire than normal trees used; a regular tree has around 250 to 300 Celsius in order to burn, and bamboo is 400 Celsius which means electric lines that short out could still be intact, but the most important factor is the cost is lower than metal, or solid plastic for use, and once it is in greater demand the cost could even go down further. Now to our other side of that sword, being these materials mentioned have such a high absorbing ability, we can now grow it with the intent of cleaning our planet from the abundance of Co2 in a very effective way by taking the bamboo and removing it from our planet to be exported to other planets as it is ground into a fine powder before transport to get the maximum amount being transported away from our world. I have researched Co2 as "The Carbon Capture Technology as air capture where Co2 is transferred into Calcium CLP pellets for post-combustion capture of Co2, and other applications are being used to make the Co2 into a liquid and is stored underground.

The problem with this technology is called: "storing it here"- the liquid is Co2 as structured is porous, and if absorbed in the future by outside elements that burn, could release the concentrated Co2 even with a greater potential of harm to our planet. So I say, let's not store it- but let's "remove it" with my Sky-trek system and air pressure launcher. Titled: ACE system launcher, the carbon materials that are stored can be transported in pellet form or taken even as in solid bamboo or other plants stated that capture Co2. Once Co2 is captured in the forms I have shown, then this could be transported, and we do not need to store anything here on our planet to potentially have a mass problem in the future that is unforeseen. So, if you wonder why our planets are gaining heat? The answer is we are burning air faster than we realize. Forest fires deplete the air, and when the height of our atmosphere decreases- now there is reduced O2 in the air space as the amount from land to our upper atmosphere is less distance; this reduced space allows the air temperature to increase faster, which leads to higher temperatures and global warming.

The greenhouses to collect the most carbon for release would be top trees of Co2 removing quality such as the "Paulownia Tomentosa" tree. Not only does this tree help the heat reduction of our planet, but it has a beautiful bloom in springtime, a lavender color, so to plant this tree is your best effort to remove CO2 as we all should make our small effort, which becomes a massive effort when every street in our world has at least two or three of these wonderful trees displayed. So do your part this year and plant one; why not plant one even on some remote side road on the edge of the road away from the highway- so it won't be cut down, we can all be like "Johnny Apple Seed" who traveled around planting apple trees is the rumor, but to plant this tree you would be called: " THE EMPRESS TREE GROWER!

This is a great way to say: "I love you planet Earth." And guess what!- It is the fastest-growing tree in the world, which means we are helping the planet's CO2 removal even faster!

15. SUBJECT: THE GLOBAL COURT

At the present time, our planet is having a hard time coping with the fact that we are increasing the temperature of our world with CO2 and people burning fires is also global. We burn much more than we realize, mind you; let me say- be ashamed of those of you who consider it a great time to burn in the backyard! Burn your air away as if we have a great deal of it (HA! Funny). We have already seen the world lose its air quality greatly in many cities throughout the world; in the early 1970s, I saw how pollution was making the sun not visible in Los Angeles till only 11:00 Am. That was a shock to me, as I was used to seeing it at sunrise in western New York where I was raised. But in L.A, the sun was a ball hidden behind a wall of the foggy mist of car fumes; the fact is now many areas are like this, including Buffalo, where at times I have stood on a high hill near Buffalo, and I've seen massive amounts of pollution, with dark, dull brown clouds hanging in the air. Too many young people in the world are aware that clouds are normally brown. However, China's pollution dwarfs the entire world's air quality, as the pollution even travels clear across the entire ocean to land in upper areas of the United States. This may not seem like a terrible action unless you live there, but the ground quality as polluted is just one of the many issues that this problem makes, and that is why I am stating once again, that a world comity needs to be established that controls the water and the air quality worldwide, and the penalty for the miss use is to block exports from the country causing the problem so that to cause a major impact on their financial status. With actual banking control, every nation's banking can be entered by the world court to accomplish fines on pollution and air control factors laws that they make a break, and of course, water pollution also stopped, even over fishing or fishing territory disputes could be addressed. The nation found at fault will be with many forms of punishments, so they will comply unless they want tariffs increased or exports to haul as to enter the nation who has made the complaint, as the areas of nations are to form alliances, each has the ability to be in a formed union that allows their complaint to have a greater impact on larger nations which have the ability to be aggressive or to flip the coin, a fund is given to the country with the best pollution reduction per year, and gives them a greater financial ability, plus nations who need funding wanting to change the condition of their areas affected. But nations who disregard the health and welfare of its neighbors, then the World Court can intervene and draft a firm and effective registration to penalize the country found at fault. This will stop nations from being thoughtless and non-complying to the request made by other countries not to cause damage to our ecosystem and the balance of nature to be maintained. To accomplish this task, a committee is formed of representatives that all travel to the areas of need, representatives made from all nations, these review all complaints. Within the group elected, a team of five people are now elected who have the highest standards as to lead the groups in action. People around the globe can submit the need factor online, with loads of video, images, and documents, if needed, that is to bring attention to this structured court and address the problem as the information is viewed daily and placed in a priority standard of the impact that affects all, then the five total group reports their findings as to the facts found by traveling to the areas of concern, taking video, and photos to present, and the court is now given all information in an open review and makes the adjustments of possible fines, and penalties for any wrongdoing found. The corrections are made and re-visited once repaired by two from the group, which observe if what was requested has been achieved; if not, fines double, and fines continue to double monthly. When the problem has been removed, the fines are now stopped from banking. The representatives who have seen all evidence will impose any fines, but all fines are held off with 60-day notice to change the problem, or the fine is implemented, and if fixed, the costs assessments for damages and re-structure are half compensated by the court. But is paid yearly to cover costs divided. The five-person committee can also address findings of world issues to close down functions as they are connected to the United Nations Police team and have the authority in every country to do what is directed by the World Court. Countries or people who do not follow the rules or break the rules can be arrested and placed in the country's jail for the term designated until the problem has been resolved.

The person held responsible is removed from the position of authority of the factory, business, or personal activity that caused the problem. This larger "representative court" can address subjects such as scams if the entity has caused severe hardship to another country, false online activities, fishing territory infringement, product theft, pollution, land taken, a business name taken, child pornography, sexual traps of theft, online gambling not approved, in short, they can address issues that cross boarders which in the past where not achievable. With this established, the door will be wide open, and penalties will be enforced not in weeks but in days. At the present time, we have seen many countries do what is called knock-off products; this is when one company or individual has invested in patents, and another country still produces the items and sells them on the open market. For example- a company has taken a product design and made it knowing that the product originated in another country. This will also be a main element of this court, to bring those businesses who practice these thefts to be given harsh penalties, including jail if needed.

16. SUBJECT: HUMAN WASTE With proper disposal of animal waste

Our waste, in general, has not been disposed of properly in many areas of the world. Even human urine and human waste could be controlled much better. Let us consider the fact that we use fresh, clean water inside our toilets to move waste from our lives. What are the best ways to dispose of it? As it collects, the mess is then dumped many times into our freshwater lakes to empty the amount that has built up because the processing to clean the waste has been overloaded. Images of these actions have been seen on Google Maps from time to time and are even more evident when signs are posted on beachfront areas: no swimming allowed, high bacteria warning. The people who decide these actions have zero regard for the damage they cause to plant life and fish life. So what we need is a way to distribute waste to growing plants that like high acid, for example, evergreen trees, blueberry bushes, rhododendron bushes, and hydrangea bushes. Evergreen bushes are to be a planet in rows that have enough space for a tanker truck to drive down the row. A light mist of urine is sprayed in the air as the truck drives at a speed of 10 miles per hour in order to not over-saturate the ground. When the truck has emptied its contents, the driver marks the area for the next truck to start down the next row available. Waste is also taken to some areas of forest pine only for urine and with the same method of distribution, and desert lands could receive solid waste that only helps the sand become potential soil. Solid waste is taken to forest areas and dumped by helicopter from distribution tanks that spray the waste between trees only on days when wind is not present. This will be more than welcome for trees, and nature will do the filtration necessary so as not to disturb groundwater. This is another reason that undergrowth will be extremely important around trees planted. The use of this waste disposal could be a great way for tree farms with areas of rows to be sprayed with human waste collected directly from homes and businesses in the same manner as septic systems that companies come and empty. The difference is the tanks are not pumped, just picked up on flatbeds, and forklifts place the containers on flatbeds to be taken to dump stations located in the center of each county; the tanks are replaced with clean tanks every year to maintain that there are no leaks and the hook up is large enough that it can be easily moved when full. There is an indicator that rises outside, which is a full stem which when the buoy inside is lifted this shows a red bar comes upward, which says "PHONE" for pick up. Both urine and solid waste can be removed in this manner, but there are two separate tanks for both. I have designed a new type of toilet for in home use which is not water based, it is a heavy paper towel that is connected from below and is clipped on an electric wheel that when the person sits down if it is urine ,which is located in the front of the waste catching toilet seat and is an oval cup like funnel for women to use , for a man only standing there is a larger cup that swings over the toilet seat to click into a lock that the person can now urinate into this funnel which both funnels have a spray that activates after use which the spray will enter the urine area and is a fragrance also that is pleasing, the funnel being slide back away for the toilet seat is how the man's spray cleaner is turned on, and when the woman of man sitting is finished and get up if seated the inside of the urine funnel is sprayed with a clean water with natural bacteria killer peppermint spray, the hard waste is caught by a toilet paper similar to the size of paper towels on a lower roller and this paper has small holes on both sides that fit into extended wheel and on the extensions that fit into the holes as to pull the paper alone to the predetermine location done by a system computer chip which designs even when the paper may be require to add more paper by a weight difference, those paper can withstand even slight urine being on it to not break but an auto slicer similar to a gelatin cuts the paper and then travels to the dump area, but this paper will decompose outdoors, also the use of toilet paper is to wipe and that is also dropped on the larger paper roll. The user can make a lower paper roll out further if needed by a push down button on the side, similar to an older design with the handle, but this now moves the paper to the next contact area. This underside paper roll travels away from the home, and the paper has a large round slicer that cuts into the paper in a grove under the paper which cuts the paper, but the activation of the paper roll movement to be cut is always when the person gets up off the toilet seat, the roll now turns by the electric motor, sending the waste to be removed and dumped into the waste hopper after the paper is cut. I realize that this has been a crappy subject. Another solution to all waste, including animals and humans, is not to allow the waste to drain into any and all areas that could carry it into our creeks, rivers, lakes, or oceans but to control the distribution of waste in a very exact way, which gives animal waste a father distance required from any water source, and areas needing vegetation, urine is a very acid-base, and ammonia is the main reason it offends and can even burn the surrounding plant growth in very large areas, so this is a very hard element to control normally, but if farmers with animals used water through with a water hose connected and placed far from creeks and any drain offs say 50 yards away and plant enough grass to absorb that waste, that would be a good distance that could improve our waters. Then the livestock would get what they need, and water from a well is healthier if the urine isn't getting down in the ground as much, and so the watering of horses and cattle, and pigs, or any other animal would not be having them urinating into our creeks and river beds, and dumping waste that you as a farmer would not drink if you were downstream!. But we are not staying with the expectations from the past as normal; let me just say that I have witnessed many times from walks down a creek to see urine foam rotating in an area where the water was slower. This is not acceptable- No, we will use drains under our cows, under our stalls inside, and these drains will allow the urine of animals to drain away from hard waste to be also carried away weekly in pick up from the local waste management, and allow it to be delivered below the ground to storage tanks that can be lifted from their containment made of concrete, These containers are allowed to fill 80% of the water and 20% of the water is added to dilute the urine and any hard waste that may have entered. The truck that picks up the filled waste tank first takes off the empty tank clean tank by backing up and pushing it to a loader area, which has a sliding metal groove and has holes in this plastic container to connect snap clips attached to chain link, and after the filled tank is loaded onto the flatbed truck , then the empty tank is then attached like this: A hydraulics lift the tank from the underside which allows the snap link chain attachment to be disconnected and now can be undone, the filled tank is able to slide onto the flatbed with ease because the tank sits on roller rails that with a slight push the tank will slide down onto the back of the flatbed onto it's conveyor rollers to be able to have more tanks loaded to the flatbed. Then the snap lock pins are connected to the new empty tank, and the filled tanks are locked down with chain hook systems to keep the tanks from moving. Released under the once suspended tank now sitting on the back of the flatbed, the truck is moved forward, and the new clean tank is now to slide outward on rails, and is ready for use. I almost forgot the open and close valve needed to be closed before the tank was changed. Boy that could have been messy!

When the truck lowered the tank onto the flatbed with braces on the sides made of rubber and attached firmly to the all- metal flatbed with conveyor rollers that actually fit to the bottom of the tank with grooves to keep the tank stable. Now our filled truck leaves with its container filled. The tanks could even use come-along chains, but a single hook holder at the top would be required, and that would be a metal hook with metal straps to maintain the potential weight. Hey, we almost missed telling you about our friend with the full waste truck! Well, he may have a serious long trip unless the farm is close to a desert. He is heading for outside areas in faraway deserts and heavy brush or slightly wooded areas or open fields or where land fires have been. His load is going to be parked near these areas to be used by large fort lifts with rotation dumping, or he may make two other applications to dump this waste. One is to drive over the land designated and with a powerful pump screwed onto the lid area , he now starts to drive over areas on land and turns on the spreader pump, and an additional hose can placed into the bottom of this tank on the side which now he can direct spray the waste around like a massive fire hose would do, and he does not just allow the waste to be in one spot, he moves his truck about three times to make sure the content is well displaced in a large area, once he is done, he now has an upper tank that was filled prior which is only a water tank which he now dumps into the waste tank to clean it as a final spread is to clean and draining inside the tank like a sprinkler that washes the inside of the tank automatically as he waits for the water tank to empty half way the water is swirling around inside the tank and washing away all contents down to the pump, he empties his entire tank and now closes the tank drain valve after he has washed the outside when removing the sprayer. He is now ready to pick it up at another farm or deliver this tank back to the waste management location. But before he leaves, he places a plastic sign at the area he started at and the area he finished; this is for the next driver or even himself to know where not to empty his or their next load; the sign also has an area with a permanent black maker places the date of the last spread, The sign at the beginning points to the right with a red arrow, the other side points to the left also with a red arrow, this signifies the area that was competed as finished and used for waste urine. The use of waste by farmers that has been a standard for many years is that the cow waste is spread on crop areas to give the soil back some nutrition and grow the crop with some natural actions that are nowhere near normal, as the farmers are also spreading a massive amount of urine to be included into this mix, so the crops fields are becoming more acid based, and they wonder why their yields are less? So, what farmers need to realize is that spreading raw mix is actually harmful over a period of time. This next subject is one that is mostly the result of forming waste, but in this case we are exploring a better way to make the needed elements that sustain life. So to avoid my embarrassment further here is:

17. SUBJECT: FARMING & FOOD

We all know that pesticides are an issue of much illness to many living creatures such as birds, and including people. But the good news is there are pesticides that actually repel insects that are not chemical in nature as manmade, and of course the big chemical companies don't want farmers to be aware of these items, let's just say it would be less profitable, so they make every effort to push their products, as if the important fact is to have a larger yield, and more product to sell- but we do have alternatives and are simply a plant grown inside other planted areas that repel certain insects. Now this may seem like a great fuss, but the FDA should have some idea that these elements are real, yet are they asleep? Maybe paid off by big chemical companies, but the farmer is looking for easy methods that are cheap so they can save money and time, oh these elements are needed, so why not have after planting the proper plants that can force insects away. This could be small plant holders on Sheppard poles placed between rows and close to the ground, with a group of bug repellant plants to be formal defender of the crop, yet takes only a flatbed to distribute with tractor the plants , one person driving, and one person putting a pot down every 50 feet, this could be placed outside the field food yielding area in some parts of the country, year round, so plants like lavender, Lemon grass, Peppermint, Rosemary, Basil could be planted as they could maintain a barrier that insects would hate, yet leave a very pleasant odor for the farmer and even the neighbors would be pleased, and I have watched the use of the pesticides being sprayed, the fact is many times the spray is coming back directly at the driver of the tractor, and he may be in a closed area , but it is coming in from the fan cooling if that is used, but I have witnessed many times the spray floats right back to the farmer, you cannot tell me this is healthy for any farmer, well, unless he likes a shorter life span, being the spray is actually a nerve agent in many cases, and a person's heart works on a nerve bases so hopefully those farmers who have been effected will change their direction for all of us who are indirectly consuming food with these harmful killing abilities, and we wonder why we have premature babies ? Skin cancer and illness after illness, could it be because what we eat is being approved but is infected with poison, I am sorry but the only answer if this is possible is "well Yeah!" (Sorry) But, let's just think for moment, if these very strong chemicals they have to be extremely strong because many bugs have become resistant to the pesticides, so isn't it only common sense that these very powerful chemicals could kill us over time? So my mission here is called "HEY wake up! This segment is to present potential solutions to our "outdoor" farmer. Needing plant vegetation to grow better and have better nutritional elements added: Have you ever heard the old saying- "If it isn't broken don't fix it?" Well we have been given a very serious way to make our crops grow very strong and very healthy, but we are going to go a bit further, in fact much further. Many years ago, one of the best groups of nature based peoples where in America and actually understood nature, they were the native Indians who I stated earlier did use fish when they planted crops and they understood that growing was only half of the need for food, but with this fish gave the food its nutritional needs for the human body. But we can make this process even better; we have a natural element that is common and extremely abundant. In a season of "fall", we have the leaves that have nutrients that enrich the trees, and other plants as they decay, so why not take these leaves to farmers, rather than to some waste dump, but "WAIT!" what is the better thing to do? We take the leaves, and branches and grind them up into a much to be made at a processing plant that mixes the shredded leaves and large shredded dead wood from swamps, and inside high growth tree areas we could revive broken tree limbs, and collect these to be processed, and these shredded materials are 30% leaves, 10% shredded raw wood from branches fallen or trees dead in woods, and 30% shredded dead fish that is also collected from ocean shores and lakes and after major storms by crews with four wheel travel and a pull behind trailer, that toss the dead fish into small trailers that they take to be given to drop off stations that will take the items to the processor plant, volunteers, retired people who have an interest in helping the farms have better nutrition, these people could be used to help the American food quality improve. The fish sent to the processing plant is mixed into the waiting leaves and mulch and water is now added to the mix, and this is now spread on the farmers lands, the farmer pays a small fee to have his crops grow with even greater yields, which is the main way to make more money, so this would be a great business for someone to start. We also have a major problem when cow waste is used to cover the growing plants such as corn, or potatoes, with this we gain,- let me rephrase, we don't gain, we have added a great amount of cow urine in the ground mix, and this high acid in cow urine also drains into ditches and these ditches are connected drain areas to creeks which now this waste reaches further away areas, and if you travel to many creeks in our nation you can observe the foam that collects from this waste around areas inside creeks which are visual reminders that we are placing pollution in massive amounts into fresh water. Not to change the subject too much, but- This new problem is not from the same substance, but our creeks and drains are seriously affected by these elements also. Oil drained , you've seen the substance on a rainy day, a colorful mix floating on the water puddle at the gas station, so colorful, yet so harmful, and we have drain ditches everywhere, do we really think that a few billion vehicles dumping just a few drops here and there wouldn't affect our water , and our food inside oceans? Our drainage ditches on the sides of roads are connect to creeks and then to our fresh water lakes, and yes oceans, is it possibility of the reason we have many of these lakes and wild life dying in massive amounts from chemical pollution, did you know that octane booster you use for your vehicle contains lead? Lead is found in the fish we eat. Big business has convinced us that eating small amounts is fine. We live in a world that allows the food we eat to be with an agreed amount of harm inside! Seriously? The reason this book is happening is to give us back some of our common sense and some value of pride of what we offer to others, so those associated to FDA, "please reconsider that lawsuits could be adjusted to those who have caused a great failure in a person's consumption of what you approve". Simply the way to stop this drainage of chemical is to have certified filters that collect the lead elements at the end of creeks and rivers.

To those who believe they are immune to legal actions, you need to look hard of all those people who depend on your ability to feed them properly. They could be the same people infected that cause a groups of individuals to unite to sue you directly, all people have to do is show a receipt they had your approved item, and it caused someone great harm, or even death. And let us go further, we witness all the time so called drugs that are the cure for some person's ailment, but can cause a multitude of other problems by using their substance.

I think it's time your standards of allowing foods to be filled with things, and drugs that can have a back lash and are even potentially even considered slightly should be removed, why on God's green earth would we want an item that could possibly kill you just because it could remove another problem? Insane people! As we see all these items approved by FDA, it has to be benefiting someone's pockets, and most likely done in gifts from large corporation manufacturers. And the financial kickbacks are pushed under the table, which is kept as low to attract attention as much as possible, but those days are ending, because this is your wake up call, and the public has been till now, sleeping, and sheep, but remember those who lead, shall be seen when followed. What you do will surface, it always does, so be mindful, the people will be at your door one day, and you are in a position of accountability that will rise. To the FDA, we need your expert ability to return to the respect to those who follow you. Hear this before a great wall that you protected yourself with- will tumble on those who built it. This is no threat; it is only what will happen if your standards remain so loose as presently. Come back home to us FDA. And to everyone else, it is time to contact your congressional representative by letter, date it and ask for return receipt, but be a voice that says I will not be silent. It is time we remove the elements of steroids, antibiotics, and harmful chemicals that infect our foods, and our health. We want clean foods, clean, and with nutrients, we do not need your fake food, artificial food flavor, and your chemicals that harm us. And a simple food treat with around 30 chemicals and every color additive known to mankind, and with preservatives that out last the Egyptian Pyramids.

18. SUBJECT: THE FOOD HARVESTER TOWERS

Some added ideas: As stated, the future farming can be with less land mass used as in a wide area of possible hundreds of acres, and this system is a more productive manner to concentrate the food development on different levels. We all have heard of , or seen a green house where plants are able to grow and if temperature control is used, in some areas they can grow year round, well this method will be the future, but let's say it is a newer design with a greater contribution to the following: There are no more seasons required, the elements grow year round and are planted to have a harvest ability weekly, and the structures are about the size of ten football fields for each product, some buildings are over one kilometer or one mile square, and there are four separate floors, not saying there couldn't be many more. But all food production structures accommodate whatever the main plant grown is, the complete project is automated and timed as when anything is done, as water is placed over plants on a timer, the timer is controlled by the computer, the program tells the overhead pipes to spray like sprinkler systems that are controlled when to turn on or off. The ground of the entire floors is just like massive trays that are plastic welded to make a perfect containment for the row after row of giant plant holders. The ground is brought in from some of the most fertile areas, and mixed into massive batch mixers with more nutrients than what ten carrots today may have, fertilizers are blended prior to the dirt brought in, and the richness is a black dirt so powerful that plants are boosted to grow larger than possibly ever recorded. The ground is tilled with an automatic set of ground tiller blades that are as wide as the entire open areas of the rows, with single push of a button, the blades are being pulled by electric motors, and chain link pulley are now lowered automatically down into the ground cutting small paths for seed to be planted, as it travels the seed is now being dropped behind each row which has just been made, and the seed is placed with a divider wheel that allows an opening briefly to drop one seed in place at a designated distance to allow the best growth, once again automatically. Now behind that area is a paddle that now covers the newly planted seed with the dirt that was opened in the tray. When the system has completed the planting process, The natural lighting is opened, this system of lighting is not like any seen, it is actual mirrors that reflect light into the building from the sides and the mirrors can point to reflection points that illuminate over the plants as each plant is surrounded from all sides in a bath of light. The outside mirrors reflect sunlight during the day but at night-the plants have what is called- a rest period which is short just six hours without light, but many growers refuse to consider that plants are just like humans , to grow best they require sleep, their sleep consists of removal of C02 which has been consumed. So yes, it is essential for these plants to make something we consider harmful. But that is why I have invented the C02 removal system for this structure I believe will be allowed to remove Co2 at these food tower plantations. Now we have in this massive greenhouse structure with another process, the timed planting, the area just planted, is in section we will call- "Planted week one" in section deemed section 1.). We now wait another week, and we now turn on the system above section two, which has been prepared as section one, with newer dirt, and nutrients added. This application is done over each section, and this is please remember for just one type of plant crop. The process can be for any type of crop, and each floor has the same ability. As the demand for food increases by world population if not controlled, then the structure is also designed to have added floors be made. Wait a minute! Who picks these plants when they are done? And what about the waste plants after they are picked? Good question; we are today with robots that can do fine technical jobs non-stop, so we will have on both sides of the rows robots along with people who may need work; these could be our pickers, but robotics have the ability to maintain a continued operation as no rest is required, so productivity would be higher. But the more I think about it AI has already taken many people's jobs, so let us use real people at least most of the time. To make jobs to pick items that are fragile to be placed in plastic containers, and set of a stainless metal conveyors that takes the items to either to packaging or to trucks to be delivered to packaging. Each row has a direct convey or exit opening that takes the items past inspectors to be packaged and placed in a cooled warehouse to be delivered. The food is taken every so often to be examined by professional equipment for mold content or any harmful content; if found, the line is shut down until the area has been removed. But could this happen? I have failed to tell you that the areas of this food manufacturing plant are built so as not to allow the following: Bugs zero. Why? Because the entry of everything is inspected, even the dirt used is inspected as samples are taken to the lab to check the content and nutrient value. We store dirt in warehousing for three weeks. Hanging bug zappers attract bugs from larvae hatched with massive amounts of highly concentrated UV light attracts the bugs to zappers located surrounding the structure and have covers to close the lights when not in use. This dirt is not just any dirt for growth; it is dirt made of a high concentration of creek sediment, dead leaf mix, chopped wood, wood taken from lake water and creeks, all wet wood, and river sediment collected, flood sediment. This dirt is possibly a combination of the richest nutrients ever produced. But made in a quantity that is industrial and speeds out, and it is allowed to ferment like fine wine. The temperature is stable enough to allow any termination; after three weeks, the dirt is then inserted with electrical prods that are attached to retracting hinged metal containment's that hold a series of electric prods in place as they are lowered into the dirt from both sides and inserted into the dirt, the dirt sits on sheets of stainless metal with no drainage as the overhead sprinkler system wets the dirt slightly, and is moist, the dirt is electrified to kill any larva that has not hatched. Inspections even include seeds used for quality and size. The personnel all wear plastic suits with air ventilators that are like a science fiction movie, why? Because people carry molds and diseases, we also protect the workers from excessive UV light and any contamination of food that needs to be perfect. So, there are no pesticides and no entry of contamination! The structure is closed with a three-chamber entering process; the first entry is a UV light bath of anything coming in, including even a fort lift. The process to kill germs from every direction. The items are inserted into this warehouse area, the operator clears the room of personnel and the lights are turned on, the operator is behind a dark screen, and a one-minute UV bath is activated. The cameras have an AI system that uses robot floor bug killers, and the system is in search mold of find and destroy flying insects; there is a wall outdoors. UV zappers and even ground zappers about 50 yards from the building, and surround the entrance areas. The floors are made at the entrances with grids that have the ability to detect crawling insects- and electricity is in the wire areas that will zap them just like a backyard hanging bug zapper. Also, there are sound deterrents that are to keep flies and other insects, including mice and rats, away from the structure and surrounding the outside areas in circled protection; even birds are not allowed near these structures, as the birds will gather to eat bugs outside the structure but a fine metal netting that is welded and surrounds the entire building top to bottom with only one area that opens, and when it does open the area that allows food containment to leave is placed on a rail car that is a short run outside the main structure so the rail has the track embedded and is inspected daily by maintenance to make sure the rail is free of cracks of any type that may allow insects to take refuge or travel under it. The rail is with white pavement painted around it to detect anything on it. The use of bugs will be the last area of defense used, and to farmers who have regular farms, this could be the answer to bug destruction around farmlands; imagine a system that detects bugs with a detection system as an identifier, and then vacuums them in as a rotational blade that chops the bug up and kills the bugs. Okay, back to our Tower Farming structure:

The contents are delivered to waiting delivery pickup trucks that sit at loading dock platforms that the trucks back up to and are lower than the fork lift operators, and automatic metal platform is extended outward to enter the truck bed, to adjust the height of a trailer bed, under the truck is a hydraulic adjuster that can make the entrance a smooth level entrance for fork lift operators, because the actual back of the trailer back wheels sits on a hydraulic lift platform and can be raised and lowered to be a perfect entry. The forklift drivers are outside under over hang structures surrounded by plexiglass walls to keep out bugs, and to keep them dry in wet conditions, and all pallets used are plastic so not to have a wood that can grow mold. The entry points to all parked trailers has a clear plexiglass sliding door that opens by movement and similar to an elevator door if anything is blocking the entry the door stays open, as soon as the fork lift driver clears the entry, the door shuts from both sides, if a fork lift happens to hit these doors by a malfunction of the doors operation system, the doors have swing hinges to allow the fork lift to still go through, but both from and back of the fork lift have fitted tips of foam on forks, and a foam cushion on the back to allow the fork lift a push entry if needed which does not damage the plexiglass doors. Truck delivery drivers are not allowed into the loading dock area, but report to the outside office to get their delivery paperwork and destinations. The electric cart rails to each truck trailer bay has a pick up point which when the cart is used there is an automatic door that opens and closes also that is an extra way to keep out bugs , so where the food was placed on this electric cart by an inside hand fork lift, it is

picked up by the loading dock forklift driver, at the end of the rail, this cart automatically moves from inside to outside with a pressure weight that makes contact to go out, and when the cart gets to the other end the cart when emptied now the weight make the cart return to pick up again, so loading is inside loader to cart, and outside loader to truck trailer. This system maintains a flow of items coming from warehouse by inside drivers is faster, as the distances are decreased, similar to a bucket brigade of the past putting out a fire. This method of harvesting is done each weekly, and areas that have been harvested and cleared of food, are then given a new ground to be mixed in from stored dirt, but this time as it moves back in place as previous there are large round blades that now chop up the remaining vegetation to return it to the soil, and this now starts the process after one week of the ground being tested, and improved with a hard fertilizer and in this fertilizer are chemicals magnesium and Zink which vitalizes and enriches the soil. That the result is to be absorbed by plants and be the better food needed for a better world. Then the new seed is planted, and the process start all over again.

The present day farmer has had a horrid playing deck dealt to him yearly, he has been told to make more for less, and less should make more! The average family farmer has issues of corporation take over, this is where a large corporation thinks they understand farming better than the farmer, and they buy up small farms with gaining of the new business they had but now consolidates the resource to make production increase as their main goal, but not the quality, and never considering the farmer who made his dream as to have a farm and be his own boss. These money minded corporations do something that was never done many years ago. But is a normal operations practice today. In many of these different food producers from milk to chickens, to beef, and many other foods run by large corporations. These corporations have convinced the FDA to give them a green light to actually give a cow artificial means to produce more milk, give chickens drugs and once again artificial means so they grow faster, but never consider the harm it could be doing to the people, or to their children; the consumer is now attaining drugs and hormones, antibiotics, growth hormones as steroids, all going in our food.

The corporations pay off the people, who make those decisions that are opposed and instead allow those chemicals and drugs chosen into the developments, but they won't tell you that, no, it would mean their job, possible jail time, and probably a long jail time, and they themselves most likely eat only natural foods because they can afford them. So what I propose is we as people get on the back of these people, make a stink, and then find out who is using these systems, find out and boycott their goods, and find natural growing food to be delivered to your home from online companies, yes the cost may be slightly higher, but the result, I think you know will be better for you and those you love. But this problem has been even driven by the very people that run our country. They can pretend not to know, but remember all things agreed must be done by someone in power, and they have "not" showed any care of our farmers who are forced to take these actions or be left with a low profit they can't make ends meet.

Farmers left in the dark who try and develop good food and good milk are now selling their lands to big industries and forcing them to sell by lower development costs. And many farmers who lived for many years and supported their children have gone to the bottom to sell and have that good farmland be taken by those who grab that land for making a housing project or some new business. Oh yes, friends- it seems like the song by Joan Biaze has come to life, "Save Paradise Put up a Parking Lot," which was a song that made jest that man today destroys the land and makes quality farmland covered by industry and business and homes, and parking lots. Farmland is being devoured away from the average good person seeking a bit of the American dream. Farmers have been selling their lands not to farmers, not- to people building new homes, now on the fertile land that could grow crops, but now have four bedroom and three car garages sitting in the middle of what was farmland. Land that once made quality food for all of the world's population, is being overrun by things that shouldn't be there. And why I am suggesting food be developed in these ways.

19. SUBJECT: FOOD PRESERVATION

I won't apologize that this subject has surfaced again, because we as a human family should care about what keeps us healthy. This said, one of the issues that reflect to farmers is the control of the insect population destroying their crops. It is a little-known fact that tobacco can be heated in water to be sprayed on plants and has been a home garden application that repels insects; this is a natural element that also is a dying product since the awareness of the ills that tobacco can bring to people and in many countries for the bad publicity that is connected to smoking. However, with this application still be applied as an effective pesticide repellent. Now this element does have potential issue, it is the rainfall that could wash the tobacco film off the plants, and then reach into the root system, but this could be tested to see if there is a problem of nicotine entering the plants. In this case, the concept would be dismissed as a hazardous chemical. But, if the plants don't absorb nicotine. Toxic bug repellent. But no worries, there are many bug repellents that are natural, we only need to use these instead of chemical compounds that effect but our water. Drain off from farms and roads: One of the major problems we have is water pollution, and the preservation of freshwater supplies is extremely essential for the future; what we need is for all roads that have drainage systems inside our city streets; let's take a closer look at exactly what these ditches do, it's a rainy day, and look at the oil film so colorful being washed down the drain and to include ditches next to the highways. These items are all to route water from rain directly to a creek or a lake; not one is filtered and is located in distances that are many times alongside of a waterfront being in the past, travel was connected to waterways, and to be close means where drinking water was also available in the past. But now the runoff from farms to close to these ditches and runoff's is everywhere; these water run offs are way too close and do not give the natural filtration ground absorption, which, if there was a wider area around growing areas, then the pesticides could be potentially reduced before entering water drainage, but let's change gears just for a moment, the pesticides they also drain off with rainfall, so this means there is an absorbing of pesticides inside the plants you eat from the roots absorbing these chemicals, and they will try and say that this is not so. But this only makes sense that anything in the ground would be absorbed by roots. So this also means that you're eating food with a chemical that can kill insects of all types, and how do the majority of these insect repellents kill the bugs, we have learned that the chemical used most actually paralyzes the insect by effecting the nerves system, so you're thinking this wouldn't effect So what is the solution? We can look at two: stop using chemicals and use natural plants planted around farms that are well known to repel insects, which would need to take more time to place around a farm, but in the long run, saves the farmer buying insect chemicals yearly, and could be a massive saving, (more profit) There are many insect-repelling plants that could be planted that return yearly as to re-grow. Second is where if the ditches had the lower end of the water run-offs, a new system is placed; this is highly effective because what it does is actually filter the insect repellent and even oil and grease runoff; it's a cradle that allows water to pass but absorbs contaminants that are in the water, these are cotton based with a heavy weave but allows water to flow through to capture 60% of the elements, but there are three not just one, and they are in sliding trays that these different sized cotton screens can be adjusted like a child's safety screen which slides back and forth to give a proper distance, then we have these units removed every six months, and we do it mainly after the crops have been picked to gain the maximum removal of pesticides. The screens are placed with large rocks that absorb oils. These could become a filtration system also, by lengthening the distance the ground itself becomes a filtration system and with this application and the reduction of fuel-based vehicles, less gas will be traveling into our water supplies, but oil is still involved in lubrication, so the use of these ideas would still need to be maintained. Now you as an official of a road crew who see's, as an official of a road crew who sees that would take some effort, I say, is it worth saving your family from eating fish infected with lead? Because this is exactly why it is happening. But let's look at the advantages rather than the cost, but wait! The cost, we could make the projects funded through tax dollars given to the forest and wild life workers, this means they are responsible and maintain these filters. Sorry guys, but the need is great. Be on board to make a difference, and it could make even your life a bit longer and healthier.

20. SUBJECT: WASTE REMOVAL-PIG FARMING AND COW FARMING

The main element that is harmful to our lakes and waterways is that many farmlands are near runoffs. Even highway run-offs the drainage from these farms causes the pollution index to soar, so in the future, waste and urine made from cows and pigs are to be drained into collected areas, which will be taken to open wooded areas where there are no trees. Still, the sewage will be taken to deserts and lands that have very little vegetation but mixed with water [20- 30%] to dilute the acids. The sewage will be sent to areas without access to creeks and five miles from any drainage areas. The reason this is happening is simply a need to revitalize our water and natural water filtration's. The removal of the acids will increase fish hatching and pure water consumption from areas that still use water from lakes. The fish will return stronger, and this will increase the number of lost species that once flourished. The cost of hauling the waste is not as bad as people may think. Each state will have its tankers with the ability to have the tanks be removed and placed on rail cars, so this means the trucks do not travel clear across the country but make short runs to what we will call DUMP STATIONS: there the rail car tanker holders are driven into the desert areas which are designated each month of what areas are next to be filled. A monthly calendar of the transactions rotates the placements and always monitors the soil and any water that is within a two-mile radius of the drop sites. All waste that is harmful as hazardous waste is not placed in these areas because these areas in the future will be used for new farming after they have been soil tested as potential growth from anything bushes to trees, the items with high resistance to acid, which can be used to build an ecosystem, and made sales of plants to people's homes, flowers-etc.

We even had an excess of food at one time as we sent food to starving nations. It's almost a thing of the past, and it will soon be no longer. And this is the most absurd thing that big businesses do to keep the cost of goods up: they give a small amount of money and tell farmers not to produce this year! Are they monetarily compensated for not growing food? This policy is the worst thing I have ever heard, so if the U.S. has two or three years, many will die of drought, and the country will not have enough food stored to even feed one State for more than a month. Then, we could watch the turmoil as people fight to get a loaf of bread. If we paid our farmers what they deserve, the federal government could develop the excess to be stored as we have learned how to freeze dry food, and dehydrate it to add water, this should be the way of the potato, and the corn, and many other things like milk, eggs, the past should tell us that to store food is many times essential. We can look at a time called the Dust Bowl of Oklahoma in 1935. People starved, and an estimated 7,000 died. Now, with global warming happening, this could be 7 million. And we have another not-so-known fact- Did you know that all nuclear fall-out storage areas are no longer supported, no food, no water, and no medical? But, stored food could have an impact on saving lives if they cared. But they have direct tunnels right under the white house to be safe in the event of an attack. Get this: they actually believe they can survive a nuclear attack if they use submarines to live in; that is very stupid, being the entire plant would be frozen if an attack ever happened from the dust blocking the sun for many, many years, but they are so smart, so freezing to death is still death. Hmm! Smart people right?

I say if you have any courage, March on Washington D.C. and insist they build in every State storage unit that can house food for the day that we have when there is no food. Even the Egyptians knew this: are we no smarter than they were over 6,000 years ago or more? The fact is we need our farmers to be helped to maintain a surplus, not paid for not growing! The people need to act now before the country has a major problem. Now, they think because we live in a global economy that, we can always gain food from some other source, this is false hope and a false view. What is to say another nation cannot have a drought in the same years? These are the small-minded men who follow these roads and need us to wake them up. And many food surplus has been taken to give away to needy nations, now some of it is given to Americans; and I have no argument that we should not help those who need help, but there also is a saying. Before I clean your house, I should clean mine first. Or maybe you've heard- Scratch my back- I'll scratch yours, but- let's be frank: when the U.S. needed help during the hurricane storm in Louisiana hit named "Katrina" with thousands of people displaced and needing help- not one nation came to our aid. Do we really think a national disaster with no food- that other nations would just give us food? I'm sorry, but you may be eating dream cookies if you think we have any back up from others; we have trouble getting funds from other nations to even help us protect their territories. We need to consider what has already happened. Not be fooled that we are loved by everyone, not many countries would love to see America standing in food lines. My feeling is we need storage buildings built that are weather resistant to high winds and moisture and made with airtight separate compartments with upper roofing to keep them cooler like an umbrella would do and have them able to store enough food for three million people per state for up to five years. This means concentrated gallons of orange juice, grape juice, and every other means of drying food shrinking the contents and keeping the most needed foods, starch-protein, vitamin C, and grains. Then, once we have these in place, we need them to help other nations build their stock and learn to cycle it. The fact is we do need to replace the older shelved items over a period of time, and this will give the farmer even more reasons to grow. FEMA is a disaster support established for the United

States, they have food support warehouses; how many, you ask? How about eight (8)? Gee that should handle a national disaster for the estimated **345,815,485** Americans. But we can at least feed the 16,000,000 illegals allowed into our country thanks to our great boarder supervisor Harris, I think we need to take what these people cost the American taxes payers, and have it deducted from her pay check weekly. She'll be living in a card board box under a bridge. (Do it.) Farmlands need to be brought back, so building structures need better design approval, and we need to increase structures that last longer and support mankind without taking valued land mass. The issue today is that the land mass being used for food production is dwindling, and commerce construction is increasing. As mankind continues to grow, a business needs to be built, or- a new home is required. But more so is the fact that is population also grows, then and so does the need for food. I hope this domino session is grasped because the curtain of our world can drop, and if we as its landlords don't pay attention, that curtain could land right on all of us, and it's made of death. Solution- First, we have made the greater skyscrapers that enter both directions, as into the ground to make foundations stronger, and that connect at points by sky walkways, or sky travel electric vehicles on the electric rail, outside street elevators that can access all floors, and these structures can include massive homes sky-high giant apartments able to be for a growing family. Still, I will get into that subject later. With a planet of ever-shifting ground mass being used for structure after structure, that has one thing in common: they are built not to last a couple of thousand years. Instead, they are often placed in areas of convenience, with land that is unstable as foundations. We have to truly worry if those structures as glorious to see from a far are practical; the truth is no, 1000% no, they are not practical because of the great harm they can cause when they fall, and will they fall? Well either by man himself, such as 911 showed, or nature, which makes man's devices look like a house of cards. When she decides to move in a wayward direction, the event can be divesting, such as the San Francisco earthquake of 1906, which took about 7,000 people, but China, in 1556, had 830,000 killed by an earthquake, so recently, we have had earthquakes in New York City, none of these structures support each other if a major earthquake happened along the east coast of America, the death toll would be millions because the structures are not designed for earthquakes in a whole. The truth is sometimes mankind goes in the right direction- but- does not see the big picture and know when enough is enough, so they build and build yet have no vision of possibilities, and when they do have that vision, they look only at one structure. It's like the child who says, please, can I have more but can't eat it all.

The concept of space being stacked is actually great, but the height should be considered for safety reasons; simply, if something is hundreds of stories tall, that means in actual feet, it has the potential of laying downward those almost exact distance and could cover a great deal of territory, which means a great number of lives are at stake below. We have wondered why the pyramids of Egypt were constructed in the manner they are and how these structures have lasted thousands of years, unlike the structures of today that last possibly 50 years and then are demolished, wasting the original materials that made them and truly wasting man's energy and cost that it once was to make the structures. Now let's take a hard look at a Pyramid; the structure resists weathering because the wind is allowed to roll away in an upward motion rather than be in direct contact. We also can now realize that in the event of an earthquake, that if the structure fell, oh my- it would fall "inward" which means that others outside the structure wouldn't be effected, t. This doesn't save, but people outside would not be harmed, and they would be alive. Even inside these structures, they have support that exceeds any structure built to date. Also use gravity to maintain structural strength if designed to lean inward. So, in fact, the design has the potential of saving multitudes of lives, and the proof that they do have longevity is what we say is in the facts or "In the pudding," my mother's quote for proof being realized. So building structures in a pyramid style should be our goal as practical and consider items of high wind reflection, heat absorption, and heat release, earthquake structures that are connected to other foundations of other structures, similar to a linked connection that now adds greater support. Also, in the past, they made them of materials that would be stronger, such as solid stone, and that is why they have resisted the wear of weather from father time greater than any structures today. So, the conclusion was to make buildings that can "only fall inward" during an earthquake. Second, make the structures movable at their base in such a way as to allow the structure to rest almost like being in a slightly curved bowl at the ends of the base. This way, if the structured has a concave design this allows the foundation movement and the bottom of the structure is made with large ball bearings that rest in steel frames that allow the structure motion. So that even during high winds such as hurricanes and even earthquakes, the structure could be stable and come to rest after the motions or the winds have subsided. Also they could have drainage flow under the buildings to allow flooding to decrease and decrease the chance of water damage to the structure. It would be a smart application to be able to maintain the bearing system from time to time with a good old-fashioned oil dip stick to monitor the oil base. To conclude, this structure could be of a cone shape or a pyramid but just wider and much taller. Being a cone stabilizes the foundation and allows wind reflection to be maximized. The future city could look very much like this for all these practical reasons. Coned sky scrapers everywhere with sky way travel centered as to go from one structure is connected to another not just below but from above an even rail system that expands to all structures right through the center middle to allow two types of travel, station stop as to each building and just enter your destination address, and the system will notify you directly where you sit that your there, and is semi-express which is to travel in any direction north, south, east west this rail is under the track of the regular rail but stops every six city blocks, and the passengers change rail to their direct needed. These cone-shaped structures would be bigger than any of the known football stadiums of today; this could be the future of high rises.

21. SUBJECT: DRUGS AND ILLEGAL DRUGS

To be aware of the problem is not the solution, but this is:

Let me go to the point directly: illegal drugs are grown; the way to stop it is to find locations where it is grown and destroy it, then you have no drug; in short, to prevent anything we need to remove the root of any problem. So who has the seed or the root source? So how do we enter countries that support these sales and exports, we hire people inside their countries who gain big money to do everything to destroy the crops, and they use drones to do it that we send them in ways they can build them, and have the use of items that they can attain even in their local hardware. Before I jump into this subject, let's consider the so-called legal drugs which have been even a bigger thorn. How many known celebrities have we seen die from prescription drug use? Way too many people. Do we have regulations on drugs that are potentially addictive? Or can kill you? No, the powers that be allow the FDA a loose rope so that many of our political protectors just get kick back after kick back from lobbyists who push a drug through as to be legal. Is this practice right for people? No, it kills people many times, yet not one person seems to care. (why?,money is made.) Personally I feel that these who have approved drugs that kill people should brought forward to be in jail for murder. I said it because if I handed a drug to you and people saw that I was the person who allowed you to took it from me, guess what - I would be convicted of murder, that would be the result. Did you know cocaine was once in Coca-Cola? Not a well-known fact, but true. Some of the same drugs today called illegal were once used in many parts of the world as normal use, the use of opium and morphine in 1895,1 in 200 people had an addiction, the drugs were sold as drug powders in many places in the United States, but they soon where outlawed. Not saying that the use was correct?, but many people have had their lives turned upside down and even imprisoned for the use or sale of drugs that once where legal and some have changed from illegal to legal! Alcohol was once unlawful during prohibition and still is a brain cell killer. Yet, is it legal. And even causes many illnesses that can even cause death, liver damage, or just an addiction that cripples the person even to fail at life. Over consumption can even cause blindness, but when you've become an alcoholic- I believe they call this a disease?" It's not a disease; it's an addiction, an addiction to alcohol. And how about smoking cigarettes? The label said for many years, could cause cancer, yet they sale makes money and yes nicotine is an addictive drug. I know, I was hooked for a long time until I got the strength to say no more! But this has the stamp of approval. The government who is supposed to be the help of the people, have given the stamp of approval because it was able to be control the development for money. ($$$). Here is an odd fact about alcohol: In Russia, there are registered, and this means people who have reported they have an addiction, but there could be another mass of people who never have reported anything, but the people who have reported as alcoholics are at 2.7 million people. (Just a few people.) But wait! You may actually be one of the 14.5 million reported in the United States! But is this legal? But we have more cases than the famous Vodka drinkers. Then, there is the issue of drug overdose deaths. The figures are estimated by the year, but around 100,000 people are dead yearly. So, marijuana or cannabis in many States in America is now allowed, and we have seen the results with mass unemployment, garbage in parks with homeless everywhere, the "I don't care" grows, and in many parts of the world now allow this use, and I can say that the use of this drug is non-addictive, yet it has some serious side effects on men's reproduction, and the studies show that the drug has been a stepping stone to harder drugs. So, are these drugs really, what people need to maintain life? The answer is very loud- "no," yet many of you right now are living under this blanket of escape. So, as a former user of these substances, yes, I once used pot as my escape from stress, but I woke up as to see the truth how the majority of long time users of these substances had a very low self-awareness, had no real goals, and where lazy in general, so drugs have a great impact on the value of how a person sees their life, and the majority of pot smokers really don't care about their lives as long as they have a drug that can numb their brain. In addition, this is based on the majority of users, which we can see on our street corners and under bridges. We have all heard of those top entertainers who died of an overdose, yet people still desire drugs as a way to escape their lives and enter a false state of mind that pollutes and yields their potential in the future. Let us not forget the billions made by drug companies to keep you so-called "normal." oh yes, they are making so much money that they can afford commercial after commercial on television to promote their latest creation. But even our children are being subjected to parents who listen to drug-pushing doctors who get kickbacks from drug companies to push their products. Just one example: "Ritalin ," "concerto ," "metadata ," and "daytrana" are just a few drugs used today for so-called ADHD children; when I was a child, we all had ADHD, we climbed trees, went outdoors in the middle of below-zero weather to sled ride, had broken arms from playing football without pads, and beat on each other with plastic baseball bats, that was the norm, I once jumped from the peak of a two-story barn just to get a $5. Bet. Yes, we had ADHD, but we were allowed to blow off steam outdoors and got sunshine.

Today, our children are kept indoors to play video games and are drugged up because mom sorry to say, can't control the kids, and chances the control is gone is because Dad left Mom and they couldn't get along, because the average home is now 43% of once married couples are now divorced. And that is something that could be fixed if divorce was expensive, but the real answer is for people to stay faithful, and work it out for the kids. But, yes, our children are being placed on drugs. Is it right? Ah, no. But I hear family after family reporting that their kids are on these drugs, and the drug companies love them. Get this: they have just reported that taking psychedelic drugs is good to "rewire the brain?" yes-- "Step right up, ladies and gentlemen." See the lady with the longest drug use ever recorded! Witness a man with no brain that, when released, walks the streets at night like a zombie! Yes people, It's a strange event, but it's real. The drug industry is making trillions, not billions; your pains, your depression, you're under the control of a medical drug. Now consider that these issues just 70 years ago where not around. Oh my! How did people couple? They use inner strength many of them, and a shot of white lighting would be in the back hills of Tennessee that could put hair on anyone's chest! But they managed, is my point. If you think that the great depression was named "The Great Fun Time," we can look up pictures of how people lived. It sure didn't look like a good time to me. So my point once again is drugs, even over-the-counter prescription drugs, can be harmful to your life, so read and learn what that drug could do before you use it and if it is something that can cause another problem or even death. Drop it like an LED balloon. I have recently had to see a skin doctor; this doctor recommended gave me a drug that can cause cancer and make me important.

The drug is called methotrexate; this drug is allowed. And at first I was convinced by this so called doctor that this was correct, after reading up on it, I decided it is bunk, garbage, and any four letter word that you can imagine. This is just one example of many times doctors have proscribed their so-called cures of medicines that I have rejected because their medications came with a baggage called potential even bigger problems that could cause death. Do you think they cared? The answer is no, not even a little, so if a drug has an added feature that says, oops! Then, I would highly recommend a different doctor who may actually want to keep you above ground. So yes, there is a massive drug problem, even with average people seeing doctors and headshrinkers. Many of these medical college degree people hand out prescriptions like candy, and let's not forget that many times, these doctors and psychiatrists receive financial kickbacks from pharmaceutical drug manufacturers, and this is a practice that should be stopped by laws made to stop the reward system given to these institutions and prevent those who support the drug sales. So this stated, let's look at what could be accomplished to get our people to stop living in a la- la- land- The fastest way to prevent drug use in any country is to have the people do a daily urine test of all employees, and could carry drug blood level testers. A recent development in the UK, has been introduced by the University of East Anglia; they have introduced a new prototype of a new method of drug detection that is found instantly by sweat in finger glands, and fingers could be handed out just like a traffic fine, but much steeper.

We all have heard of the countries that have drug problems, and it once seemed like a way of life that we would not allow, the attack of our people and our youth today live daily in this activity of drug use and the lie that says today- "Oh it's okay to do a drug. " Let's call it what it really is: The action that allows my body to have a foreign element inside my bloodstream that allows my brain to become dim, or foggy, or high so I cannot feel pain or be polluted with substances that kill my brain cells, like legal substances called alcohol, and let's all embrace marijuana which enough of it consumed can make you into a walking vegetable, not able to have the ability for care or even discernment. Under the influence of marijuana has been proven that "yes" an overdose of cannabis can kill and has been recorded after tests as autopsy diagnosis found firm conclusions that people have died from this substance, we can also consider the low-grade sales of this drug, which have been laced with other drugs such as "PCP" phencyclidine, and cocaine, and crack can all be added, and the unsuspecting buyer is now possibly taking their final ride on the roller coaster. I fully admit that in my youth, I tested cannabis, what we called pot, and as a former drug user in my youth, the pressure from so-called friends who thought to act like an idiot and not have the common sense to see the world as it is, this drug taken gave a numb way to live, I observed grown men massive men die from overdose of pills, and people who smoked pot, as a daily activity, I watched their family life dissolve, because they had no real care of their situation, because, to them, everything was fine, as lost their loved ones, friends, and homes to massive negligence. As they became obsessed with the drug of choice-pot, they watched the life they once had become a horror of waste. One person I remember well had a cat; it was allowed to do it's poo anywhere and urine inside the house everywhere; the roof of his home began to leak inside his living room, so he moved his bed to a couch in another room, then the front porch fell into ruin, then the back porch, then he found himself living in a bathroom and sleeping on the floor! He lost everything! This was because his life became numb to real thinking; real observation was not important to him and many like him; In a few years, he was living in a dump; you may be one who believes doing drugs is a normal way of life, let me ask you this question: "Where you born with drugs inside you?" Sadly, today, some people would proclaim – "YES" because they had a mother who did drugs. But, in a normal situation prior to drug users, would you be with a drug inside your body? The answer, without any doubt, is "NO!" You were not meant to be with these substances inside, and the people of the past did not even have Aspirin; they lived drug-free. So to those of you addicted either mentally or physically too- pain pills, pills to calm you, pills to so call keep you balanced, Smoking Cigarettes? Drinking Alcohol, Smoking cannabis, and even coffee, which has been in my life, yes, sorry it is a drug, and is it harmful? (Yes) It also has a backlash; try not drinking it for two days and see what happens. We can look at any of these elements and see the so-called positives, but a natural body will survive with proper food, proper nutrition, and proper exercise. These were the way people lived for thousands of years, and died yes as we all do, but they had something called : Clear Minds, and they had the ability to see things as they were. So to you users, try something called- LIFE as it is, and stop making life out to be the boogie man that you have to escape from. Rather than do this, my advice: "face it, and make it better, fight for you! Fight for what you really are. I did, at age 38, I found my real self again, and mess that started when I was sixteen! I gave up everything stated here, and only one remains coffee and tea. And yes, they have caused my bladder harm and many sleepless nights, but sadly, I use them to wake up faster and be more alert. Is it bad? Yes, so no excuses here, and to you, dear reader, my hope is that before this book is finished, That you have broken one of those addictions one day. And I hope that all who have these problems re-think their lives and come back to us as one who sees the world as it should be, not as being numb, but being aware and clear minded with purpose. And to say it firmly- please notice that drug users are highest percent of life's losers who have no idea they are.

Now to the drug world and its corporate rulers: Money, Money, Money, the ruler of the trade of drugs; the players are many people you may know; what happens the day someone you actually care about dies from the same thing you created, will money bring them back? Sadly, the methods that have been the way to stop drugs have been given an open door as our borders are now made as a joke; the users walk our streets in many cities as zombies, and the people who see them simply walk around them as if they were a tin can on the street, with no regard, and no care. But we need to see, my friends; we need to be aware that the people who desire our lands, our future, only love to see the greatest nations dissolve around them, so this makes their takeover much easier, very easy. The world has always had those who see with greedy eyes, so we must unite to make their mission fail. We all need to help say no be regained, and we need to get our borders back. And sorry, I love all people who are good, but those who come to this country, USA, and any other country illegally need to be stopped and come in properly, and be what I call one who wants to care about the country they live in, not sending money away as if they are on vacation. Addressing the drug commercial- The people of the past used herbs and devices we would consider prehistoric or ill-conceived, as once doctors would drain a person's blood to relieve pain in the head, and the results of many so-called remedies actually caused death. Today we have the same exact thing happening: professional licensed doctors who proscribe medication without direct supervision and give these medications with the greatest confidence that it is the fix for that person, and over 50% of people who have this happen to have complications, and we have all have seen the advertisements on television of the so-called cures of some illness, yet then after they say what the medication is good for, then here comes a massive list that it is now proclaimed as possible side effects, such as liver damage, or heart failure! Sometimes, that list is much longer than the actual cure. The fact is we first should consider anything that has been proven to harm a banded item- period! I know- I said this before, but it doesn't hurt to try again.

A fix should be just that- a fix. Not a maybe, or oops! we did that?, and we are sorry. But you're dead now? I have always researched the items given to me, and I find that many times, a doctor will provide something that does exactly what I have just said; it gives a later complication or problem. So, I stopped using their so- called investigative methods. My advice is not to allow compromise when it comes to your actual health. And so the real thrown is the drug companies who convince the FDA and medical communities to pass the use of many items without the proper research given, and even with appropriate research, they still allow the company's item to pass inspection, case and point: The so-called fix for the " Covid " illness, the shots that had been approved, which killed people, and many did not work, or the person got Covid again and again!? What are the facts of "we need to add a booster shot?" This is just one example of thousands. So let's look at THE FIX: We need lawmakers involved directly with the FDA with a backup of top picked people who are qualified to check the so called experts, and when they find a company making a garbage. Then a clean house is used of chemicals in lives and yes, our foods and meats that can actually harm people like the overuse of steroids and growth hormones in chicken and beef, and the injection of these fast growth animals which are now infecting our families, but the FDA turns a blind eye, why? Because here we go again- Big money is once again at stake. (This sounds like a broken record). Some corporate payback is being made somewhere, and money is king over caring-isn't that right, FDA? You and the "Better Business Burial" should team up; you're both worthless! Oh, and these are labeled as legal drug applications!, and all drugs should be taken off the use as-directed list that can potentially cause "any harm" or have been proven not to work for "everyone." I realize the food and drug people will be upset when I say we do not need sugar substitutes that can cause cancer; we don't need drugs that can cause another problem to a person as it is supposed to be helping someone but have a list of issues it "could" cause, what kind of B.S. it that!. Now, back to Illegal drugs: The world has been dragged into a false ideal that drugs of any kind are safe to use that distorts reality is okay. Okay? I remember in my youth; I had experienced the use of marijuana (pot) as a pleasant numbness to things around you;

you can watch a man on fire and say: Wow! Look at that! "Cool", you can be delighted with a simple potato chip seems incredible, the flavor is strangely changed and is enhanced, and all food that you once ate, looks so different. The quality of these doses of the chemical THC can be on many different levels and can even cause a person to hallucinate and see traces of trails behind movements as they move their hands or arms. And yes, it causes massive laughter or the dumbest acts. In short, a person can become desensitized to trouble or danger. The drug can distort even a person's awareness; under the influence of this drug and driving, I was with another person, and this person reported, who was not high on the drug, that I had just gone directly through a red light! I was very lucky because I did not get us killed, obviously, but this gave me a real idea of the danger this had to my ability to have a continued life. But we are now seeing the use openly, and people walk around in a daze and act like everything is fine. A perfect application to dull the senses of the population so that now you can perform anything with the approval of the majority because they are not EVEN AWARE, OR CARE, a perfect application for a new application toward control of mankind to be implemented without any concern, and we now can introduce something called communism, ring a bell?, for the openness of socialist ideals are the stepping stone to the next level communism. But hey, folks! Who cares? You're higher than a kite, so does anything really matter? The answer is if you ask the drug users what they think of anything that affects their lives as who should run the country, they will reply- "I don't care." I have asked many, many times people in our youth from ages 15 to 45 what they think about politics, and they reply almost every single time," I don't care, and nothing I can do." It is a defeated attitude because they are numb to the ability to care; why? Because they do drugs that are designed and cause escape from the real world they live in, no different than alcohol, and we have, over time, allowed our leadership to cave into ideals and give in to pressure or the profitable applications now did give permission to the public to use the once illegal drug or substance to now be allowed as long as "they" controlled the sale. But who is the main root of these approvals? And let's look at some folks that love to make our nation a bunch of weak minded zombies so they can do their worst. Let's look at today's political action: They are supposed to work for the public as a servant of the people's interest, and we give them control to be paid to be aware that a certain thing can be allowed, which is supposed to be good for the public, YET!, They will enable the item to be approved as a law because by the support of this item is backed by corporations who lean on them, but the real facts are, they have abused their positions, and many have abandoned the higher standards of the past. Instead, we can see tax increases, and they are in support of getting a raise associated with that public office, all in favor, say yea? The enemies of the free world are firmly behind the drug use, and China delivers the drugs through Mexico into United States, yet what is our Governments resolve? Let's continue to support a country that wants to kill us and destroy liberty, does this ring true? But the American people who could stand up and say , we need American made products and we need tighter manufacturing costs so we can compete global, instead have tossed up their wallets to China and any other country that offers the item they want cheaper, 50 years ago the city of Detroit has the world's greatest car manufacturing. Today, it is in shambles, gone and nobody cared. Years ago Chrysler Corporation was failing, In 1979 the United States government and with republican leadership helped them with a 1.5 billion dollar loan and they recovered and paid back the debt. The Democrats could have helped Detroit car manufacturing not to die, but oh my, my, not much we can do to save American jobs. Yet they claim to be our help? You need to see that they can lie like a man accused of murder, just to save his skin. But facts are real- their agenda is to trick the youth and the black community into their bad dream of socialist / communist control and joining China and Russia, this once great party, and one of my most valued Presidents John F. Kennedy who to this day I set his standards for my life, and for years when I had my business, his black and white picture was behind my desk, but if alive today he would be crying foul play and in disgust of what the party now embraces as saying democracy but achieving control, and destruction of the nation! But his once great party now allows socialist ideals and communist infiltrators to enter their party, the takeover was slow with false claims and funds under the table, and the puppet master was China and Russia, but they made claim after claim that President Trump was in cohesion, when the facts came out that their leaders where on the take from those nations, our enemy to freedom. And the woman running for President if she wins, will turn the country over to China.

And because these political issues given permission to dismantle our country, they support actions drug business's to grow that are linked to the legal drugs, and many have greater income to support their greedy agenda (with more taxes gained). We hear of people who drive and drink alcohol and die in some car wreck or kill others from the negligence, and these people many times are harmed so hospitals have more income attained because what I approved. Insurances well improved in value because of the increase of people driving under the use of drugs now having car wrecks; this is called the domino effect. With the corrupt use of corporations using lobbyists, the money one again flows under the table, the Democrats and some republicans gain bank accounts seemly overnight, why because they don't care about the country anymore, they care of the profits they gain. So they support the many areas of business that actually harm the nation. The legal fees gained in billions year after year, a long trail of caregivers to support those who have been disabled, the medical community is with a green light on open costs to patients, thousands per day just for a bed, and we see equipment sales of wheelchairs, walker, neck braces, the list is longer than a football field, it goes on and on, so my support for "business" financial growth is not my beef, it is to the approval of the drugs that caused these domino's of disaster, and our leader's tax even on items that people need, they tax even our way to get to work, yes gas is now taxed, so we are not gaining people who work for us, but work for corporations who give them perks and far from the real job they were hired to do. They should be considering the public they serve, but they say inside their minds this as an excuse: "I can't get richer caring about people; the public wouldn't care about my safety." Harsh, yes, but exact.

So- the next time you choose a candidate, you might want to put down the drug or ask the question- does that person support drugs of any kind. But I will now reach out to the ones who say- well, it's okay to have a drink of wine or do some pot; I will say this that there is a time and place for all things if our maker had not wanted these items, he would not have made them, but abuse is when we use these items daily, we cause our lives to be lacking in care or even wanting to be improved as individuals or our appearance to others, as a drug user standing on the street if he or she cares of the fact they are standing in full view that they are not able to function! This is my yell to you who say- it's okay? The world needs us all; we all have special gifts; many do not even know the great potential that lies in their hearts and minds and the structure of what they "should be as a person," so they become one of the sheep, and they have no motivation or as said-care is gone but only for what that drug provides- The escape, and what are they escaping from? Cruelness, yes; poverty, yes; people who place great pressure on their lives as "stress," yes; and if it does take a clear thought, a vision of self to change bad habits, and these habits are connected to us like leaches sucking our good away, sucking our lives like blood draining from our brains, and we get controlled by a "substance!" This substance is easy to get, and we have allowed ourselves to believe the media promotion of these substances as "not harmful." No- people, these items are not safe to use. And so we fool ourselves into a false idea that what we do is all good. Many people say: "I need this, or my life will be in a mess!" The real mess is that you're in use and standing in the poo for a long time, and the smell is now normal. You have convinced yourself that this is normal- I need my coffee, I need my cigarettes, I need my booze, I need my drug, I need, I need, I need. That is the real world; the real issue is you now "NEEDING" something that controls "YOU." You are no longer in control of you. So, what is the real cure for drugs? Here it is: STOP- (I did over 40-plus years ago. I stopped smoking, stopped smoking pot, stopped drinking, and I am clean, strong, and 71, still playing basketball like a 25-year-old. I gained weight, but now I am even doing better than ever with my awareness that not everything we do is for our good. My weight is much better-I am 6' 1" and should be 220; I am now 240, and was 265. Life is a battle, but being aware is half of that battle. Be aware and become the person you know is really inside you. Do this- and good things will follow. And if you're someone who really wants to change, talk to people who don't do drugs, and this next section may be your path.

22. SUBJECT: HARMFUL DRUGS REMOVED

I know, I'm kicking a dead horse, but- I am not one to be without ideas of possible resolve when I see an injustice. As I stated before to get to the root is the real answer, but what do we do in the mean time? The answer is we don't sit back and wait for the problem to arrive, we become aggressive and go into the other nations who profit and hire their own people , with ($) to bring down the developers.

We have people who fight the drug problem daily; the police in some areas have a battle that can many times be compared to the front lines of a war. They die for those of you who should be grateful that many of them are actually dutiful and that many of them actually care about you. And to those who know this, stay on the right side, don't give in to bribes, and remember when you had the desire to help people. The job has made you, in many cases, can show you the dark side of humanity, but please know that we out here traveling through this world need you; we need to have officers who are the shinning nights of old and yet still have compassion and give us kindness to all, yet you need to be supported by the public, and many cases have tarnished the badge, but never think that badge can't be cleaned. Now the next sections to follow will give a rough idea of how some actions could be accomplished, I am sure law enforcement have many tactics already, but these are only suggested actions. So the first thing I feel you as a police officer needs which may never be given in your training. [Compassion] Here is my example: I was drinking and driving around one night after a party, two girls sat in the front seat with me when I ran out of gas, suddenly a police car appeared seemly from nowhere. The police officer used caution as he yelled to put my hands out the window holding a flashlight, and his hand was on his pistol. He then approached the car I was driving, and he asked for my driver's license and registration; he then asked me a direct question. Have you been drinking? The obvious was in the back of the car; empty beer bottles were all over the floor. As he shined his flashlight inside everywhere, the girls became silent and sat like deer caught in headlights. He then asked me why I was parked there, and I replied," I ran out of gas." he kind of smiled and, nodding his head, said: "Sir, get out of the car." I was convinced that I was being taken to jail for drunk driving. he then said, get in the car, and then when I started to walk to the back side, he said," No, get in the front seat." Then he told the girls to sit tight, that we were going for a ride, that he would be back, and that they were not to leave. They agreed, and we drove away; once again, I thought he was taking me to jail and then would return to get the girls. Instead, he said: "I remember having some wild days in my past at college; just know if I ever catch you doing this again, you will be off to jail. We then went to every gas station closed, and back then, the hose was many times still with gas inside, so we tapped into gas from three closed gas stations. In a way, we had taken gas for free! He then drove me back to my car, and it was about 2 am when we got back; I used the small amount of gas to get to an all- night gas station further away, and I took the girls home as the police officer followed me. When we got to the gas station, he then repeated his desire for me. He said: "Look, I am letting you go, but don't think if you ever do this again, you will be sitting behind bars the next time." I thanked him, and it had sobered me up pretty fast. But I have never forgotten that night when a police officer, a name I never got, made me a real believer that there are those out there who actually care about helping the people they protect. Now, this statement has only one real purpose: It is to police who think everyone is a bad egg; we need more police who help our people rather than be willing to empty their pockets of money they may need for bills or family. And the reward of letting a guy go home can make the difference in someone choosing a life of crime or caring about others. You see, you are on the front lines, but even in war, we need to have a clear mind of who is the bad and who is the good. That night made my life want to follow good because a good cop gave me hope. Now let's address the stopping of crime, the real way to stop it. The real way is to place police what I call Temp-stations closer to troubled areas for greater response time, these temp-stations are just that , they are make shift trailers which park in a central trouble area as to be with minutes of a call, and also be able to attain back up much faster, but even better is to make groups of police who are using regular cars which have no special lights, except an install of a light system to pull over suspects, the lights will rise upward from the trunk and from the roof, (so a new invention), and the use of a siren, not something you should ever use unless your miles far away, it only warns the bad guy your coming. The cars are normal in appearance and no black vehicles to easy to spot, and the cars are American made, and modified. The lights should only be used in heavy traffic, and the use of a speaker to move people should be priority over sirens, sirens just say hey I am coming! And this give bad people a warning you're on the way. The flashing lights are only used when you arrive. Then they will be with the mind set of: Oh-Oh! But now it's too late. The cars also have and a rear flashing light that when others pass by, the name police is in large letters that when the lights rise up then you see very clearly who is there, so on police cars the name is no longer on the sides and the cars are not all fancy graphics that display the car like a skunk strip, easy to identify. But once again, only on back of any car is the words police, so the bad guys don't get warned but people passing by are seeing who is there. Okay now you arrive at the crime, but your entrance is not a single vehicle, no you have minimum of about three cars are almost instantly on the scene because they have been with you about two or three blocks away, and even if your closer your cars are seen as no threat. The front car is always a single person who has a cell phone to call others and this looks 100% normal, and all the police cars are built with doors that have bulletproof doors and with bullet proof glass, a metal rear plate 2 inches thick behind seat and one inch metal plate in roof, and a front plate hidden in front with layer slots to allow air flow to radiator but bullets reflect back, so by parking directly at a crime you can exit with a protective cover, the future cop needs armor and there needs to be lighter and thin layer protection which could be placed directly into pants and shirts so when your dressed the protection is there also, consider it like putting on a football team uniform that covers more than just a chest, arm, and shoulder armor, stomach armor that covers the lower parts, and a football style pads that are worn on legs, and cover knees and the last armor covers the leg and the ankle. The technology is here, but we need to invest in better to keep our police better protected, and still look cool, hey I like football (lol), so why are the people who need this protection not using it? Funding, and I hear rumors that funding should be cut? Seriously, why not just say – crime ah go ahead! So, if we protect our police we now have an advantage over drug dealers with handguns. But let's go further; when traveling in your designated territory, you're not alone anymore, you have been followed all night by two other cars on your route, and the action has just been sent that an area has had shots fired; you're not alone as stated before; you have your back up already with you, six officers armed and a team. The drug dealers use lookouts to watch for you, so uniforms are covered in travel, but when you go into action the outside covering thin jacket comes off. So that front car is our "proof" of drugs, when this officer is driving, he can drive up to any drug dealer and ask for drugs waving money, he doesn't arrest the guy, no, he makes sure the guy is videotaped by a mini camera placed directly from the mirror, and from the hood ornament with microphones as to hear his request, the recorder is in the glove compartment, now the surprising action is, you actually buy the drug, why? Because no the sale is recorded and by the video you have the person 100%, then you return a few days later, a new guy is there, he is doing the same thing, he also is recording the crime, and sale is closed. This action is done over and over by different police from departments further away so they are not known as police; and why are we doing this over and over because we are collecting groups who are all potentially connected, and then, the next step is the tracking begins, and the images of the people are placed in the system. A.I. face recognition is used, and the address of the person is verified by the further away van that is parked with a long-range drone to watch the operations from above. The drone is used to make the I.D. as the individual selling to be followed by the drone high above the person to verify the drug dealer's residence. Once the people involved have made sales and are recorded over ten or more involved, the local police group with help of swat team now are sent to do the big bust. They all make a move around 5am once again in unmarked vehicles; the difference is the vans have signage saying Detroit NEWS or New York Times; the reason is these are to look like delivery vans; the delivery with a pickup van just a few blocks from the arrests. Now, the idea is not to draw a great amount of attention to the main supplier. All these people are given a jail cell at first but a judge comes to this to see the evidence recorded for each person and sentence is held as five years hard labor and if anyone died from a drug an execution by electric chair ,with good behavior a three year term, but the purpose of this arrest is to make them believe they are to be in living hell, they are not staying there; they are taken to a holding station outside the city on a bus, this bus with all windows blacked out enters a very large warehouse , and inside the warehouse unknown to the public is the inside barbwire fence with two sections, one with hard wire doors like boxes with bars , and a single bed that is hard as a rock, this is not a picnic area , no it's a hard labor area that provides the person found guilty to work at farms to shovel cow poo all day. They are bused from one cow farm to another, and the dump trucks are also following the direction of poo shovel workers. BREAKING NEWS! {SEE. This is where they are told that because of new laws passed-which we will all vote yes for anyone selling drugs is automatically reduced to no rights and no legal phone call, considered a threat to national security, and their actions of distribution of harmful drugs may have killed someone with drugs given; and that they promoted, these are not confirmed but are possible as all drug sellers have this potential with certain drugs sold. So they have lost all rights; this means abuse is possible if they refuse to move; they can be forced to move by stun guns, they will be forced to work, the work is always connected to hard labor, another work project to make our friends comfortable is the project flood sand bags.no less than 80-pound bags of sand are dug and placed on trucks that will be used for making any area where flood water area could be needing these sandbags. So they are forced to lift and carry these bags over and over, and after each day they are bused back to the warehouse jail, and sleep on hard mats inside very closed in metal screens of heavy screen and welded to the bars , this makes the stay an open forum and everyone can see everyone, so sleeping is hard, and metal toilet is in the back with a flap that can only be opened by the guards when a rod is pulled by a hydraulic system that opens all the toilet flaps all at the same time to drop human waste into the septic system, so the smell of human waste and urine is all night as the use of this toilet is actually a smell that will linger inside the complex, and they get to eat there last meal on the bus as they are in motion , and very little food is given which is always finger food, so no spoons, no forks, no knives, and is in a paper bag, food is prepared in the back of the complex and is always a bag lunch to be taken to the job site; a sandwich and metal ladle attached on a chain for water to be given, are all they get, enough to keep them alive. After about a week of this activity, they have been lined up to be able to have a porta potty visit at lunch time, and every day after eating super, they are called by name to line up in two separate lines to go back on the bus to the waiting cells, but there are always two outdoor toilets , , the type of plastic porta potty is next to another porta potty, and is able to be attached to the back of the bus, and one is for guards and the other is for prisoners, with a vent on the side that does not show anyone there is a vent , so it's similar to a confessional in a church, but the person in the next porta potty who is a police detective asks each person doing his or her business a single question and must be someone who can speak Spanish or English, the question goes like this: "Look you have been watched and I could help you " how would you like to be rid of this treatment? If they agree, then the person is told- okay, after lunch, we will talk, but keep your mouth shut, or you will remain right where you are; I may be able to help you tomorrow; what is your name? And what is your I.D. number on your shirt? The info is recorded and unknowingly the person if cooperation was positive is now to be marked on a chart with everyone's name as potential to give information. And is told -Now, remember to keep your mouth shut because now you are being monitored, and that means you cannot say anything; if you do, you stay here to work forever. If they agree, then at about 5:00, the people who stated they wanted out are told to line up when their name is called. This time, one of the lines is told to follow the policeman in front, and they are each to get on the other bus, there are always two bus's one for prisoners and one for guards, the guard bus has a room in the back that is empty with recording devices, two way mirror that is used in regular police stations today, and the prisoner is told to sit down; The person sits alone for about ten minutes and the bus stays at the location for 20 minutes. A voice is spoken over a microphone into the room that says," I understand you have a desire to leave this place. Do you want to stay or leave? If they say stay, then the conversation is done. But if they say leave, then the voice says- You need to know that this conversation is recorded, but the information is not shared by anyone who you may know connected to your operation, so you're never going to have to worry about being a person who told us information, do you understand?, you're never going to be disclosed to anyone, and if you tell us what we need, you will walk freely and we will give you a sum of money a new identity and will be moved to a place where they could never find you, and if your found to have top information that is correct, you will be a rich person to be taken care of for the rest of your life, does this interest you? If the person agrees to cooperate then they are told that this may take up to three days to have them processed, so they will be traveling to work area for just three days to confirm the information given. And when they return they are no longer in the same cell, they tell the person to cause a disruption so guards will look like they are taking the person down from anger, then the person is removed to detention holding area, this is done right in front of all prisoners, and is every two days of every five days, depending on the amount of people willing to give information. This statement is used to attain information about the people who are in charge, their names, the contacts, and where they are. The fact is, many of you may say, well, this info is now open to the public. Won't the criminals know this and be able to do it? The answer is criminals are not people who read unless it gives them directions to a home to rob. Therefore, the person is not set free, info is received, and the individual is not placed back with the others but is taken to a prison work camp, which has an easier work detail and better food. The suppliers of drugs are the target, not the street sellers. To arrest them is the way we reach those in charge, and we work with other countries to get these illegal sellers no longer placed behind bars; no —street sellers arrested are given a new choice of life, which I call- productive applications. The truth is we need full inspection weekly or even daily urine tests to see if someone is using their abilities or if they smoke a joint during lunch or take down enough beer to choke a horse. My strongest recommendation is that business owners make a two-day inspection of their employees, one as they enter and one as they return from lunch. This will give authorities the ability to be notified, and rather than arrest them, users of illegal drugs are placed in the workforce as a cleaning up from any hard drug use and given community duty if they can tell police where they got the drug from, but if they don't tell them, they are sent to a place to clean up their behavior, And now the reward of being a drug user, or if found selling drugs is to be rewarded as follows:

23. SUBJECT: WELCOME TO FARM LIFE PRISONS

The guilty person, male or female, is now taken to a farm camp for men or women. This is a camp with no fences and no housing; the people all live in single one-person tents, this is similar to warehouse prisons, except they never travel, and both programs stated here have no possessions but the two sets of clothes they wear and two sets of shoes, which one is rubber boots which are numbered for each person's use. These are colored in bright yellow, and the residents are only fed good meals when good work is made, and okay meals with work not done well. They also gain TV-nights rewards in outside large forums that display the TV, or a movie in evening, but they do not have permission to leave the staked area around the camp,- you see prior to being placed inside this camp the person is given three very important elements-

A necklace or neck brace that wears around the neck and is attached and locked with a very special key that is never seen by the visitor who is blind folded when it is put on. The reason is the device never is taken off and the person can't leave an area on they can be shocked, or if they are starting a problem, also can be shocked, so it is a dual purpose device.

I mentioned that the person is a visitor because unlike other prisons the only people who are in these projects are people who have a release date, this is nothing like those for many reason to be explained. The neck collar as stated has a contact that works similar to a stun gun, but with remote controlled, and directional signal with voice-activated sequence; each prisoner has a number big and bold on their shirt and pants similar to football uniforms so the numbers can be seen easy. The transmitter has a series of buttons with numbers on each button that is connected to each person's collar, which the guard controls and is attached to his arm and secured , and the receiver is connected to each person's neck brace is installed in the system and tested on low frequency when the device has been applied to a person prior to being placed on their neck; the device is presented to the individual after it has been installed to show the person why they should not make any attempt to leave the complex area marked with bright red markers and red lights at night. The device is live on an actual dummy which then the person is to stand facing the dummy about two feet away, the device is activated to show what happens if they leave the compound area. The device has a comfort of foam and is very light, but a steal rod closes like an adjustable house shoe and is worn every day and night; the camp has an area where if a person tries to run away, they will be shocked past that is the same as a dog invisible fence, but if they approach the marker area within ten feet an alarm will tell guards that an escape is in progress, but the sound gets louder much stronger as they now have passed the marker area now the electric current is activated on the neck, the further they travel beyond the red markers the signal increases, the further they go, the more the electric run by solar power increases even more if the person stops running the device is linked to a GPS locator and the device shuts down, only when motion stops from that spot but a loud beeping will still be on when a device is activated it also sends a signal to the camp foreman and crew to go and get the visitor who has most likely passed out. Or has decided not move further from intense pain, or may have died from over exposure. The second application is the device that can save a person from possible death in an escape attempt with the neck brace activated. Another device is also activated, the two braces that are on the bottom of the feet, these are magnets that activate when turned on by the system that now pull together when electric signal is turned on, the magnets will lock the persons legs together, and they will not be able to walk or run. What does a person do at this camp? They work for farmers and for cities, and any job that requires heavy lifting, repairs made, painting, crops hand-picked, trucks unloaded, train cars unloaded, food bags stacked, hay harvested, sewage cleaned in tunnels, creeks cleaned from metal and fallen trees, dead wood removed from forests, the tasks are many and road crew garbage pickup, and lastly "garbage separation plants" our next category. But before I exit this, allow me to state that if someone does not want to work at these prison farms, they will have one way not to be there longer, and this is similar to the warehouse prisons but these are not the hard core criminals who worked for drugs, but never killed anyone, but can attain the following if they tell where the drugs started from and their contacts, this will bring their sentence down to months rather than potential years, and they attain something called the transfer and name change, as a citizen they receive a new social security number with a new name they pick is established, and they get a free apartment for two years to be employed by a state-run work program that helps their every need. This means they attain food, electric, water, heat, even free entertainment T.V. and they get a bonus , that if their information leads to a massive drug bust worth over one million, ($1,ooo,ooo.) they attain a check for $10,000. To be paid directly into their bank account for their use. For every additional million taken off the streets, the reward increases by $5,000. Per million. The incentive to know that the bigger the fish fried, the bigger the reward- is the reason the program will succeed. To bring down a massive drug lord, will be a life of luxury, never to need anything again, and with a completely new identity, and even a slight plastic surgery could be applied, so they can't ever be found. To be able to keep prisons in areas without massive structures and massive costs to maintain these people. The collar device to be considered is a way that all costs of prisons can now be cheap as no guns are needed, and attempts to leave a work area designated is a work section which can achieve the portable system to have to advance past the markers in a collar that can prevent the exit with a greater sting if they continue forward, and even a greater possible of electric shock so great it could kill a person. But these devices would allow prisoners freedom to Rome and travel inside the compound area.

24. SUBJECT: ILLEGAL ALIENS WORK PRISON

This subject deals with illegal aliens, and sadly the need of this is becoming a real issue, We have an opportunity to use illegal immigrants , yes use their efforts but today they crash our party and they are doing very good at it, because they get assistance from others who are already here, so they can blend in very easy, so first we need to do the old style census reports as to who lives where, this means a door to door unannounced and a random street to street teams that can use their authority to stop a person on the street and ask where they reside, if they refuse to comply with proper I.D. , they are placed in a very large truck with seats on both sides and in the center, as they are now to be taken to homeland security and the person is detained to see if what they say is true, if it is not true they are then delivered to a cell, where they will be transferred to "WORK JAILS" As manufacturer workers , jobs to industry manufacturers to do assembly and warehouse working for no pay, and the no pay means they do get pay but only if they have been in the prison work camp doing quality work as assembly, so a total of 1 year must pass and the sentence to be caught as an illegal alien entering the United states is no less than five years imprisoned , and they only get $3.00 a day so after five years they attain a check for $5,475.00 , the money earned is placed in a personal bank account designed for the person, and when the person has done over 5 years , prior to release they are offered a chance to become a citizen with rights if they have been a good worker, they are given citizenship, but they only receive the type of job they have been doing, and only if they have done the work well, but they now attain free rent , and free food outside the prison work program Here is a device that could be worn to maintain control over prisoners as so they could be put down without killing anyone, and have a restricted distance that can be set for any type of work project. If they have been a problem at any time when they have been released, it is on the other side of the border in Mexico, and they only receive a payment of $3,000. To start a new life in Mexico.

I know what some of you may be thinking, why should we be giving money to people who obviously broke our laws. The reason and all you illegal folks listen up, it's because we will like that if you break our laws, you'll be back in jail "again" doing almost free labor. Which yes for the fact your actions have cause great hardships for people seeking jobs, and for people in the U.S. who are paying for your free ride as gaining perks like free food, and free medical, all from U.S. tax payers! This program pretty much puts a period on your free lives.

The fact is that our leadership has failed to see is that these people that come into our country uninvited and sneak into our land, they are actually invaders of our territory and country, because they come here at present with needs that our bleeding heart liberals who see that their way of treating these people as our friends of need is costing the country trillions of American taxes to support their illegal actions. I am not anti- immigrant people; I am saying that the majority of these people care more about their former nation and send billions of American dollars back to their country per year; if the former people that came to form Italy or Ireland had not wanted to be American citizens they would have sent money away back then, but instead they wanted to become true Americans to help their lives and support the country they lived in with pride that they were Americans now. So why are we not giving these illegal's what they really deserve to be prisoners of the United States, not getting free housing, and making their stay in comfortable hotels? This should be considered stupid to do that for people who break laws. And all these people are breaking laws of entry, which should not be rewarded! The term illegal's is here not as "legal people" to be welcomed, we must realize that the majority of these people are leaving their homes to find a better life, and they do not understand the great amount of billions spent on their entering the country in the way they do, cause a massive drain on the United States economic structure, and these same services required for the American citizens. These program established a growth of illegal's who are being convinced to join the Democrat socialist which will be the future (Communist Party),But they are gaining millions into states who are the highest number of electoral votes, why do you think that the illegals are staying in California , but being shipped to N.Y.? , if we look at the electoral vote number, they are the highest to make the Democratic party win an election, and another four years of these people, they may not affect the present vote tally, but if the socialist Democrat's win the Presidential platform, in four years those millions will be new voters who will only help the socialist who allow them here to stay. This means they will control America as socialist, and America will join Russia and China as allies, (this sounds impossible I know) but they who are already controlling many parts of the United States in many farm lands, and forestry sales, and buying the United States in a slow methodical way, as not to make any attention to their actions as much as possible, all so they can claim it, but the ability to gain votes for the change is needed and obvious that is why they allowed all the illegals to enter our boarders. So they smile and laugh, and when asked why they laugh and joke, oh we never did that Ha! The Socialist Democrats who many choose to support don't seem to see that many people coming into America are running away from socialist countries and communist countries, why? Because the system of government is nice and kind? No!, because they have been involved and seen the terrible way their people live is under a full domination of controlling people, so these who are now in control are leading us to the end of liberty, the end of freedom of speech, and the full control of you as soon as they remove the guns, because the guns are being taken away more and more, by new laws they make, and by building the resentment for anyone to own a gun. And why are guns for freedom to remain so important? It is because with them the people have a way to fight back a repressive government and have the ability to rebel if your freedoms are taken away. So we all hear our leaders who say how terrible guns are, but the fact is without them we are doomed to be imprisoned by their control. So research socialism, before you join them, and ask yourself this question, if I have a desire to better my income, how can I achieve that when they refuse to allow me to be in control of my business? You can forget having a hope of achieving something, The socialists take control of all business's and the owners get pennies. It will be controlled and your wages will be monitored, but the real end is when they arrest you, and your family slowly disappears just like socialist Nazi Germany did, it was all in the plan a plan so well devised that you're sitting so comfortable and you kind of see it, yet you feel helpless. You're monitored, you're watched, your words are listened to, yet you think it is to protect you? Funny.) It is so when they achieve full control after 80% of guns are gone, then- the hammer falls. But we could turn this around by arresting these traitors, and arrest the media who has also been bought by the socialist /communist agenda. Sorry, got off the track a bit, but it's a link, oh yes a very big link to why the illegals are here. And we could change the illegal aliens who enter a super big penalty for undermining our system of governing, with making them work for nothing, this could help this country and gives our manufacturers a better price for their goods to compete against China, so all the need is food and a place to sleep and wash their clothes and showers, but the savings to industry is potential trillions as the labor is used "free" for a period of up to five years minimum for any illegal found, and the week these illegals are set free , the group is now loaded on to a massive enclosed semi-truck with a porta potty inside and a spot container of water, and air fans on roof, and seating both sides and in center, and they are taken across the border of Mexico and un-loaded, no fuss , no problem if they return, they will be caught and two more years of labor will be done, and the fact is many people will prefer to live like this because the life they had prior was much worse, with drug cartels and no work available. This program will allow the United States and any country to build lower compete in the global market as quality at reduced development. Now, who pays the people to be working for just food, and bed, and a shower? We can look to the manufacturers to give back a profit percentage from their actual sales, and the State they are in gives a small income up to five years to whatever that amount is from the use of the workers as any job that they do for roads, painting, construction, all the wages which would be for them if citizens are now used for everyone in the prison work program. And any amount that is left over yearly goes to new startup businesses that need capital if they develop a new product that can be sold as an export only. This also helps the country as income is now returning to the country, which allows the economy. For families with children, the children are not separated from the family member of the woman, but men are placed in separate areas away from the family. However, they can have visitation from the wife and child every three months. The idea here is not to reward their lives, but make them realize that coming into America illegally is no way to enter, and if you come here, be prepared to work for the country to be awarded as a citizen. And yes, a person in jail can request citizenship as long as they have cooperated with the authorities. So, to conclude, these people stop being a strain on the country and actually help the economy and can be welcomed if they agree to work over a five-year term. Let's say they have paid their dues. But if they are ever found sending money that is not for relatives to visit, they can be stripped of citizenship and sent back over the border. No country can afford money to be drifting away from the internal needs of its people. This is just the reason trade laws are made, so the countries doing trade are fair with their profits made. This issue is not over:

25. SUBJECT: GARBAGE

I have made many references to this because it is a major problem, and garbage seems to be a very wide variety so that these subjects will be just as wide to solve). The stinky business has everyone wishing it was not an issue but should just be buried out of sight like a cat digs holes to bury its poo and cover it. The old saying" Out of sight, out of mind! HA! Funny. That is not happening by any means; I have seen firsthand the results of out of sight out of, out-of-hand garbage results. I had my business across from waste management for over seven years; I watched the hills of garbage across the street rise up into the mountains! And they continue to grow and gain more hills every year. I watched as massive trucks dumped their garbage bags made of black plastic; the stench was horrid! And contents would break out when pushed around by massive bulldozers and reveal the metal, glass, or plastics everywhere, as the birds feasted by the thousands on routing polluted food that could only be a hazard waste to their bodies, so the so-called plastic bags which hide the contents allowed many different items which would be considered harmful to the environment the items like paint cans, and paint covered wood, many items have rust and could drain and drain into areas away from the garbage dump. But mostly the gases they make as a result of piles of garbage roughing on top of garbage some over 100 feet deep, now make a gas that they try and capture to run their gas-fueled trucks but only captures less than 30% because just down the road in a gully and road called next to this massive dump-is called: "Hand Road" You drive down this road, be prepared to breath heavy contents of these gases which are flammable, called methane gas.

And is this gas harmful? Not to the owners of the garbage dump, these are what garbage dump owners; these are called "acceptable amounts of release." Tell that to my stepsister who died from bad lungs and water formed around her heart at this same location. The water is tested by the local country health department and is approved, but peoples pets who have been given ground well water near this dump all died from intake of the approved water, which a pond which is exactly across the street and drainage comes from the waste dump has a zero life of fish, showed me their efforts to contain this mess were just that- a mess!, but I am understanding that the task of making safe methods to dispose of man's mess, is no easy task. Allow me to give you all the hope of tomorrow that we get this under control, and those who work in the garbage, I respect you greatly. You are the front line, but your leaders are not seeing clearly; they are only making the best of a bad method, which is no way to handle a problem, and the solution is not to have garbage, but to have it processed 100% back for use.

26. SUBJECT: GARBAGE REMOVAL

In the old way, we all put our garbage in bags and garbage cans, we separated them, and we thought we were doing good, but that is not true. Many dumpsters have been used to mix these elements in abuse for many years, and I have seen these actions firsthand numerous times from many different places. So what we need is garbage that is placed in proper containers that are automatically sorted into separate containers and then sent in separate truck pickups to be sent to recycling plants that take all the garbage that we create as separated and make it reusable; if you're saying to yourself they do this already, no it is done on a very limited scale, about less than 5% of garbage is recycled, and yes to appoint, but not to the extent that I believe is necessary. Garbage is metal, and metal is made with plastics, like refrigerators, computers, and air conditioners; plastic bags are discarded when they could be recycled plastic bags, and all plastic bottles yet, because some have no deposit return, are tossed in with regular garbage, some are recycled, but the majority is not. This means all types of containers, metals, glass, paper, and glass bottles could also be recycle and products that are assembled could be taken apart or disassembled and with metal detectors the individual parts could be recycle. But these disassembly plants would be costly, you say? The cost actually makes our hunt for more resources lower, which means we are saving money in the long run. I have a good idea of how it would be done. First, as we said, the home needs a system to separate without all the fuss, so an automatic distinguish-er that when you drop the garbage inside, it is marked to what is required to enter, you drop it in, and it is still shut means you didn't drop it in the right area. Garbage will one day be worth money to you, but that isn't right; for any home found returning garbage correctly every year, they get a reduction on their State Taxes by 5%. If this garbage is returned properly, it will only benefit many other things, like no garbage dump yards. And job developments for people as inspectors and disassembly mechanics, you see all items in, are taken apart just like they were assembled, and many items will return to the selling floors just outside the garbage factories for re-sale, simply some things will be repaired if in a repairable state. There is basically four components of garbage biodegradable which is your banana peel or a peel from an onion. This element can be collected separately and placed in our forests, which would be beneficial to plant life. Metal can be re-melted and reused. Plastic can be chopped and mixed with virgin plastic and also be reused. Lumber can also be chopped up and used as paper or plywood and other elements that can be reused for building new buildings. Chemicals this is our biggest concern and should be because these chemicals today are more harmful to many more than all that was previously stated. Some chemicals can be reduced in their effective nature by direct sunlight. This could be done in some of our areas of the planet, like in deserts. But the harmful materials need to be changed, what I am saying is that we need to create natural elements that can still do what we need. The NFD needs to enforce the use of new safe soaps and detergents that protect our water and need to become natural to protect our children from chemicals that harm them. Petroleum products must be recycled and we need new regulations to check applications to see if they are functioning properly. They call it checks and balances. Financial fines need to be implemented two companies that pollute the water or the air and laws of deep fines to be enforced for these companies to comply with regulations. If they do not, we should have the right to close them until they have resolved their problem. All factories should be required to have drains and sewage monitored of the substances daily that they are sending out of their company drains or factory smokestacks, and this checks and balances a regulatory institution is not to allow accessible bribes and other financial gain, which can be given to people in charge. Anyone again. Imprisonment with work detail as community duty for the rest of their life. A hard real fact, that they may have killed people for their negligence of allowing pollution.

27. SUBJECT: JUNK YARD & PARTS BLUES

I know it sounds like a blues band or blues tune. But most of us have seen the massive amounts of cars and trucks at these places. We seem to have millions of different parts that people will go and try to find the part that is only on their particular model of vehicle, or they will go to a parts store. Maybe the item is still in stock, and if not, it could be ordered, but the warehouses for these parts are the size of football stadiums. The fact is the parts to build an engine of any type should be regulated so that when you need a part, well, guess what? Everything under the hood is exact. That is not what you may think, as an impossible action. The truth is you could have different styles outside, and even different interiors, but the bolts and the parts needed to run and the function of the actual parts that get you from point A to point B are the same in every car. This means all car manufacturers would have to make their internal parts all the same and this saves our planet from massive junk metal being everywhere rusting, and getting into our water supplies, and harming our environment. Why is this really important? First, We have junk pilled up in some junkyards that flow over because they can't get rid of certain engines or certain drive shafts, or a particularly used disk brake, But if all the manufacturers got together and said- Hey let's join forces as to have the parts the same for the functions, we can now save money because we can all contribute to the development standard and saves us billions in tooling. Now, the features of cars can be different, and yes, the demand for body parts will still be, but when a person needs parts to make the car run, wala! The parts will be easy to attain. This also makes the Junkers and the auto parts job much better because the parts are now easy for them to find, the need to collect hundreds of different engines will be gone, and the need for used items from car wrecks. The only differences will be exterior, interior, like seats, interior parts, and body parts stated. The electronics could even be the same, such as wiring and lighting inside the housing. So if the car companies unite in development tooling as to all inside components to run a vehicle, now they can also share the replacement parts sold as a united sale for each company and divide the sales worldwide equally and yearly. The sales could all go into a joint account that their accountants supervise, and at year's end, the total earnings tally is divided. This could be because my company divided the tooling cost for all items. Equally, I now can attain funds at year's end, even from models I have not been in my inventory, but sorry, that is not how the money is divided. Instead, when a customer comes in, the particular car model is also logged, and that is the part bought. The profit goes to the company, which is the part in that particular car or truck model that is to be given to the car company that supplied the design, the interiors, and the exteriors. Still, the real savings are two real elements. Each car company now saves in development because the newer, improved cars are introduced as parts change by improve performance or newer tech. Still, every ten years, a group evaluation is done to see what is needed to improve the car's running parts, and any tech is presented. This action reduces the massive mess in our lives and reduces the time wasted getting parts. So bolts, screws, and everything that puts the car together is the same, and people are much happier because they don't hear these words ever: "Where sorry, they stopped making that part."

28. SUBJECT: GARBAGE MANUFACTURING

We make all containers of clear bags and clear containers as to see contents always, never hide garbage with other items not accepted in the specific containment, and if repeated offenses could be a fine. We do not allow metal with the wood, wood with the tin, and so on; instead, we once again place the first responsibility on you. The household person now owns a given container that sits outside your window, and you can use the garbage intake, which detects the type of garbage it is- and separate it right as soon as you place it in the holder; it reads the contents and won't allow it to drop until scanned, and if you placed a content inside not welcomed, a buzzer sounds and lights up the area that has a problem, once you have fixed the issue and now the container opens and drops the proper item into the proper chamber. The garbage man has the ability to take the entire unit on your garbage day and he retrieved the holder of separated garbage and now wheels the fresh unit to your location, and removes the filled one, and then wheels it to the waiting long garbage crush truck, This truck is designed similar to the garbage trucks that press garbage together, but has four separate compartments , glass, cans, Paper, wood, food compose , Metal and large wood items like electronics, and old furniture a stove or any item not listed as garbage is considered hard items a separate truck goes around a massive flat bed with an extending elevator lift as the items have two men crews take these items from truck to Garbage Manufacturing work station and dis-assembly this includes car tires, car parts, and any hard metal objects, the workers at these parts removing are trained to take apart all items including the nuts and screws that assembled the items and place them on conveyors that dump them and also will ID the item as the correct metal to the correct dump area for recycle melt down. This plant has the ability to make metals clean and ready for use by manufacturers needing different types of metal shapes and components; a manufacturer can even have molded parts made on- site. The process per item is as follows:

First, let us address the type of truck for garbage pickup, It works very similar to the trucks used today, but because of the need for more people making more garbage, these trucks are longer and have five separate containment holders, each with compression mechanism the same as today in the garbage trucks used with the ability to press garbage with hydraulic power. The difference is that each area has its own separate area for the different types of garbage. The households are supplied the proper containers which they actually use to keep the convince of being able to have garbage taken away, but is separated right at the entrance of the container with marked entry and even detection to stay closed if not the scanned item as proper area for disposal, These containers that are used at home when filled have fill signal (WIFI) , that is sent to the garbage pickup that will be schedule pick up automatically in truck pick up, but when they come they pick up all garbage so it is best that people are aware to not fill one area faster than another as much as possible but these containers are able to be detached if say one of the containers had no items inside, in the process of a very large pick up need such as a construction being done at a home, then a phone number to the garbage pickup office must be made to make any added pickups, and the unit garbage bin is five separated compartments that have the ability to lock together , or in the case that one of the units has very little garbage , the bin can be detached very easy with a turn of the side lock mechanism clamp, and all the bins are the same size, and they also sit in a tray and on the bottom their connected into a slide groves that keep them in place during dumping, and the carry tray has large rubber wheels so the entire five bins can be detached from the home and rolled out to the dump truck, when detached the inside garbage doors inside the home will not open until the system is back in place as emptied and reconnected. The garbage separator container is moved to the truck with a forklift slide in the bottom for loading onto a hydraulic lift. Then with a simple press of a large button, the fort lift forks rise upward with the full containers, and then rotate backward toward the opens on the back of the long truck. They now empty the containers, which are now directly over the proper separators as square slides that dump each item in its proper area. This truck is twice as long as today's garbage pick-up, but the compactor press's all containment and garbage crew then latch the garbage separator can onto the lift holders before the items are dumped, then is crushed to compact that area. Another container dump option is an operator can detach a bin or bins and now take only what is needed at the present moment to discard the trash. The bins are taken to the truck, and the truck helper dumps the properly selected bin into the proper containment. The compactor is used for each item placed in the section. When the truck is full, it returns to the GARBAGE MANUFACTURING PLANT. Now, to change the subject

slightly, I have included an additional subject of space travel to help our planet with garbage removal, but we must remember that time is an enemy of gathering bad elements such as hazardous waste, and if not disposed of properly, will only come back to bite a person in the but-x in the future.

At the plant, the garbage is taken to each separate workstation for recycling, and the applications are as follows:

A. GLASS- placed on a conveyor, and some workers separate the glass by type, Clear, White, and colored, and each is placed on a conveyor. It is then crushed, and then the heat process melts the glass and is poured into molds per customer request; if there are no customers, the extra glass is used for windows (clear) and storage jars and bowls- Cookware/dishware (white) (Color glass) is made into small broken glass to be sold to glass manufacturers USA only. (Profits are made, and the buyers save money for recycled items, so they make more money for their items.

B. Cans- These are dumped from the truck onto a conveyor with a small crew that separates wire or large items of metal using a hand- held metal distinguisher that is the same as a metal detector, but any item that is too large goes on a pallet to be delivered to disassembly station, this is possible when a customer has placed items not considering the need of the item being taken apart, but this can happen so this is the reason the workers will have to remove certain items from time to time, now the majority of these items are cans from cooking food, but because they can vary in size, the conveyor has its own metal detector that now has a channel size separation ability by the size of the item and the cans are allowed to enter the proper size first, then the automatic metal detector now separates types of metal, TIN, from Iron, copper, or brass, (even gold-but doubtful)-One can only hope right? Lol.

The metals have one last two-person crew that looks for any item that has slipped by the separation people and the mechanical system, and these items are either placed in the proper conveyor channel or taken off the belt. Once the items are closely separated and have their own conveyor, as they are now heading to melt down, the items are melted and recycled into products once again in demand by food processors. (Profits are made)

C.PAPER, The same application is made with workers separating types of paper. Still, they mostly look for hard items that need a larger machine to chop the wood or need to be disassembled as a wood and combo item, so it is placed on pallets to be taken to that station. They are also looking for binders with metal clips. These are taken off as much as possible before reaching the paper shredder. The work crew consists of eight product separators, and they take the items one by one from the dump bin and toss the items on the conveyor. And each item is inspected very closely, and item found with metal attached to paper is tossed in a separate conveyor that will take those items to a paper work group that only takes these items apart, and then places them back on the conveyor. The paper is now in the chopper and is made into a mix for the making of: cardboard boxes, customer food inserts that may hold food, and customer cardboard sheets for separation layers for pallets. And any customer-molded items ordered from mold developments. And last is the separation of the first part also has another conveyor, and the third is for use. Each conveyor has a sign above that conveyor that says: COLOR PAPER / CARDBOARD- the largest conveyor, PAPER REMOVE ITEMS, and last- WHITE PAPER ONLY. This is for paper clear of ink and will be remade into paper for writing again with press rollers and cutting machines; any waste cut from any process is placed in bins to be placed back on the proper line. (For Profits made.) Regular wood is a separate pick-up for mulch making and will be sent to make factory pressed boards.

D. FOOD COMPOSE, As I stated earlier in items needed for deserts to flourish, we will need everything from turkey and chicken bones to your banana skins or potato skins, will be here, and yes, this is a smelly job, so each worker has a face mask with a tube to their and oxygen tank with an air indicator, and they breath clean air. At the same time, they check for any elements that are not welcomed. But the process all starts with the proper garbage removal, and separation done in the homes of each individual. However, the process all starts with the proper garbage removal and separation in each individual's home, and these items also required would be wood items, such as branches, to even large cut wood, or wood that is being degraded, but when they arrive at the separation plant to be exported to deserts requesting the compose , the deposit will not include items that are not biodegradable , so we have help of a metal detector which will beep and stop the line , then once the metal is found with a handheld unit, the metal item is removed and disposed of to the type of metal detected and the line is restarted. The waste is going to a chopper and goes to a chopper and becomes a farmer's dream come true. Once the items are well chopped, they go to two separate areas that divide the items as they travel on a conveyor that reaches a metal " Y" as a divider, and the divider allows the waste to go down a very steep decline slide to a very large dryer oven-air blown heat powered by electric coils will start the drying process, and this will be chopped again into a fine powder to be bagged into home garden fertilizer, The second metal slide takes the mix to a very large vat, the all-aluminum mixing tank when full to the marker indicated-will have water added, for farm customers wanting to nurture their soil with the best nutrients and best growth available. But is this mix just this waste? No! It's dead fish because when this mix is completed, 25% of this mix is containers retrieved near waterfront areas, ocean beaches-or very large lakes like Lake Erie for the added boost in this mix, and the mix is pumped to even larger holding tanks where the mix is now added with the chopped fish. The fish are gotten daily from pick up crews from beaches and lakes will be added from tanks filled with chopped dead fish. In short, it is the world's best fertilizer without urine acid, as applied today, and without chemicals added. Farmers will love the mix because it will bring back the soil with added nutrition and yields.

(Profits Made)

29. SUBJECT: THE DISASSEMBLY STATION

This requires a once-a-month pick-up from flatbed trucks to customers who have placed items on a curb. If someone is over age 60, the person will place a sign on the front of their property with a handy cap emblem; the crew will now knock on the person's door and request the item, if it is too large, then the call goes back for an evaluation by manager as to how to proceed, but if it is a single item, then the work crew will use a large dolly to remove the item and take it to the truck. All items are picked up, and the back of the truck has a drop-down hydraulic-powered platform. When the truck is full, they mark on their route the area last picked up as to return to get more item stops required to finish their route; the items are now all taken to the disassembly plant and separated by- (size and content, Metal items such as a refrigerator or stove go to house appliances but will be looked at for evaluation of refurbish. A piece of furniture will go to another evaluation as a possible refurbishment if the structure is intact and the style is sell-able; fabric cloth from the furniture removal is placed in bins for shredding for recycled carpet mix, and dye is added for colors. Furniture without the ability to be repaired is taken apart, all wood is removed from nail pins, and any metal is placed in a proper metal bin. To be exact- The entire function of this station is to place items for repair and re-sell as refurbished or taken apart completely, and all parts placed in proper metal containment, and plastic containment bins full take to melt down or chopping stations, the bin will be marked with a delivery order placed inside a clear plastic holder on the bin as to where the items are to be delivered for a meltdown, or recycle.

EX-LARGE- DISASSEMBLY- Just imagine a full old train, an old jet, an industrial lift truck, all massive items being re-done for metals and components to be reused down to the smallest bolt. Once again, the evaluation is made, and the concept is always team made, as one. Materials available? 2. Cost to restore? 3. Sell-able? 4. Profit made? But many items will go to auto parts as inspected and working, and to be sold at the lower process than retail outlets, and all items are taken apart from vehicles top to bottom, aluminum is separated from iron with magnet testing, items that are whole as body parts to newer cars are removed and the items resold to body repair shops at half cost but customers are given the cost difference decreased half of all latest. And all items redone as refurbished are posted online for the customers' review, the doors, the glass, and every part is labeled as to make, and the parts destroyed by a collision are placed in large outside drop-offs that are taken to our largest meltdown. The warehouse has bins of every size and can house up to the top 50 cars on the market for up to twenty years old. Items over twenty years old go to our final resting place warehouse and stay in the bin they were placed in but were removed from online. Now, we can only be called to check an item over 20 years old. The ware houses for these items are by year and model of vehicles now sent to disassemble two stations, and all parts are melted down according to the type of metal. This workstation is for everything from cars, trucks, and boats, and the size has no bearing. The parts are collected and taken one by one to the housing elements warehouse; this is a warehouse that we can refurbish even- TIRES: This is a process that has been one which many believe is impossible, but the metal cord is recyclable and requires a machine to remove the cord by ripping it out, and then the tire is ready for chopping into a fine rubber that now we can place in a new rubber mix process as a rubber mat, or rubber can be added to road pavement to give roads a better ability to retract or expand during cold months or hot months, this idea has been given by myself many years ago and has not been applied as much as it could be. But another development of rubber could be rubber added to sidewalks in concrete as the small rubber pieces would prevent people from slipping and possibly help even if a thin layer of ice has formed; some of the tips of the rubber could prevent someone from falling, and if riding a bike ever is allowed on sidewalks again, which only makes people safer so they are not hit by cars, which I have seen happen more than once. But if the rubber is applied, could help the bike to stop faster because rubber against rubber has a greater resistance. The conclusion: Safer walking and safer riders.

30. SUBJECT: THE TIME ARK

The process of information and even re-development of life as a faster information and stricter of "how to" and the potential of starting humankind over on this planet, or the ability to use this same system to achieve space travel to new worlds as to start life a new on a new world. The need for preservation is greater than at any former time in history. In the past, people did not live under the threat of total devastation or removal of humanity as never to be again as we do today. So what I have devised is a way for mankind not only to survive but also to be able to retain all forms of information that could even be shared with the life forms that may replace mankind. And that thought is not welcomed, but we have to consider the possibilities. But the main element of this structure is to be able to maintain it's ability to withstand some very serious events, such as tornadoes, earth quake, flood, and even meteor storm. And nuclear disasters. The device containment is designed with a sensor system that can do things as detected. The structure is like a living adapter which uses computer programming to achieve actions needed to maintain the visibility as to future generations to be able to be able to either find the structure to retrieve the information from outside, but this structure is designed to evaluate the intellect and the ability to prevent entrance if the outside individual is hostile or not with proper ability to enter, the outside can detect vibrations, and the eyes and ears of the computer can do a proper evaluation of those seeking entry, the structure sits on an elevator that can retract the building downward slightly to reflect any heavy activity such as a dust storm, or a nuclear explosive device. Radiation detection is another feature that will retract the containment; these shock devices and detection devices are sent by solar-powered transmitters to the controlling functions of the structure; the board mother control system is an A.I. active responding to the outer areas and will do what is programmed as the proper response to the actions made outside.

INTRODUCING THE WORLDS-ONLY KNOWLEDGE PRESERVATION AUTOMATED LIFE RETRIEVAL SYSTEM:

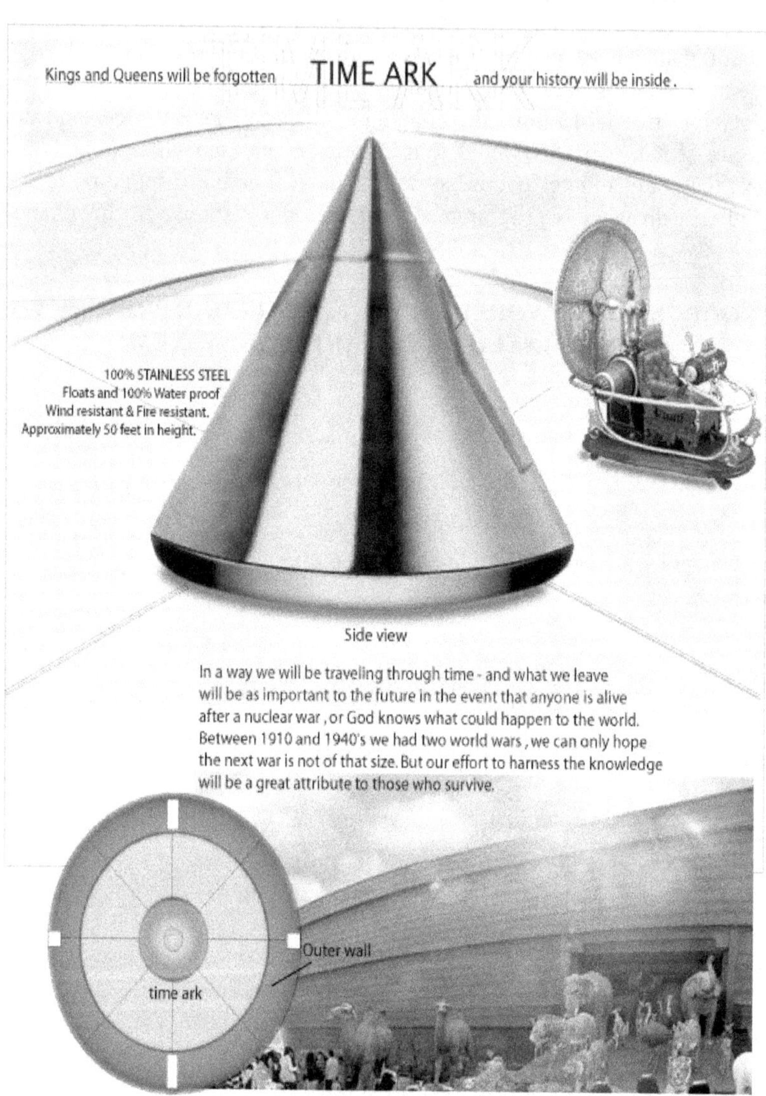

In a way we will be traveling through time - and what we leave
will be as important to the future in the event that anyone is alive
after a nuclear war , or God knows what could happen to the world.
Between 1910 and 1940's we had two world wars , we can only hope
the next war is not of that size. But our effort to harness the knowledge
will be a great attribute to those who survive.

The first " Time Ark " shall be to preserve
as many technological achievements as possible.

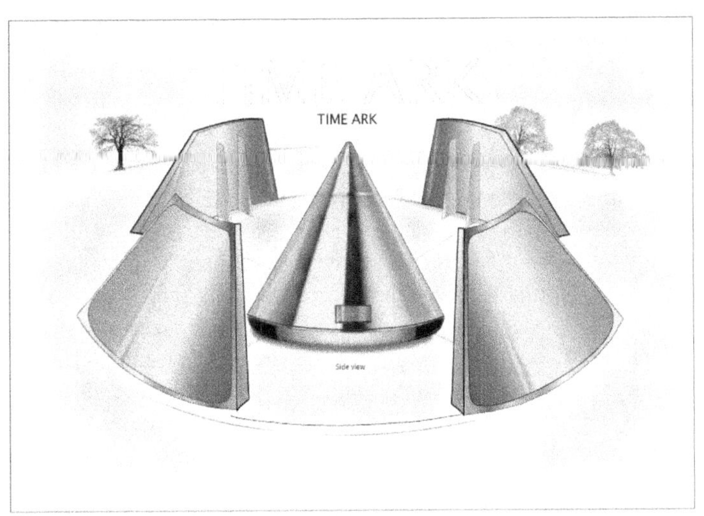

The TIME ARK structure also has the ability to float in the event of a large flood or even planet shift; it has retractable connections that allow the structure to be able to rise above the water level or be like massive hydraulics that can extend the structure above a contaminated area, or a flood, or disruption of ground, the reason this need is maintained is that over time a structure will settle, and also be covered by dust particles yearly, so the structure needs to make adjustments to this fact of being able to stay above the ground level. And so the hydraulics are reflective indirect light ports that maintain solar powered batteries and back up separate electric powered generators for energy use is located inside the structure to maintain all electrical features which include but not limited to- Lighting, storage lighting, artificial human development, with robotic parenting to bring life back as within an artificial womb, and the parents are now the robots who give the children their lessons as a step by step is applied in a time framed method no different than what regular family life is as 3 meals per day, defecation and internal waste management,, and the children are taught around the clock as information is even being delivered into the artificial womb, at four months arrival, and feed and care for the children as they mature the lessons increase, but learning starts from conception from artificial womb opening and umbilical cord cut, and child is now placed in proper seating and structure that allow it to form as required, and also be given the same attributes as any child needing exercise and nutrition, the lessons are done with a programmed step by step lessons that insure the child is gaining the information correctly as the tests are defined as the lessons progress the student now has more freedoms to move within the structure to areas that are open upon we can say graduation or completion of the child's growth being seen as correct. They will eventually learn all the needed survival abilities even to grow food, and yes, the ability to have other animals will emerge with the understanding that these creatures are to populate before use for food, the seeds of animals will be preserved, and the seeds of plants will be stored, the equipment needed to build and even maintain will be provided inside. Shovels, hammers, nails, and many devices will make the ability to build, and the step by step methods of use of all devices. In short, our students will be able to bring a world evaporated back to life. But there will be one main teaching, and the program will be made never to show violent action, and that all people are to respect and never harm another human to fight, but maintain its distance relative to its original placement. And A.I. will bring the lessons with the ability to have the information be restarted in the event of a first failed attempt , the system has three backups which will bring perfect DNA humans to life so that later procreation can be achieved, and disks of learning or instruction will be protected from UV and climate, and the equipment with the ability to present these lessons will be a giant round player with actual thousands of disks, and hard drives that also retract the information as the students can now move step by step, through any category. They start in alphabetical order, and the information is in detail in any and all areas, For example, how to make a computer would be the advanced system, but the equipment needed would be maintained, and the need for plastic development, metals required, and how to find them, the lists of each item will be long, and extremely precise. The fact is the structure is timed to start life over after 10,000 years, so equipment must be maintained as water and moisture-protected like no other structure ever conceived, and the equipment inside is also protected by plastic constraints that seal the devices that are essential for the startup. This means a single A.I. robot is protected from elements, and as you know, I am against the use of A.I. being mobile, this is the only exception because of the need of the robots to nurture the new born children, but every area has sealed entrances, that to open them requires a lesson learned by the new born, and many things required to build will be inside many of the containment areas. This structure is made of a stainless steel outer hull and is wider than any naval ship ever made, it has six levels, each with the ability to be locked down as sealed waterproof, and prior to the structure being closed, the entire structure is removed of O2 and a vacuum to prevent rust and moisture, but the air tanks are massive and activate when process of life begins, the O2 is maintained in tanks until a clear outside presence is found as no radiation or air contaminants found by systems lab testing equipment that takes air samples from outside, so once again the inside does not have the ability to rust, there is even equipment that shows emergency exits which are only accessed if a danger inside like fire starts when the system is turned back on from proper air being presented, then if a fire was started from moisture or rust in the electrical system, the onboard system would close those areas as locked areas, and the air would be vacuumed out to destroy the fire, and once taken back is then to be opened, but all areas inside show the exit area and the information of how to open it manually. But a diagram would also present the directions to restore a break down process. But the robot programming will be the main fix in many cases because the entire diagnostics of the craft is logged into their systems to cross check any malfunction. Even if the entire system of learning was destroyed, the robots have the ability to teach from their logged memory banks as a backup application, and are solar powered to maintain their battery recharge which that is also programmed for operation as a charging station is located inside. But when they exit they can use sun light to maintain their power. The top of the structure has a camera structure of 180 degrees, and the cameras have three backups per area of a rotational ability in the event a camera break downs, the entrance areas also have monitors, and the motion sensors allow the camera areas to open above entrances as they are protected from weather elements with sliding panels that even have an automatic oil use to maintain the movement as to not be rusting, all doors and equipment of movement have a yearly small amount of oil displaced on moving parts as timed actions. The other viewer is also near the top of the structure, which allows metal panels to open to see out through glass that is also covered on the outside but has a removable plate that can be done from inside with screw connections of plastic screws and stainless threading, which has heavy sealed covers over the exterior to prevent water damage of the threads. The inside tower will have the ability to even be retracted or extended as this is the control tower for the new students that will one day emerge from the mother craft. But only after graduation are all allowed to reach their learning peak level of understanding from the systems robot teachers, concepts as to potential dangers and what could be required as to maintain life, which once again each compartment can only open upon completed studies, and the key to the next equipment and understanding its usefulness. I also said that these types of structures could be used to build life on other planets, which means the systems are exactly as to be what is needed to determine a good environment, but when reaching a potential planet probes are sent from the space craft to first to see the potential to have the proper atmosphere. If a planet has the proper elements are transmitted as sent back to the mother ship, then the release is made to send the Time Ark life craft to a designated landing area also predetermined. It is also tested for land strength for a proper landing area by a drill to bring back soil samples and test for bacteria levels and harmful components. If approved, the system starts to deliver Time Ark. Once landed; the process is no different than the land-based units back on Earth. The structure will descend only and open 100% once the last studies required are completed and tests are final, as then the outer door is made available for exit. All this time was to prepare the growth and the ability of each new life made in the event that the outside atmosphere does not have the proper composition for breathing; this is once again determined by A.I. sensors, which were automatically sent out from the mother craft before landing, as a test is made prior to child conception being activated. Again, if the test results show harmful amounts of gases, the process will not start until a proper atmosphere has been found to maintain life. This means the mission would instead be aborted, and the ship would travel forward to the potential next nearest planet to be observed by a long-range Hubble-style monitor system. The A.I. will go into sleep mode for 12 months and then test the air quality again. The process will continue until a day when the air quality is correct and can support life. The dangers that may or may not be outside the TIME ARK would be unknown, but the equipment would be given for self-protection, and at the right time, the proper use for protection will be given and trained. The Time Ark is a potential way for mankind to start again, not by scratch, but faster and more effective, as inventions are now based on priority and the ability to build fast, but with one main element included. Man is not to harm man in any way or form, and the protection devices will only be opened or explained if needed because prior to the onboard mother A.I. making the final exit, It will send out seek and find mission search drowns to fly outward in mass detection systems all programmed to show any form of life, record its size and observe it, as something that could potentially harm a human will be sent back to A.I. mother ship to consider it and then take appropriate action, as the drowns view the entire planet, and attain life forms reported as threats or non-threats.

If threats are real, then and only then is the protective equipment taught, and its purpose is to protect others and self-handed to our new crew. The drown observation process could be applied even before the child's development. So, monitoring from space would be extensive prior to launch to the surface. So, I go into a bit more detail and hope you enjoy the way mankind can be planted across the universe. The next application for the survival of mankind I call:

31. SUBJECT: THE TIME ARK, WHY BOTHER?

I am revisiting this subject because I know that if this is made, then humanity will survive, and all the millions of souls that once lived will not be in vain. So, let's look hard at this concept. I am hoping that maybe you will wish to help support this worthy cause and help me attain funding from those who see the great need to keep our future intact to survive. Suppose you can imagine this device that allows all mankind to jump-start up in the future from any world disaster, a massive flood, a moving of the poles, a massive virus, and, the worst, a nuclear war. All of these actions would not affect "Time Ark." The reason is that the entire structure is designed to take massive impacts, massive weather changes, and even be able to float. The structure stands twice the height of the largest pyramid in our world. It has massive outside storm panels that reflect high winds such as tornadoes and hurricanes. The structure also has these attributes: the entire surface is made of stainless steel, with an under-coating metal of titanium for added strength. There is a white plastic covering over the outside that reflects heat and holds solar power sheeting under bulletproof glass. The entrance to the Time Ark is a manual door with heavy latches that have a diagram of the direction the wheels turn to unlock the sealed doors. Doors that are with three metals to maintain the perfect seal and are also stainless to prevent rust. However, the door entrance has a large cover outside that is similar to a shield. In order to open the door, the shield must drop down like a draw bridge that is connected at the top and makes a protective cover to stop all weather and outside conditions, as stated previously. But we must consider that some people in the future may try and enter or destroy this structure, that is why the entrances are like bank vaults with massive lock systems, and the only people who can gain entrance is the people that are inside waiting to be developed, and that system only activates if the structure does not detect human life for over 1000 years and if radiation, or any harmful gas, or virus is present outside it will also be delayed in activation. Now this device can open to outsiders if human DNA is detected on the exterior indentations, but it is like a learning process to attain full access. So why this is structure so large, and what is it for? The Time Ark holds every piece of knowledge ever made, every device, and every book that was with learning is now on an air-tight UV-protected digital disk. The equipment to read these disks is also protected in sealed plastic to prevent moisture and temperature destruction. But before I go too far ahead, Let's take a walk into the year 6396 if to date with our present calendar, mankind has been gone for thousands of years, the survivors will eventually stumble across one of these structures, and they may no longer be able to understand our English language, so as we teach them and their children a video program, this emerges when DNA is matched, and is not an entry point, but will be emerged from a thick metal slide plate a video screen protected by clear plastic as to not be broken, and what we can call the teacher program will start, and this program eventually after the programs have been correctly answered , the computer will ask questions that need to be answered, but first the person will need to learn English, with the first soundings of the alphabet, and each sound as they are presented together, this introduction is with images of objects, and the video has a large play button and rewind to review the content over and over, just two large buttons with arrows- rewind and play, the screen is activated at first with a motion sensor and a floor plate that can also start the device, and a book will be opened from underneath and ejects, the book is made of plastic coated pages and is sealed in a clear plastic cover air tight, and has a pull string to take the cover off and release it from it's contents, the book will show images of things we see today in our world, and the letters of the alphabet will be with pictures, but this book will have a solar panel which allows the sounds of the pictures to be sounded out by chip prerecorded messages. The book will be "Introduction to the English Language" and will teach the user from start to finish every word and sound and how it is represented. Then, the second book emerges with a second video. Now, to be exact, there are more than one of these structures worldwide. Six locations in upper mountain areas are still of good air quality height and on land that has a good level grade to be flat with good rain and winter snow drainage. The entire inside device is made with chambers that seal just like a submarine; this includes the seal of the front entrance. The chambers all carry devices with instructions on how to use them with step-by-step images. The videos are in each sealed room, and the devices all have videos that can explain the use of any devices- The instructions of metals and materials required to make any object start with primitive designs but advance as they make tools that require the next level of development. Every tool is shown how to be made, every part required to build is shown "how to be made," and step-by-step applications show even assembly. The information is with every manufacturer worldwide, every mechanical and electrical manufacturer, and yes, even the development of electronics and how they are made. The list goes on and on and on, and each category only has essential information. Then, we have the history recorded on the top floors and the warning of many ills of humanity, from sexual misconduct to crime prevention, legal actions, and recorded penalties to make a better future. In consideration of the potential of nuclear war, the inner chambers are with an LED covering to stop radiation and covered with water resistant and fire retardant foam. Lastly, in this wonder of information is the "rebirth of mankind." This is a system that activates after thousands of years have passed, and robots are involved in the system, which was discussed in an earlier chapter of Time Ark. The fact is this information and application could be implemented into Time Ark. Some other features of Time Ark are the ability to detect motion outside with long-range radar and a set of loud beep sounds to attract anyone who may be near its doors. The entry points are at the top and the bottom, as over time, potential volcanic eruptions and land erosion with dust sedimentation could bury the device, so the top is also accessible. Both the top and bottom doors can detect human DNA to allow the doors to open. All disk information is sealed in cases with LED covering inside and a covering of plastic outside to seal the containers; each disk has a separate case and is also covered in plastic; the cases have a player in every five cases that can have a disk inserted into the case with image instructions showing the disk being inserted and the insert causes the instant paying of the disk, it will stop and eject when finished. This system is yelling out to you, it is saying "help me" and if you contact this inventor online, maybe you can be that helper to save our human existence.

32. SUBJECT: TIME ARK- NEW WORLD EDEN SPACE CRAFT

This is very similar in conception to the Time Ark, but more like the origins stated as Adam and Eve. These will be the names of the first two to be created, and thanks to technology to add to my creation, in Germany, as recent as February 19th, 2020, a child was born in a lab, and this was considered many years ago by myself, as I explained the concept to a friend who said "That's impossible! But today, the artificial womb has been established. And we can have these devices run by A.I. as to have the proper care in travel, but more important upon arrival an A.I. robot will be on board, and actually is not involved with the travel unless a repair is required, and A.I. onboard system would cover all areas of supervision and directional travel. But once the destinations are confirmed as a potential planet, then the robot would become the developer and the nanny to the newborn once all tests are final as to a landing to be made, and the facts are these robots would be more than able to even protect the children if deemed needed. But our children are with DNA cells that are primed and protected from UV and harmful radiation, and there are many on board, actually hundreds of unborn , that will be developed over time from the original group born, but our first two new potential humans, - just for fun we will call Adam & Eve, with perfection designed in DNA, will be the beginning of any world evaluated by A.I. in every possible concept of the word, but with some very different features or applications for this craft does hold all of man kinds completed ideas of creation and how it is done, which is very much exactly like TIME ARK Earth, but the equipment which could be stored inside "Time Ark Space Travel" is not able to be presented but only in size and how it is made. But I don't want to get ahead of myself. First, these are spacecraft sent in every possible direction outward away from our planet but pointed to stars and planets we have considered as possible life-based, and let's remember that the past is not gone from the memory banks as to the origin of the mission and why it happened, and each unit sent down to a planet has the ability to rebuild other systems, exactly as the system used, in short, the "Time Ark Space" can be a way for mankind to spread across the universe and bring wonderful never imagined. But unlike a manufactured craft, the ship can be taken away from the original course by A.I-.if the ability for life to exist is not possible. These hundreds of crafts are driven by A.I. not people, ; there are no people onboard, but there is potential for people as the system is based on the artificial conception and development exactly as stated prior in the Time Ark, but here is the main difference: the crafts will be evaluated by A.I. as to the size and the condition observed as the proper size for gravity, the proper sunlight that is present, these will be the two starting structures that say to the commanding A.I. pilot- "This is a planet we will investigate", so the craft now goes into the proper orbit, and the A.I. releases three probes to the surface, which now are hopeful of making an entrance with detection sensors to relay all activity and all life forms found, and most important air quality, and acceptable for human development. Once the reports come back as positive and dangers are minimized as even viruses and germs are tested by the drones sent down to the surface , and separate area that no contact is brought up to the craft but , a robot works in the remote system to test and gain visuals of bacteria and virus to see what is able to destroy any particular item found by send vaccine fluid down to the drone probe, and to the area to do all testings, and any virus found even after landing will not able to be allowed to entire inside of the craft. The next action once the quality control has been established is a release of a satellite to observe the planet up close in observation. The probes sent are run by this solar-powered transmission. This device also has is a Hubble telescope but is on the satellite craft to observe items extremely up close, and extends outward only after a clear path of no space debris like rocks, or items which may be surrounding the planet has been found. And if these elements are found, then this planet would not be a potential for life because of the dangers of rocks traveling into the lower atmosphere would increase potentially as a major danger. But let's say that the devices report back the activity is not harmful, or has potential to be inhabited. Our ship does not yet descend to an area predetermined as solid ground, but now the planet is to be observed even up close, now smaller groups of probes to retrieve dirt samples as scout crafts are sent and land on the surfaces to take samples and detect radiation or harmful compositions known, all monitor the same applications, some things will not be known, so we will have simple creatures developed inside our craft prior to see the potential of dangerous species. With the news of a planet being a proper life force for humans, the process starts first the child development, and all food is dry food with water mix for nutrients to development. The entire crew being developed uses these nutrients, all based on an abundance per individual, and the contents are stored as to be for each person. After the children have reached an age of understanding and live up to right at age 9, they are now taught about food production that has already started outside by the robots. They will now begin to learn how to take care of animals, and how to develop their own harvests, and how to maintain food growth. The seeds of animals, which will also be formed artificially, are now started by the robots, as the robots programmed now begin to build the structures with tools provided as to size and a step-by-step step how to make the cages and the proper feed which is to be grown first and stored, the ship has a group of prefab units that make this activity easier, as the systems can actually be taken off the ship remotely and then placed by robots as to the proper distances required for structures, this includes housing, electric, heating, cooling, insect control, territory observation, security, and yes deterrent protection elements if needed. However, to be a second step is to determine if life made for food nutrition can be accomplished. So a detached ship with two robots are sent down to begin test for seed life needed , corn, wheat, rice, grass, potatoes, cabbage, carrots, beets, to mention just a few, then if the items have developed in full, and are then tested for any contamination, then our animals seeds to see if certain living animal elements can now survive, Chickens, turkey, pigs, cows, , and let's not forget fruit trees, and fish , yes if there is water , the fish could be developed in small man-made lakes and fenced to prevent any possible predators for all living elements, and the feed for these animals is to be for a ten year period each to start with six animals with three male three female, with recorded progress observed as to infections or potential harm to the development, the robots are equipped with programs that could save the lives of animals, and have a series of antibiotics that may be effective, they also have the ability to develop antibiotics for everyone and all animals have a system security locator placed in ear and a camera monitor in the event an animal has been killed. So any potential danger will be compared to life forms of our planet; all this information will be delivered from the probe scouts sent as they will be equipped even with diseases to be identified as possible elements that have already been worked on by Earth science. What has been proven is that viruses and germs have been found in meteors that have been opened, and the components prove that many viruses come from other places throughout our universe. The probes will be able to test the elements with an onboard analysis and do antibiotic testing to a degree of if they work, then the problem can possibly be controlled. So the process of the landing has the potential of years upon years prior to landing, but let's say the process pass's go, and then the information relayed back to A.I. pilot, we have now made the new home for human life. Of course if the reports are found as positive, and all evaluations show the planet has the desired components , but we can consider that there may be actual creatures that are similar to our planets food sources, and we must have the animals that are grown to survive before a landing can be done. If there are animals that could be used for food source this would be tested also for the elements must be close to what is eatable. But are, say a fruit that has vitamin C in the fruit, this is a long-term research. But once the findings are completed. We then have "The process for life." And down we travel to a new world and descend to the new world. But there is one major problem, if this planet already has intelligent life on the planet, or items found are extremely dangerous to human life. The A.I. Pilot will retrieve the scouts as much as possible, then remove them from orbit and continue forward to another possible planet.

33. SUBJECT: SCAMS AND FRAUD

Today, we live with a massive amount of people worldwide making false claims to gain our money, and they use every tactic to get your funds, as they will even pretend to be Godly! I hate to say it, but hell- bound is their destination! To think of using God as a device to steal, not well advised, but he is laughing; well, if he is not, the Devil is! But now, to a more serious tone- These actions have gone on far too long, so let's look at the real view of these issues and what they cause. Well, the obvious is- "Financial loss" for anyone who says, gee, I get 1 million dollars for just $350. What a deal! Well, they go even further with many people who have had their bank accounts cleaned out, the access coming from areas of the world where the parties who lost their money-have never had their funds recovered. Okay, so what is the second result- we are a people seeing an opportunity now have been so aware that the people who had the damage done to their funds taken, are now extremely leery for anything which resembles a risk, or opportunity. These people have been causing a division unknown, a way that other countries that hate America can say- we are winning! You see, they have turned you into the American inventor, the American willing to help another to become cynical and downright resistant to any opportunity that may actually be real. So, how do we weed out these stupid fakes? The answer is simple; we need verification of all business opportunities online, a signature emblem that says click here, and a live individual on a chat line can now take the I.D. number as verified business, and the name of the business is a scan code you take that offline to send to the fraud department. So, the entire Internet of Business must have a business approval code. This code will be even for non-profits; the goal is to stop all scams that contact people in emails. The email must have the link to the official scam verification if it is to be taken seriously. The job of the scam finder is to give a report of any actions that have to be investigated. Now, many of you may say, many of you may ask, what is wrong with this issue? Well, as I said, we have a nation that has stopped trusting each other; if we fail to have the most valued possession, there is to be halted; it's called growth! You see, without trust, we fail to accomplish the smallest tasks that may take trust; we have to worry because we have no security against theft! And the real issue is we need each other to build, and when we cannot build we fail to grow. As we have seen a massive decline in growth, here lies the foundation being torn down, as people refuse to listen, refuse to act, refuse to look at the little guy who may just need a small financial lift to establish something worthwhile but distrust has been built, and the powers that be have no desire to find the scammers, so we need an international online investigator who can arrest people who do these terrible acts. The main issue is for us to recover and start being open to those who say they have a real opportunity. So, I have decided to say we need a forum that does not allow fakes because the idea must first have an NDA a (non-disclosure Agreement) signed by the reviewer and this should be set as law. This protects the idea person, and it has no time limit. Then the idea is presented on an online video such as "WhatsApp", and we can now present it to live investors, but wait, we need to verify the investor also as actually someone who has the investment funds, so before the investor can see the idea, he must show proof of income, once again set by law, and from as screenshot from his or her bank, and all account numbers are cleared, so we are able to keep the balance of idea and investment as a way for people to grow again. , and more importantly, to place faith back into our nation's people. So here is the real, even more serious solution:

34. SUBJECT: THE PARASITE ACT

Our first major change is any business email to be taken serious as a legit offer, or sale of any item, must have a posted in header I.D. clearance number, this is something anyone can go to the check online website to verify if the company or person is registered, if they are not your looking at a potential scam. A law could be placed in the future that all countries are to have the IP addresses of their nation's populations connected to an international hub storage internet cloud, and they should be with teams of investigators who follow the IP address trail to the parties who are at these locations. The emails of people who send false information can now be traced, and they can be arrested. The fines against scams are divided with the nation as 20% for recovery, 10% to each nation, and any parties with claims can now have access to the remaining funds confiscated. If a single case, they would attain 80% of the funds taken, and the 20% is for the efforts of those nations involved in the arrest to be divided. The future of false claims will also be in this agreement, whereas if a person says that something is "free," it better be exactly that without signing up or making some effort to attain the item called free. The term will be held as a responsible action, and those companies or persons who decide to make false advertising claims will be held responsible as potentially arrested and will be fined. It's called fraud, and when advertisers say something called free, it should be penalized if it turns out the offer is a fake, and you're now forced to fill in items and buy items to be able to get the so-called items that are not right up front. So the amount of the offer is the penalty. Take that, scammers!

35. SUBJECT: PRESS ONE, PRESS 2, PRESS 3, LEAVE A MESSAGE

Yes, the wonderful answering services we get from our phone company, and other business companies, and there are many businesses we need to contact in our daily lives when we try and contact our bill collectors, or V.A., or hospitals, or any business that we are dealing with, and let's not forget the long dragged out on hold music!

Gee, a wonderful way to spend an afternoon just sitting and hearing a song repeated so much the only invader is some voice that interrupts the music just long enough to repeat an online way to get what you want to be done, and of course, if your calling about your internet down, gee wonder how that can be done online, and let's not forget the over 50 crowd that never wanted complex phone systems, and we just wanted to make a phone call! And let's not forget the so called smart phone which is designed to have gadget after gadget that you never use 90% and this is okay because they can do facial recognition technology in picture taking which the so called up load for your phone is actually to send back to the government of your phone information and who you speak with, and what you looked at, all to invade your privacy which they insist is okay? But if it is okay, why are these devices on our phones? Because they use them would be the answer and they do it even if you don't want it. The fact is, people we have never realized that these so called devices that are supposed to help you and business but help government control, and your tax dollars help watch you and actually harm the nation in ways never ever considered.

It's called removal of your rights of living without invasion of your privacy and not able to revolt against a government which wishes to take your freedoms, this my friends is in our constitution that we have the right to re-gain our freedoms, but if the enemies of freedom have your every move, with GPS, and your Wi-Fi location, and every word you say is recorded as said on your phone, which now has voice identification so they know if it's you speaking, looks like privacy may be more important than you thought? Now, let's return to the press one phone issue. If you have ever had a company that you had to contact, maybe a bill that was wrong, or you needed to speak to an office personnel for some reason. Perhaps you want to actually pay a bill, but they want to give the customer or the person needing information or to attain information a thing we all know well- it's called DELAY! And the frustration of sitting on a phone, for even hours, has been done by my contacts made as to sit there and keep hearing over and over and over, "We will be with you shortly, and then they go into this long information as I stated before, that if you need to resolve your issue go online to www.no-thanks! Because they never consider that-oh! My reason for contacting you on my phone is because my internet is down! And that is why I am calling you! Okay, the reasons to be on hold can be hundreds of reasons.

But we now need to look at just one example that could bring some light to these so-called leave message saviors, or the wait to be actually taken to a real person, the press one if you need this or press 2, if you need this other department, and the list can be very long, so you now are sent to the wrong department, and then that person finally answers and transfers you after you have sat explaining for five minutes, but guess what, they send you to the wrong place, or your now sitting there and the phone drops the call! So you have to start all over again! These devices that delay our businesses and homes need to get rid of these time wasters, and very soon, you will all know exactly the real reason they need to be gone, and yes, the good old operator picks up calls to send you to the correct department, but even they need to be better trained. I have actually had a secretary stop contacts to the people I was supposed to talk to, that is absurd, but that also happens. We in business need an insert on our cell phones that have a business verification I.D. that says, "I am real" but back to this time waster. The computer A.I. Press systems used daily by millions actually cost American billions per year in failure. The fact is when it comes to business, delay can make the difference between failure, and success! Why? Well let's imagine I have just invented this new device, and I am reaching out to find some help with development. I am calling patent people, being placed on hold, and told to leave a message, and they get back to me a few days later. I missed the call for many reasons, maybe my cell phone powered out, I tried again, only to find that they were not available -"again", finally the next day I was in contact but told I had reached the wrong department, the-person who gave me this new bad news, now sends me to another person who is supposed to be available, but what do I get? An answering machine!

These are just a few examples of how the instruments we consider so helpful actually delay our progress. But, before I stop just one real example that happened to me as I am a businessman needing to contact an artist to do a packaging design idea, and once again, the company sends me online to a chat box; the chat box is another device that says, "we'll get back to you, leave your email and your name, and so you're not even talking to a person, but- "DELAY" has happened again, this project for one single person now has delayed days of the progress, and we have not even touched the surface yet, why? Because one day, the project may need a "BUYER!" and I call the company, and they have press one, press two on and on and on, and I never get the department, because what I want isn't even listed! What is the result? MORE DELAY? Yeah, because people never even bothered to call me, back. Why did they never even listen to the answering machine, and maybe it was erased? Or the message was the message cut off by mistake? My point here is if we counted in just one person's week of how many times a business is delayed or a personal issue is delayed, or a payment is delayed, we have accumulated hour upon hour of wasted time, and a massive amount of delay. Now, just for a projection and the example of what this impact has on our country, let's make a calculated lowest possible conclusion of how this may or may not affect the progress of a nation and, in truth, can cause a nation to fail using this system, that's correct fail as a productive nation.

THE FACTS: So let's say for giggles- that a single person in a week has been delayed-oh, - three hours at their business, this is just one day, 3 hours, placed on hold with the cell phone company, or had to leave a message, but a list of different things that all caused their business a small delay. Now, let's just look at how many hours our single businessperson has totaled per month and year. With just three hours per week, we have in a year 144 hours. Now I am sorry to report but I am a businessman, and I have had three-hour delay in many days in a week, sometimes I won't get a reply for weeks. So, this 144 hours in a year is really dirt low.

But, now as Google reports that in year 2022, we had a total of 33.2 million small businesses not even including the large business's at all which are: 33.2 MILLION businesses! Just a few, but all are having one very important thing happening really daily- DELAY and inside an office with thousands of employees having these same issues, so how many hours do you think they are at a minimum- lowest possible for the nation in hours wasted? How about 4,636,800,000. Hours. Now big business. So, how do we fix this problem? We make sure our employees are no longer connected to these devices, the country will run smoother, faster, and effective. An operator will listen to your request, and operators need to know exactly who handles what, to include people seeking to present an investment opportunity, at least a mailing address, but we need to function respectfully as proficient and speedy applications that get us much faster communication results that don't cause delay! Plus, why do people who pick up phones not know who does what? Training, training, training needs to be every week, as to who is working, and a list of who's who for each department with a brief explanation that tells the caller this is who handles that request and be on the money. And if this is applied will be able to interject into the call to see if your being helped and getting an accomplish call? That question not answered correctly then that caller will be directed to possibly someone who knows the facts. This will be faster, and much more effective, and saves business's all throughout businesses all throughout the world trillions of dollars of valued time.

You get the correct answers and are directed to the proper person who will pick up the phone, or the call goes to an open phone, but if the parties are already on the phone, then the operator, after a minute of seeing that there is no answer- then will take your number and have them return your call as next in line. The country needs to back up, as we can compare this to a truck stuck in the mud and a very large tree is sitting in your path ahead; reverse is the only solution; reverse to operators who direct customers, direct the buyers, direct the idea, direct the production, direct the growth faster and effective rather than frustrate the public, and the company seeking to do business. If we did a research of the actual hours wasted per American from these devices, it would be more than what has been projected here; it would stun the companies to finally realize they have lost productivity! It is obvious "MADE IN CHINA" they do not waste time, they act ASAP, they are motivated to win, and they are winning because time is being used effectively to even develop. Now lastly, I spoke about the delay factors earlier and what is the real motive to be the one FIRST TO MARKET; while we are all not working effectively and lazy about how we communicate, we have lost just within the month, so many ideas to overseas developers, not being first means your second, third and last, and your market is smashed if you're not in the front row as the first company to be showing a new idea, a new product, you're a loser if your to believe that being last to the market place is okay. First to the idea, first to the development, first to the money. It is high time we are first again.

36. SUBJECT: FLOODING

The idea to stop flooding is basically a "relocation" by allowing the overflow to supply water to areas that have no water. The Romans had aqueducts that took water from mountain ranges miles away and brought the water in a downward slope so it was used for many purposes including Roman baths, so the concept of transposing water to another location was developed before we were all born, and some have considered it but not to the point of what I propose, but the water used cannot just be a simple transfer, it must be a filtered water, and made usable, and the storage must be sufficient to maintain the areas needing water as required consumption, this must be obtained to make the water consumable.

So my idea is to build large pipelines from areas that are known as floodwater areas, but before the water is sent is allowed to settle any sedimentation, and this sedimentation would be used for farmers and growers. Including drinking water. The pipelines are to travel to underground storage tanks that have the ability to even travel, being they are all tankers that can be loaded on rail or flatbed truck, that when each are full, the water is now sent even further away but always stored underground to not allow water evaporation, so there are underground warehouses that the tankers are parked in rows with container attachments that have a solar power electric pump that can pump water to the surface when needed if there is an excess, the excess is used for desert renewal to make new farms also delivered in water tanks above land so water when released can spread by water sprinklers.

The problem we have in dry areas is the water will evaporate too quickly to be ablest the water will evaporate too quickly to sink deep into the root system. And the cost to make massive greenhouses need to be a progression as to grow a crop with this water supply, give the farmer the seed and the overhead heavy canvas with poles, and shade the growth area , this the water to stay in the soil longer. A heavy canvas similar to a circus tent crops to keep water from being evaporated as fast and still allow enough light through to allow growth. The poles are braced no different than circus tents, but the ends are with metal covers so that a large hydraulic hammer can pound the stakes that are twice as long into the ground to make sure that high winds don't blow the tent covers away, and during high winds, the canvas could be closed to protect the crops.

This could be a basic way to start new crops in desert conditions, now we will need rocks removed and fertilizers so that process is first. Now back to the distribution of river water to desert lands is to the new design I have made for tree growth and food growth can also be for a whole new investment I mentioned earlier as "stack farming" methods could be developed to even grow small trees that could be moved later when mature to a good height, the system water is stored the same exact way, the difference is we can build massive underground tank structures for abundant water levels the size of reservoirs that have over flow, could be a means of distribution that can move water from further areas in other states, but storing water is the secret, because when there is droughts, the water is there ready to take on the problem.

But we have never considered that where water has gathered a small lake can evaporate, but even this could be slowed down by large shading high above the lakes, it used to be trees that did this, but mankind has cleared the trees away from many water sources and the reducing sun light reduces water temperature which will slow down evaporation but this method takes time to replant trees large enough to keep water cooler, so the sky shade windmill is the answer.

Not saying that this is a perfect concept, but could be with visitor towers as to overlook a lake, and the windmill is made of a colorful heavy replaceable fabric that when in motion actually turns a generator and produces energy but also shades the sun side of any lake to keep water cooler, that stops the sunrise and the sun set as to gain more shade, and they could be made to look like blue sky, which would not look so bad. So we have step one, the development of pipe lines that are located near the edge of the rivers say Mississippi, or Nile river in Africa , or the Ganga river in India. What you will be seeing is design that shows how the future water runoff is made.

This depiction is for a new withdrawal of floodwaters method not strictly for rivers but also for mid-size streams or even large creeks that may flood from heavy rainfall. But in each case, this extra water can be kept in underground tanks that are massive in size, and when I say massive, these are different than the water tankers, they are like previously designed containment's made of stainless steel and a whole the size of twenty soccer fields is the actual size of this man made under water storage tank, which when completed is then covered with a plastic coated cover , that is sealed together and makes a water runoff for storms, and has ground cover of 6' (feet) deep now placed over the entire tank to keep the water cool but the water is first allowed.

The pipeline for the extraction of water is located in a slant facing up stream and is only used when the water level reaches into the entry point; the water taken from a river or lake is extracted from a pump station lower to be pulled up by the pumps to leave mud and debris. But to reduce the mud and any potential debris such as tree limbs, the water near the actual entrance of the intake is filtered as it passes through a series of large bared screens that prevent the trees and scrubs from blocking the pipeline entrance, and have an alarm that sounds when the water intake is being blocked a pressure plate will be the activation alarm to let someone know that section B of the water intake needs attention to remove a blockage, so they then un-block the line. The first screen is very large, then a second screen is used for smaller branches, and then a third is used for twigs, then a leaf screen size of 1 inch, which is essential that at least four people monitor these screens in the event of a flood. Each has an alarm and has to be cleaned out when not in use. But the lines have manholes located every 50 yards that have elevators to be able to enter the lower water entering and this area is like a massive underground tank that, when the water is all removed, leaves a great amount of dirt behind, so there are automated scrapers that push sludge into conveyors that carry the sludge upward which will be used for farmers to have new top soil.

The ladder shafts extend just over the water intakes so that workers can pull off large trees with a bucket to grab any large trees that passed over the filters from water pressure to clear the entrances of water. The water now travels to a series of holding tanks and is pumped into the tanks there are twice as many pumps for each line in the event that a pump breaks down and a backup gas generator system in the event of electrical failure to prevent the use of pumps.

The pumps move the water into holding tanks, a series of 20 very large tanks located outside each tank has an over flow line that when the water reaches a height of 150 feet the water will now travel into the over flow located at the top gravity takes over from there. And the water travels to the next station where the process is done again this method insures each area of excess water supply is cleaned effective and perfect before transfer to main pipelines or to tankers waiting for delivery.

When flooding is done, the water can now be pumped into the final water sediment trays that the over flow is now inside these massive trays. But once again, before the water ever reaches tanks, the water has to go into a basin.

First, a large man-made tank; I am repeating this because of it is essential to remove the sediment, and this is the first of three-step removal, so the water has a pump that enters into this lower pool, and the filter has a cleaning mechanism that actually uses a clean water spray to exit and be cleaned whenever the system filter says it is being clogged, so the pipeline to extract the water is also drained after each use, and once the flood has stopped, and the water has settled inside the massive holding tanks, then the mud is removed to become topsoil to be trucked to farmers.

Because this mud is filled with nutrients of dead life, but the outside filters, which are near the water entrance, are used with a conveyor that moves the items to waiting trucks that also use the elements and remove garbage in the water like plastic bottles or large debris is all separated from the screens prior to the cargo being re-processed, but as the items are placed on the conveyors toward the end of the conveyor is a metal distinguisher for items that cannot be chopped, such as foam and metal cooler, or any car metal, but these items are collected first and never gain entrance to water being removed.

But they could be sent to a garbage dump for reprocessing; all items are directed to the proper truck waiting on the scale which will tell the operator the truck is full by an alarm signal to the main office. This means plastic goes with plastic, metal with metal if found floating on a structure, or wood with wood.

The final water destination is now ready to be sent to farms by a high entrance pipe line, similar to the Alaskan pipe lines, and water to desert to be distributed by smaller lines to holding tanks in closed located underground and water is pumped to needed area. Smaller Section 2. These are a series of ten backup pipelines per water intake area that are activated only in emergency flows.

No filtration is done at this point because the holding tanks became over-clogged, and now the system is in Pearl but is more than able to handle the overflow as to direct this water to another water holding system, which has no filters but can take the excess, and if needed can have tankers line up to take the water directly into tankers, this water will be needed to be transferred to cleaning after the flood waters reside.

Also, we have emergency pipelines large enough for men to walk through that can take water to manmade lakes that are actually hundreds of, miles away from the storm area. This system will require that a cleaning truck enter the pipeline to move dirt out of a cleaning truck to enter the pipeline to remove dirt after the storm has subsided.

This run off is above the original entrance but at a higher placement next to the river's edge once again above the other system, and this is strictly an over flow pipe line which allows water semi filtration as to have metal bars stop large debris water is done till later. The same in effect except now the water is used in the local areas storage of 20 miles, 100 miles and storage 500 miles but can be directed to the desert lines if needed by runoff lines. The desert lines do have holding tanks that can store water for large cities that are in the same direction as the line.

But are not to be used by the community when lakes or reservoirs already establish sufficient water; the majority of this water is for areas needing water from drought conditions and already in an emergency as needed; 70% of the water storage must be used each year to allow any overrun storage to be available space, all massive storage tanks that are not transportable are located below ground level and will automatically turn off by themselves when full as the water will now continue forward by gravity by the same line connected but is now a bypass to allow the water to travel to the next tank waiting potentially hundreds of miles away. So each tanker is now being filled by the continued over flow from floods, and allowing a drop off water to each underground storage.

Let's say an area which has been considered a drought area, if this water is needed badly let's say in Nevada desert area these underground tanks drop can be closed. This water delivery system could even be used for regular lakes that have been drained for many years by drought, which could allow the development of fish and help the wildlife population.

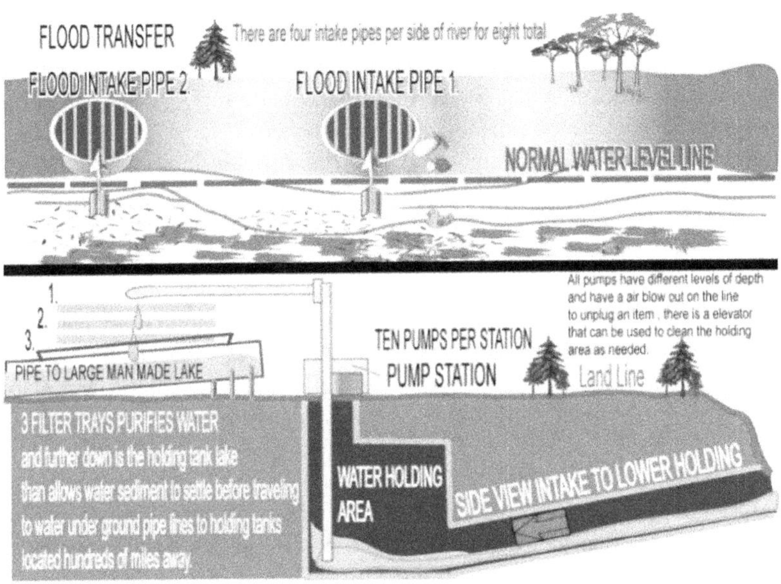

37. SUBJECT: FLOODING CONTINUED

The main filtration plants located in massive rivers are around 50 per side of large rivers situated every 20 miles from each other and use bridges to maintain the water flow degree conveyor screens. This filtration system is a device like a massive filter, to another massive filter to another massive filter, which by the time the water reaches the delivery pools is already cleaned in many ways and could even be transported at that point of receiving the water as final conditions are approved before water is released. The main and largest of the water flood control system is the debris removal, as stated before, but let's look at the system a bit closer; the entry pipes into the pool water capture tank, which sits about 60 feet below the river about 60 yards from the banks is where the water will start to come in, the water is to pass but the level float device which is floating on top of the back of the intake tank, this is to tell if the flow is being maintained as to not be disrupted as the water is being cleaned, over 20 pumps are making the water travel upward to the upper screens that are stainless screens, that can take the elements of sediment out so fine that it leaves the water pure and clean, the water being sent is also bombarded by UV light inside clear pipes that will reach the final filtration systems, and a dose of a 1/2 bottle of chlorine is added every 5,000 gallons to kill bacteria.

The lines submerged in the water almost to the bottom but raise when the sediment increases that blocks the water flow, and on top of the water also about eight feet above is a float to make sure the water entry does not break the lines , they are secured by connection rails that allow the pipes to be lowered and lifted when needed by the float system a detection level indicator is about half way to tell operators the tanks are needing extra run off to lower the intake , but this also helps prevent the excess water that is possibly filling the entire tank a way to prevent any items from blocking the next water flood control system which is above in emergency conditions, and the water gates as to remove heavy logs, limbs and very small items in the water that is causing, The outside top entry port has very large metal ladders with heavy connections that are welded large swivel ladders, which can swing outward to collect any and all items that are big, possibly blocking the device, because of the possible water current being so intense, these structures have a very long base connection deep into the ground and are made to with stand a great amount of water pressure if being hit by a massive over flow, but with all the services in place the chances of this happening is practically zero, there just to many run offs that can be added to decrease the water retraction intakes. They are run by electric engines, which are run by battery storage, and the device has a pressure sensor that can detect a large item as a full-size tree and sends a silent alarm to the local highway department if the device is jammed. They will have the ability to make the device go back and forth to dislodge an item stuck and a walkway over screens on the rivers, but the screens can be lowered down into the water if a boat needs to go through the screen area. But the device normally can handle almost any size item because the areas that the grab ladder is moving through have very little resistance, but the water pressure and the items are with powerful scrapers as the cables holding the bars turn upside down as it goes ashore this can dislodge any item stuck and the scraper devices that are to dislodge any limbs is extremely strong metal alloy to maintain the integrity that the lines move properly. And make sure the large debris chain stays free of trees and logs and brush, or any other item that could be dragged down stream , the dislodge is actually a chain saw mechanism that can be used to cut away a wood that has caught. Let's say a boat is coming toward the cleaning screens, any boat traveling will have the ability to see signs far upstream in bright neon colors that they are coming upon a screen system, and to use contact to chain operator by a cell phone, and a CB as to open the areas, boats traveling must also have schedules that are linked to operators, this system could be operated by A.I. but to be perfectly run the system should have workforce and make some very important jobs of supplying fresh water and cleaning our rivers of items that should never reach our oceans. So, an alarm if anything ever gets stuck or someone needs to have the cleaning filters lowered temporarily, and he will have an overhead lift that he can use for hard-to-reach items that reverse does not allow the item to be detached. In the illustration provided, we see the over-the-river or large creek filters as actually three filters that allow water to travel as once again, these filters catch debris in smaller sizes; they rest on a concrete slanted area which is covered with stainless steel thin sheets that prevent the concrete from erosion, and this is also our be converted to a water overflow screens that can fill tankers with an attachment cone- like intake for additional " flood water collection." The water removal on this river that now all the water is now flowing over this as the water flow they removed nutrients from the water and sediment, which is being transported out of the water in the same manner as the large filter ladder removes trees and large limbs. This application is to retrieve topsoil erosion.

As we have stated extensively, the water will be used to transport water to farms and communities that have no water, but at one station, which is before each of the six lines, the water is placed for drinking purification plant that is for a 50% drain off to be bottled, and transported by rail, or tankers. In short, the water will be used for farms, people, and farm animals, and the water tanks in drought areas will be underground. The need for water is now gone, and the investment is around 20 billion, but it will save the future for thousands of years. Now lastly having water return to deserts means the ground temperatures decrease, this mean over time the planets temperatures will return to normal activity. A win-win for everyone, everyone that is thirsty, of course.

38. SUBJECT: FRESHWATER LAKE POLLUTION REMOVAL

In a Google map application where we can see high images of land and water areas, in my observations, one day, looking for the change in tree cover, which was older pictures and the newer same areas which showed an amazing thing called decrease in trees, there also became a mess seen high above from the shore of lake Erie, there a massive ugly dark brown stream a stream about ten miles long, there it was-, a massive dumping from a shoreline industrial business. A gigantic brown mess was recorded in a photo image, an image that made my skin crawl, knowing that this was the same Lake, Lake Erie, that when I was a child, our school had many summer events with lifeguards taking us to go swimming in the freshwater Lake, this mess was heading just off those shorelines. To see it being destroyed and many beaches are even closed in many areas, or the signs say, "Life Guard must be on duty, and there is no lifeguard. Simply, the authorities have tossed up their arms and said, "We will do nothing to stop pollution!" I propose that the days when factories were near waterways should be relocated miles inward, but being close to water travel, we don't use it anymore, and it should be a thing of the past. As stated before, I am determined to get people to listen, which, guess what, why do you think T.V. commercials state a product about ten times before the ad closes? The reason people don't listen unless it's being dumped on their brains over and over.

So yes farms which pesticides have used be "moved" yes moved further from creeks, and from drainage systems, and with this application, our waterways could see a greater improvement. But we can do our part by having lake water purifier systems like this sun evaporation system, which purifies the water and returns the clean water back to the lake, and the sediments and chemicals are later scraped off after the large tray areas dry completely. This could be applied in summer months, around every two weeks, to the air temperature and amount of sunlight predicted by weather detection stations. This may actually help people who use lake water to be a place to attain water again, and here is some bad news for all those fishing boats and water crafts that use oil engines: a law will be passed that only electric motors can be used for water sports, and battery-powered, and a fine of $1,000. It is for anyone who doesn't follow the law of keeping these boats off freshwater lakes. Let me know if you don't like this idea because I have an invention you can use that won't need gas power to move you. It will be available when the law passes, so this means all of you who actually care about drinking water (unless you're a camel), then do something called:- "I'm going to call my congressman and tell all my friends to do the same, and tell him this: "Hey there is a guy who says he can provide water fun without pollution being added !" he's a nice person, and he likes to drink water that is "clean": how about meeting him with a nice cold water bottle filled with fresh, clean water? How does that sound?

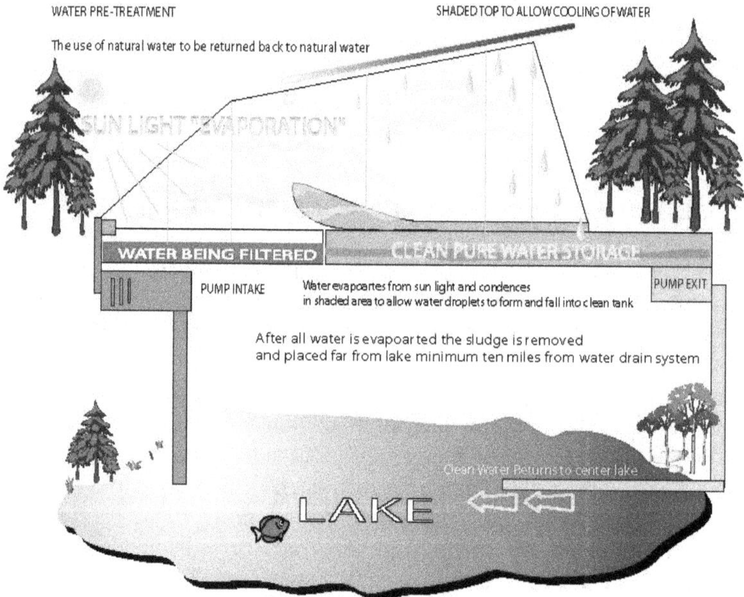

39. SUBJECT: ELECTRIC CAR AND TRUCK STATIONS

It is the hope of this writer that all automobile companies will be transferring there interest to electric cars and the creation of electric stations. And for anyone who has an interest in making electric stations more profitable, then, please contact me. I have a perfect way to do this in a much better way. And surprise- guess what- here it is free! The future operations of these facilities is the same as the locations we have now where we have to stop and refuel. The big difference it will take half the time and much less from our wallet. Because electric runs by line not by trucking as stations are filled today into tanks. There are no tanks at an electric station. Only battery cells are mounted on a carrier basis that operates by being hooked up to a connection" recharge station," and there are over 50 recharge stations per electric station, which are in actual ability to charge a battery when one has been brought in, and the recharge station is similar to a medium size warehouse and in the sides, and the back is connection station. Now, your car doesn't hook up to them, these stations are for recharging batteries only. When a customer needs to refuel, they pull up to a connection area outside the building. An attendant takes an empty transport cart and removes the batteries from the vehicles simply by opening the truck area, and the entire battery set has one connection release, which can be disconnected safely with an electric pre-shut off for the electricity that is still left in the batteries as little as it is, it could still cause physical harm, or possible death if not handled properly and safeguards would be made as to prevent contact. But let's face it so can gasoline. The customer never comes in contact with the battery transfer. In fact, you could smoke around these station without worry; the attendant doesn't need to know how much fuel you need as all replacements are the same amount and cost worry. The refuel attendant now goes inside and goes to a" battery set" that shows that it is "full" by a meter that displays the electric capacity or amount of electricity stored. And It has a light above it that shows it's ready. Then, they turn off the connection grid to that battery that was charged at the station area only if it's still on , but an automatic full capacity could be above the battery system that turns on a green light when it is full, and these recharge power grids have an automatic shut off when battery are completely recharged and they sit on long travel carts that the forklift hand held controller can now pick up a battery needing to be removed, and now place it on the cart, and the next battery in line taken off the train travel cart, and that new battery is installed in seconds. The contacts are turned back on after being connected, and the trunk is closed, and the payment is made. Bingo! We have just made a complete electric transfer in less than three minutes, especially if the loaders are waiting for customers; this could be done with a rotational conveyor, which rolls the needed batteries to be recharged indoors, as another roller conveyor rolls out the new battery to be installed. Now, the battery operator he or she takes a connection cart and unloads the old battery set just taken from the vehicle onto the inward recharging line as stated, and the inside operator connects the needed battery to a recharger and leaves it in a connection recharge station until the indicator shows full, each refill station is about the size of a small closet, the cart now is placed under a newly recharged batteries set lifted out and wheeled to the vehicle and connected, and you're on your way. All battery connections have an exact battery connection slot to be able to load or unload a forklift- style base to be entered into the holding rack, which can also lift a used battery from the charger cradle. The cart has wheels that allow it to transport the batteries safely. The load area of the battery replacement installation can be in front or in the rear, which is preferred by the manufacturer. The area of transfers is kept dry, and wind tunnels are used to dry cars on rainy days. They are loaded on rubber mates beneath the wheelbase to prevent any electric transfers to the ground and harm to attendants and customers. Customers are asked to stay in the vehicle until the transfer is complete. This transfer time will average 3 minutes. The time it takes today is 10 minutes and a wait to pay either at clerk or even transaction. The payment for a new set recharged will be $5.00 only, and can never increase because of the new methods of making electric will be at no cost to the developer. Wind and solar will be the only means of producing electricity, not by coal, a fossil fuel with limited potential and hazard to water and the natural environment, and also is mined and strip-mined, leaving major damage to the balance of the planet. With the newer designs of electric-batteries I can only say to investors, what are you waiting for?

40. SUBJECT: THE BALANCE OF THE PLANET

Very little is considered about this issue; the planet is a revolving large ball, rotating on its axis because of its size. The inhabitants of this planet in the past could not move mountain range after mountain range, the displacement of rock, minerals taken from the planet, and dirt in the number of billions of tons displaced, let's not forget the billions of tons made by massive cities with high rise after high rise, oh we can agree this seems so unimportant. Still, the equipment today can make land where there was no land, and man has left massive holes in the ground, removing diamonds and coal, and so much more. Research should be done on areas displaced and removed from it's natural environment where bridges and railroads have been established or material was removed should be replaced with counterweight. This may sound absurd to the average person. Still, mankind has never considered that these actions of displacement, such as the Pyramids in Egypt, have been moving the planet's stability for centuries.

Let's consider a small ball; if we placed a rubber ball on the floor and then gave it a spin, the ball will mostly stay where it was spun, but if we placed, say, a weight like a small fishing lead weight, and taped it on the ball, then let's spin the ball on the floor when we spin this ball, the ball will wobble, the action is a simple fact that if mankind has no consideration for the planets rotation and the displacement over time would make our orbit change no different than this ball used. This action would be an instant freezing of our planet if the orbit had a disruption that could send our world outward from the sun, and all life would become ice, including our oceans.

The space observatory equipment should be equipped with a yearly depth of land mass, a visual calculation and tracking of the areas already affected to rebuild the mass, and image displacement areas balance should be observed by space of what is a potential hot spot if the land is displaced, areas that have unbalanced can now be filled with areas that have been removed in the past. Materials should be highly considered as the almost exactness of what was removed previously needs to be filled in by the same people who did the damage for their profits.

So the best way to do this is when any material is removed and it is taken out as to be stored outside. A material of similar weight and ability to maintain the consistency of what was taken is now replaced with a similar material of the same descent and weight volume.

For example, coal is burned, so it changes its weight, but to replace it would need a substance similar in weight and able to fill the mine; the substance that could be used is a combination of hardwood chips from fallen trees, and this element added into a prior coal mine could become a decay and become like coal many hundreds of thousands of years from now.

41. SUBJECT: OZONE DEPLETION & PROTECTION

By theory, which means unproven, the ozone has been depleted by numerous attacks from everything from CO2 and rockets shot from many angles from many directions into space. The exit of our world should be maintained at an exact exit point at all times so as not to punch holes in this ozone layer. Also, it's possible that if carbon based products and oil-based emissions were not being used to the degree used today, the ozone might repair it's self as nature has the ability repair it if left undisturbed as much as possible. My reference to this is limited, but common sense says that if these are holes made from exits, we must respond to the future, as the need for the ozone layer is a very important protection of our planet. We must be more mindful of where we use our exit to the stars. The equipment I am introducing as an air pressure lift device to launch into space could be with areas predetermined as launch times that would use the same exit location in the Ozone. Let me add that a new method of traveling into space has been considered by another inventor, one that is not stable; allow me to present the way the item is used; it has a ribbon that will enable an elevator to travel up this ribbon, which is attached to the ground and to an upper state where payloads can now be picked up, sounds wonderful, except! Such an extremely long device has a major flaw. The idea means that in the event of natural storms that can rise in just minutes, storms that could disrupt this ribbon, grab it, and destroy it 100% as it is pulled downward and pulls the space satellite with it. It sounds impossible; anyone who has seen high winds knows that even aircraft have crashed due to unexpected upper turbulence. So this is a total bust. But good news, my system allows me to travel non-stop into space with the ability to launch payload after payload in relatively minutes apart. Plus, after the crafts are emptied they are simply dropped back into waiting pick up cargo ships in the ocean to a GPS location, the craft is none manned and A.I. controls the drop point with onboard computers, with backup system. The system is only used when weather permits, as the device is like a massive peashooter. It makes the ride easier and much faster as directed to be caught by massive magnetic attachments to slow the item's travel and latch on to move it into cargo bays. We have one major problem: we now have many nations making holes in these very sensitive areas in our ozone, such as Japan, China, and other countries seeking to add their communication technology or war devices into space. The need for a global agreement should be established that only a small group of countries can use the devices of space launch to enter space, and the areas must be of the same use of location as to be monitored by those countries so as not to make new holes. What should be done to avoid the destruction of the ozone is for nations to unify to gain access to space, ship the materials and ability to stay in space with a space city able to produce plant life, and with radiation-protective metals like copper, silver, and brass, to include metals that reflect the visible spectrum of light such as

Au. But the space city would be with an exact number of crew with the skills to build in space. These men and women would stay in space for periods of three years as the materials have already been docked in space for the construction, a minimum of materials is required to maintain the crews working to build the predesigned structure prefabricated on Earth, and the materials to build in space, the process done this way will give our ozone time to heal which is essential for ultraviolet rays to be reduced to our lives and our planet. The world community must be more aware that allowing any form of chemicals to be released into our atmosphere, which can cause our ozone to dissipate, is a crime that must be dealt with harshly. The global community I propose is to handle these issues as mentioned in this book, but the need for this exclusive team of world judges is essential to enforce the laws to protect us all and the future of humankind. As "Jean Luc Picard of Star Trek "would say: "Make it so."

42. SUBJECT: FUTURE MASS TRANSIT TRAVEL

Recently, it has been reported that tube travel with magnetic propulsion is being considered. The tube concept is fine, but the rest of the application-- well, we need to look closer to this popular opinion to be shredded as a form of travel in the future. Yes, it could be done, but as soon as it became a large travel frame, we would see that our planet would no longer be around; the reason I say this is magnetic travel was designed by myself back in the early 1990s,s in the early 1990s, and I logged the invention and did some art drafts. And unlike many inventors, I look at the potential- an example of someone who did not look at the potential was a man called "Mr. Ford," the inventor of the combustible engine. He never considered the number of cars today globally and the impact that could cause a major shift in air quality or harm the planet directly; we can assume that all those fumes inhaled are not a good thing for the human condition, so after looking at the possible of magnetic travel.

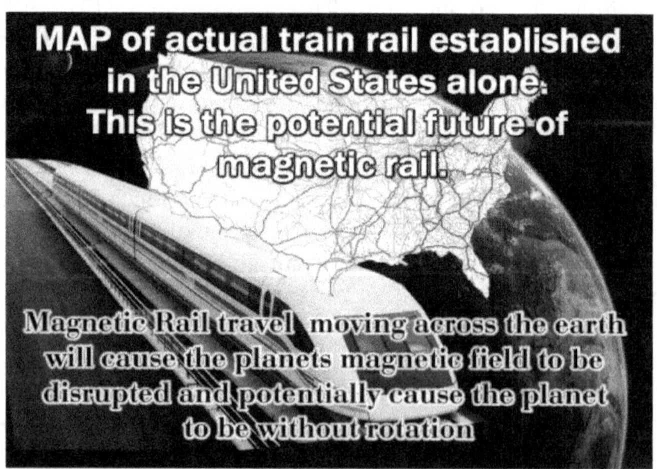

It hit me like a hard like a sledge hammer and the real thought that our planet is stabilized by a magnet field which is the main reason our planet stabilizes its rotation with this field along with many other applications we may not even yet be aware of for just one possibility,-without the rotation could we lose orbit around the sun? And drift away?- which instantly would lower temperatures to massive freezing globally, and if these people are allowed to continue to build these magnetic trains, that sound all so wonderful- Their only interest is- " Oh my, we can travel so fast !". And people want to get there faster $$$! Well, we should know that we have seen an increase in massive storms, and it is possible the devices now being used could, over time, be used to nullify our planet's magnetic field in relatively no time. Let me rephrase that last statement," I have zero doubt that this is a terrible idea!" As I stated, we could all be sailing away from our sun. I heard about the train in Japan, and I tried to send a message to the White House as in a letter reporting that this form of travel is dangerous to our environment, and the future safety of our planet being destroyed. But my message was ignored. But if we look at today's events as weather disturbances, people can't seem to see the direct link to many massive events of weather change ever since these trains have been traveling. I firmly acknowledge that the link and the time of these trains are the main contributors to the disruptions we have witnessed on scales never seen in history. So before we go crazy about how fast we can go, let's consider the actual results established on record after the record of new weather damage and floods, droughts, and new temperatures, the list of massive storm surges, and destruction in the trillions globally. So, to be exact, anything magnetic should be banned if it deals with magnets in motion traveling across our planet. In truth, any form of large magnet should be prohibited.

There is an old saying that it is better to be safe than sorry. It can even be changed over time, so make things stronger with protective elements. Let's place one improvement that car mechanics will thank me for, and even home car mechanics, all bolts that hold parts not heated as to screw on with Allen wrench entry and plastic. Wake up car companies- make mechanics happy, and they may recommend your vehicle, and they will say: "Hey, I don't need to fight to get a bolt out!".

43. SUBJECT: FIRES

Fires have been global, but now home fires and mid-size homes can be made to have true fire protection, and with this fire protection and these new inventions used, the need for high fire insurance rates drops like a lead balloon. So what is this new system? As most of us know, it takes air to make fires; without air, there can be no fire. So, the future homes built and, yes, even older homes can be tested and made as improvements to remove fires.

The first step is 1. Is the house is made with a firm plastic that is sealed inside the inner walls, any home can be tested with an inserting of a color smoke that fills the rooms and then air pressure is pulled in from any openings such as chimney , a bathroom vent, and they are sealed with heavy plastic bags with rubber straps that tighten with a set screws, to keep them attached to any item it is covering, this prevents air from escaping a the air pressure increases in the home , all areas of air release is seen by the orange smoke that is odorless, but highly visible, and where ever there are air openings around doors, and any area, these are sealed from inside , testing is done again in about an hour to see if the air pockets have all been found, then the main systems are put in place, windows are given extra sealing with new window frames if needed, and the electronic sensors are put in place, the concept is simple yet an exactness must be made, no air can be able to enter this home when the system turns on, the system activates by smoke alarms that detect any smoke will now turn on the system, a voice is now stating that you have 2 minutes to leave the home and a countdown starts, when the system turns on, two very large vacuums will now deplete the air from a normal size home, and all fire will be removed instantly. Now the only downfall to this system is someone cooking, so to prevent cooking smoke from activation as a fire alarm, the user of the home must have a stove that has two functions added, one an emergency fire connection to the system; this means, I am cooking somehow my grease caught on fire, I push the emergency button located above inside a small clear box screwed into the wall, and I hit the button, the system turns on, no more fire, but also the second needed application is an overhead fan must turn on automatically when cooking. Hence, the stove knobs also turn on a fan always.

This prevents smoke alarm mistakes from happening. Now, I have included a system that does not need any of these applications but is to cover a standard home. The system works similarly, yet is not as fast. Yet, a fire will stop in more than half the time of a fire system today. No added damage will be done, such as windows broken and firemen with axes making entry areas, but the device is driven to the house fire and parked over the top of the house that is on fire, it must be cleared first and recorded as a home with the proper access and dimensions before it can be used at that residence. But it will cover the entire structure in seconds. Once the fire distinguisher unit is set in place as completed, the entire house is now covered with a fire retardant cover bag. As soon as the determination that all residing are out of the home, the vacuum of air is now turned on, and within a time frame of minutes the fire is out.

Let's take a look at the in-home air vac system for a moment for it's construction. All windows are inserted with a heavy rubber seal into the areas around the edges so that air cannot enter. The windows also have a pressure latch that, when a window is locked, the pressure lock actually presses the window to make an airtight seal. This means the window, when closed, must have an indicator that shows that the window is sealed. This solar-powered L.E.D. indicator will be lit when the window is closed completely. The house's doors also have this same pressure-sealed ability as in different rubber edge forms. When the door is closed, the rubber secures the door from all air intake and works to prevent insects from entering under and around doors.

Once these applications are now completed, the last areas of the heat furnace entries and any outside chimney, all openings as air exhaust fans or any system of air intake or exit, such as the kitchen exhaust fan and air exhaust fan in the bathroom, will be installed with a new close cover activation system with heavy plastic plates that close by electronic connection sent by the system in the event of a fire either detected by fire alarms hooked up to the system or by an actual fire level manually opened and pulled. These devices are designed to be high away from children, and a code entry is entered to test the system.

A message is placed inside the system on and off that gives access to the system by a phone call to the fire department, which has the house on record by address. The user will be asked for their birthday and age to verify; once this is done, the system can be also turned on by fire department, and the system is now activated to test all entry areas of air intake. All in the house are to leave in less than two minutes, and let's say the fire is outside a door exit. You are trapped, no problem; every room has a mounted air tank that has a face cover to breath fresh air while the air is being removed for the home or office, and yes, even high rises.

But in most cases, people will leave, and the exit doors will seal in two minutes; there is no need to gather anything, but people run outside, as the down being sent loudly; as soon as the system is on, the recorded message is announced in every room with solar powered speakers mounted to windows, and when the detection smoke alarms have no smoke detected, an "ALL CLEAR" is announced, message alarms goes off. But when smoke is detected, the message is ATTENTION! And a extremely loud alarm is activated- Then a voice will announce, "A fire has been detected; please calmly exit the ("Home"/ "Office") to any available to exit door."- You have 20 seconds before the air is removed, EXIT NOW! You now hear a countdown before the air is removed; The voice countdown states: "Countdown commencing as 30 seconds before air removal and lock down, then the air pumps turn on all needed locations, in the attic, the second-floor closets, and the basement, the air is pulled out so fast it only takes less than one minute to have the entire house air removed, now as stated before for say someone elderly caught inside as to have fallen, the emergency air tank station would be located next to the bed, and the emergency on the system would be in reach also, this means the user only needs to break the seal.

The air mask is placed over the face, and access to air is available to the person breathing the air from the air tube inside the home or office, because the air tube is actually getting air from outside, whereas the air being removed from inside the building is now put out the fire, and once the fire has been determined as completely out, the all clear is made, and the doors are unlocked which allows the air back into the building. This entire process took exactly 1 minute; any normal fire detected will not be able to cause any serious damage in such a short time.

Now I have included a concept drawing of the portable home AIR VAC drive system which could be used from moderate to somewhat larger homes, again, the difference is this firefighting equipment will take longer to arrive, but once in place the air vacuum system is much stronger and the fire can be removed with a cleaner result of fire damage, the tarp is a fire proof is rolled upward in travel, with weighted sand bags that when the unit is in place around the home they descend to make a seal on the ground with the sand bags surrounding the structure on the ground, and fire fighters can easily adjusted the sand to any ground level differences, and even stakes can be pushed in place to make even stronger sealing, the hook releases to this massive bag can be engaged as manual or electronic, with a chain elevator ladder on each end, and the tarp descends to all areas surrounding the home or building, and then once the gaps are secured to the ground as the sand bags will form a heavy displacement and added flaps are now placed down to secure any air gaps, and where there may be an opening cause by a pipe, or opening which allows the air to enter the fire department use rolls of heavy plastic to seal off those areas with duct tape to seal them. This action will not be required most of the time, but homes with trees close may require this.

To avoid this, the fire department has strict rules that homes must not have trees grown too close to a house to prevent using the tarp air vac system. All homes will be measured as potential air vac users and placed in the system as recorded as user-friendly. Other homes will use the traditional systems today, but here is the concept design for using a transportation Air-Vac mobile unit. Now this design is in the rough, but the principle is to have a tent like structure that can suck out the air extremely fast to remove fires. This said- (Enjoy.)

Title: AIRVAC AIR REMOVAL FIRE PREVENTION

Down view travel mode

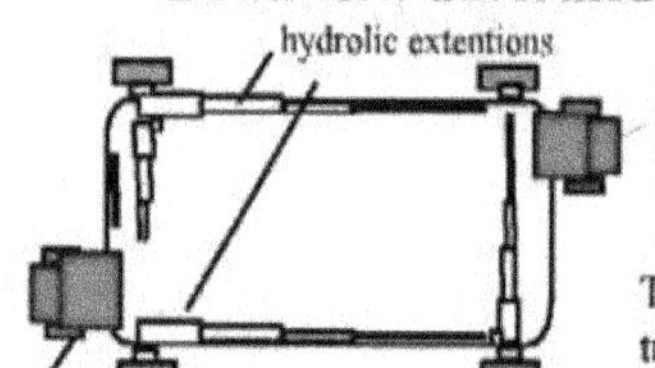

back opperator for complete visual aid., and opperates
the east west extention , and the front opperator controls north south.

Travels just slightly wider than one lane road , but on coming
traffic is to pull over to make clearance.

front opperator and driver. This is a designed fire proof house tent to remove fires faster.

law passes that states that no tree can be planted to close to home that prevents
the use of this device, and property owners who have trees this close must be informed.

duel tires to prevent sinking in wet mudd , and four wheel drive, and has hooks
to pull the unit into place in events of very poor ground , and if stuck - four
electric pullys with hook ropes that can rap around anything solid like tree trunk
or a pole can be driven into the ground by hammer to make a instant pull area.

Frame adjusts outward in every direction, laser beams help the placement at night
and drop ropes real to ground with drag weight to keep it straight to guild during day light.

Air vac

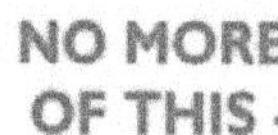

Air vac

hydrolics raise the unit up to 70 feet

Drop cloth fire resistant, when it hits ground
as a large tent , the air is sucked out. And the
fire consumes any air left and puts itself out.

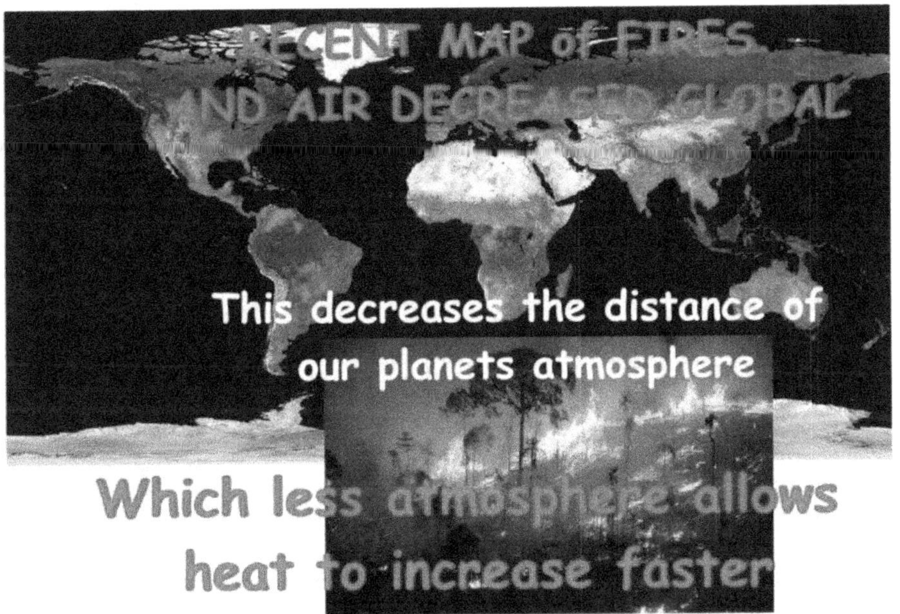

44. SUBJECT: SOLUTIONS FOR FOREST FIRES

The prevention of forest fires has been a long struggle that man always loses, especially when the wind is active, to encourage the flames even more. The systems used today are non-productive as a plane over the fire fans the fire, and so do helicopters, which carry water or chemicals. The chemicals are not something that shouldn't be used because they harm wide life and humans who could come in contact with these chemicals.

Here are some designs that can help stop forest fires and some equipment that could be used in tough-to-reach areas- see illustrations) I have included a concept of a water firewall that, when the plastic melts, releases a massive water tank to spray outward some twenty feet or more. These are water tanks that allow water to fill during the rainy season, but when filled, actually keep water inside from evaporation because there is a seal that closes the tank's upper entry when the water level reaches the top; the tank lid is closed- the seal can only open again if there is rain present to cause the rocker panel to re-open and will then allow water inside again until full the lid then closes again. These very long and large water tanks attach to the tree line, and they are mounted clearly around the tree to be three massive tanks that, when the heat appears, the lock pin will dissolve and open the release drain; the natural gravity now shoots water from a sprinkler bottom. 2nd, this tree-hanging system surrounds a tree with a similar application to the much larger tanks- these tanks made of heavy plastic bags hold around 90 gals of water each when full. They have a float inside the water intake that works to close the intake during rain, but these hang around the tree trunk, and more than one can be applied to any tree, depending on the size of the tree trunk. The device uses two means of water distribution, and one can be a manual pull rope that, if used, sprays the water in an outward and downward flow, spraying the water in every direction. These units can be used as firewalls of water, as each tree is given this water tank device. Another way to prepare for a fire that is moving fast from high winds is with a solar-powered smoke detector. This detector can open the valve release, but in the event the solar battery inside the device failed due to possible rust or electronic failure. The system is still useful because the bags are made of plastic, and the adjustable frame is aluminum, which channels the water to spray. Now, the fire itself can cause the water bag to open as it melts, and the water is now released into the tray and is sent outward and downward along the trunk to keep it from burning. Making the tree now covered in water at the lower areas, this system can be placed for miles and cover areas very wide, and we can add something I call the auto water tanks. These are similar in application but very different in location as they can be placed between trees and open field areas as a water wall. Let's call these Goliath tanks because they work very similar to the mounted water tanks, but these are massive in size with major water to gain fire control. Still, they store around 6,000 gallons of water that use pressure plate gravity that can spread water across a wooded forest every 50 yards. These are used only in areas known for high wind fires, and they should be lined up in the main direction of oncoming winds recorded as normal predetermined wind direction. They are stainless steel tanks that look like a tanker truck standing on its head, such as long tube-like tanks with stands that hold them upright, and they have water-fill areas just like the two items mentioned to store even rainwater for months upon months. When a fire is approaching, the tanks have two sensors that can activate the tank, and these can be manually activated by remote control to individual areas, by section number, or by a manual chain that can be pulled at the source. The tank's heat sensor is the method of automatic release of water. The water pressure can travel around fifty yards away as stated before, and here is why, just below the water intake is a massive metal weight that looks like a round washer but sits at the top of the tank when filled, and when the connectors which hold it at the top are removed by an electronic screw system that turns the pins that allow the massive metal plate to start downward, once the system is activated the water ejection opens, pulls open a rubber and metal gate valve, and this now starts the water to shoot outward from the massive pressure plate descending downward all at one time on slide tracks that guide the plate, this baby will destroy any oncoming fire in minutes. And makes a massive area of water over the tops of trees and on the ground. The designs of these water tanks are made with water collection in a unique way, but the plastic units, when used, will need to be replaced as the tree hanging systems make a water wall. The more expensive the stainless tanks will be, the better use because they can be moved, can weather the storms, and do not melt, so they are reusable when reset for use. The metal tanks must be re-loaded for the plates to be in place and the water intake opened to allow rainwater to enter. Still, if a drought condition was in effect, the tanks could be reloaded by fire trucks carrying water to them to refill. This metal monster invention will enable water to save our forest areas. They can even have a hose connection to direct water to be used like a water tower spray gun, in the same manner, that is used at sea for ships on fire and can also have hoses unrolled a group of six per tanker can be used from the water tanks, all stored on site, and a locked storage holder, the best areas to be placed on a hillside so that if a water tank has run low, a hose can be run from a higher tank on the same hill to switched the upper tank water into the lower tank operators that may have the fire now almost beat. Today, forest fire individuals are asked to go into an area and cut a dirt pathway to contain the fire. Instead, let's make a mandatory dirt fire path, a farmer tractor-style plow used between the rows, and a 20 ft. spacing between each forest could be maintained in large squares of five miles each. Instead of having an area that burns beyond this, you've only burnt one area and possibly saved thousands of acres. This application would prevent the fire from spreading across ground areas. The air resources of our planet are reduced yearly by massive fires worldwide, and we refuse to see that the protection of our resources can be maintained if we look at the full scope of our global community and that every nation should be involved in fire prevention with these additions.

45. SUBJECT: FIREPILLAR FEATURES

1. The tanks can be up to six pulled in at a time depending on whether the

Turret is level enough, but this design is for traveling through wooded areas directly into the fire.

2. Changeable water tank inserts can reload in seconds, and more than one water tank can be connected.

3. Changeable water tanks are also behind water cannon stations.

4. Fire pillar can be driven in from either side, as pulled and pushed. All-wheel drive is a six-wheel drive on each command drive system, which can connect as many command drive stations as a full fire may need to be stopped to allow transport into the deepest areas of the fire.

5. Water tanks inside 100 gal. Per side, and water connectors that can redirect the water flow from one side to the other side for water cannons. To include massive portable water tanks on wheels.

6. This unit is for close combat, and the interior walls are refrigerated, and the inside is refrigerated to cool inside airflow. There is an emergency air tube that drags like a massive tail. This tail is a fireproof hose that can attain air from areas where the fire is not present. Being low to the ground, the air is filtered with fine screens that can remove smoke before air intake into all compartments by flexible metal tubes.

7. A backup electrical drive system is available in the event of a shutdown, and it can be used if the vehicles are stuck. The system also has hook connections that can connect to trees to pull the vehicle from being stuck. There are two on each end for double pull strength. A hook system can also be linked to a free-moving drive to pull the stuck Fire Pillar.

8. Electric motors on all wheels power the unit solar panels on roofs, which can be used to restore power.

9. A (3") glass visual ports have a metal cover to be pulled away, a wiper in the metal cover, and water spray to clean it. Replacement glass can be used by unlatching the broken glass, pulling out the window, and inserting a new pane stored below.

10. Retractable power saws at both ends, front and rear, able to cut a tree in seconds, can be lowered and extended or just extended between rams

11. Front rams can pull the unit out if ever stuck with a power cable or push a fallen tree.

12. Rotation cameras at both ends for infrared vision to see through smoke can change to regular vision viewing.

13. Water cannons, operators control from inside and sit in swivel seating that can direct the water stream to any point and any height.

14. Front and rear view ports protected by upper solid steel bars and dual window cleaner with wiper scrubber to remove dust and smoke build-up.

15. Top of all units have tree crash covers that reflect fallen trees or branches. Solid steel roll bars on top protect tanks

16. All sides have changeable water tanks strapped in by metal brackets and locked 100-gal containers.

17. Air tanks can be in thick smoke; the users inside can place masks that connect to air tanks, and the wearer has clear goggles to prevent eyes from smoke that may enter through air filtration breakdown.

18. Retractable power saws at both ends can cut a tree in seconds and can be lowered and extended for the lowest cutting of a brush or taking the tree down. The

ram has extension arms that can clamp onto a tree by an inside operator and act as a holder to direct the tree's fall so that it does not land on the fire pillar. The hydraulics make a large two-pin style closure that inserts into the tree bottom before the tree is cut to allow the fall to travel to one side only, away from the Fire-pillar.

19. Each wheel has its power, and the front and back can go forward together or in reverse, as the operator drivers can communicate to each end. If the Fire pillar is going toward forward at one end, the other side is in reverse.

This allows the power to be 12 total wheels with traction in normal use.

20.Water tanks, depending on the terrain, can be up to six water tanks, and if there is room to maneuver, the system can have what is called as long as there is a way to travel in full circle. With the wagon train, the water able to be pulled and water able to attack is doubled and requires the FORWARD and REVERSE water cannon compartments to be more, so if we have what is shown here as three of these, then we have the ability to pull 12 water tanks 1000 gal each, and have six cannons and drivers added to the fire.

21. The wheels can be retracted inward when traveling through narrow woods. When on a steep hill, the wheels can extend outward by hydraulic extension bars.

22. Wheels are made of high-tempered steel with replaceable rubber Snap-On rims that wrap around the wheels and are attached to all wheels with a hook latch, and then the fire pillar is moved forward to secure all rubber outer housing on wheels. These can be put on if traveling on highways but are removed by reversing the process when reaching the ground that needs traction.

46. SUBJECT: AIRVAC HIGH RISE SYSTEM

As I explained earlier, the new concept of airbags removing fires can be done if installed correctly to remove fires instantly. However, with some slight modifications, this application can be used even for the world's largest high rises. The system secures an area with the same air-tight sealing of all outside air entrances, such as doors, windows, or exhaust systems. Now, the area has a vacuum system, which, when a fire is detected by a smoke or high-temperature increase, the AIRVAC will activate a warning and begin the countdown to alert anyone to use an air station tank, and this is where the difference starts. The air stations are on each floor, and elevators are all open to stop at all floors so that people can exit, not by the elevator, which could be descending into a fire area, but the people are directed to two locations, the floors outside each elevator and inside all rooms have an air station, but the majority are in the hallway, and these are with separate sterile face masks that a person simply opens the compartment and removes the face mask, applies it to their face as the air is removed, but let's say this fire is one caused by an act of a terrorist, an explosive device, which. Still, the people are directed to two locations: the floors outside each elevator and inside all rooms have an air station, but most are in the hallway. These are with separate sterile face masks that a person simply opens the compartment, removes the face mask, and applies it to their face as the air is removed, but let's say this fire is one caused by an act of a terrorist. This explosive device has now made the system non-workable because air cannot seal on a particular floor; this is where the human slide tubes come into use. Located toward the center of the structure, they are protected by massive solid steel beams that support the entire tube from top to bottom. The escape device is easy to see because it is highlighted in bright yellow and orange instructions of fire to identify and shows a person holding a fireproof mask over their face which can be retrieved next to tube entrances and have a plastic air tank for breathing, if tube entrance floor is on fire entrances would still be open even on a floor with fire to allow someone to descend downward. The tube is designed to allow a person to slide to the bottom of the high rise, and there are four such tubes per building from each corner of the structure. It is a smooth ride to the bottom as it loops around and around to keep you moving at a proper speed and level as you descend. A person's size has no bearing because the size can handle people of almost any size. When entering from the top or at any entrance opening, you will find a seat that extends to the side of the tube with a few steps to sit down in this grooved spoon like seat, when the light is not on as an indicator and turns from red to green, you are to push off onto the slide, and this is allowed by a motion sensors which are located higher above the floor your on to detect anyone who is inside the slide moving downward, as soon as the light turns green-you go. Then you enjoy a fun trip to the waiting rotational wheel that slows you down as you slide out like a piece of luggage, and you are now told to exit the building as directed by firefighters. But firefighters need to be able to get to the floor to put out this massive fire, don't they? No. Because on top of all high rises are two main elements, a fire sprinkler system and the top entire floor, which is not a floor at all, it is a massive water tank that looks like it's part of the building and even has a small bench park with tree shading sitting on top, and helicopter pad to land. Still, this area is a massive water storage compartment filled with millions of gallons of water and can store rainwater also, which can even be directed by separate water lines to every floor in the entire building or be opened 100% to any location as the entire structure is about to get an inside and outside bath. With the AIR-VAC system installed, the escape slide would not be needed because a fire would be systematically be removed, as the system would do the proper air removal. However, the slide is being mentioned because it could be made on all high rises today as an exterior or new interior design addition to save lives if the air removal system fails. Let's call it a solid backup. The slide tube system if installed with the Air-Vac system, the Air-Vac would have priority in a fire event, but if the system failed for any reason, the backup slide could be used. The slide is closed and sealed when the Air-Vac is turned on, but if the fire has air available from an event like the former "Twin Towers" high rise structures when damaged, then the slide tubes would be opened or use by hitting a red button that would open the sealed entrances to all the tubes. And, of course, there will always be skeptics, people who own these high-rise structures who do not want the fire AIR VAC system.

47. SUBJECT: NEW PROTECTIVE FIRE GEAR,

With cooling ability through added hose connection attached to main water hose. So, each firefighter has a CO2 remover hose with a cooled air intake.

The design below is for firefighters who are involved in extreme danger, such as large wildfires with high winds, forest fires, or a large structure that has never been installed and requires old means of putting out the fire but with a better ability from smoke and heat protection, these outfits secure an air-cooled intake to the suit and have temperature control, even can make a firm stand with high-pressure hose adjustments that can be a support to hold the firefighter from water pressure being over strength, to include all fire retardant wear and fire retardant knee pads, shoulders are encased with air- conditioned inside, which resists burning. The suit has a clean shield that allows water to spray with a simple press of the wrist water cleaning fluid directly on the face shield. The water transfer hand holder can be body-supported to support the water pressure being sent outward.

The feet of the user with fire retardant inserts air cooled from the suit, and metal bottoms with air exit from the suit to cool feet. The metal is highly resistant aluminum with a tube connector that travels down the inside of the pant legs. These tubes are connected to a powered high-pressure fan belt attached to the user's belt. With a simple flip of the switch, the air is pulled up from suction cups located below the metal feet housings, and now a high-pressure suction will begin. A proper stance now gives the user greater control of the extended water pressure, and any increase in water pressure enables the user to stand firmly while releasing the water.

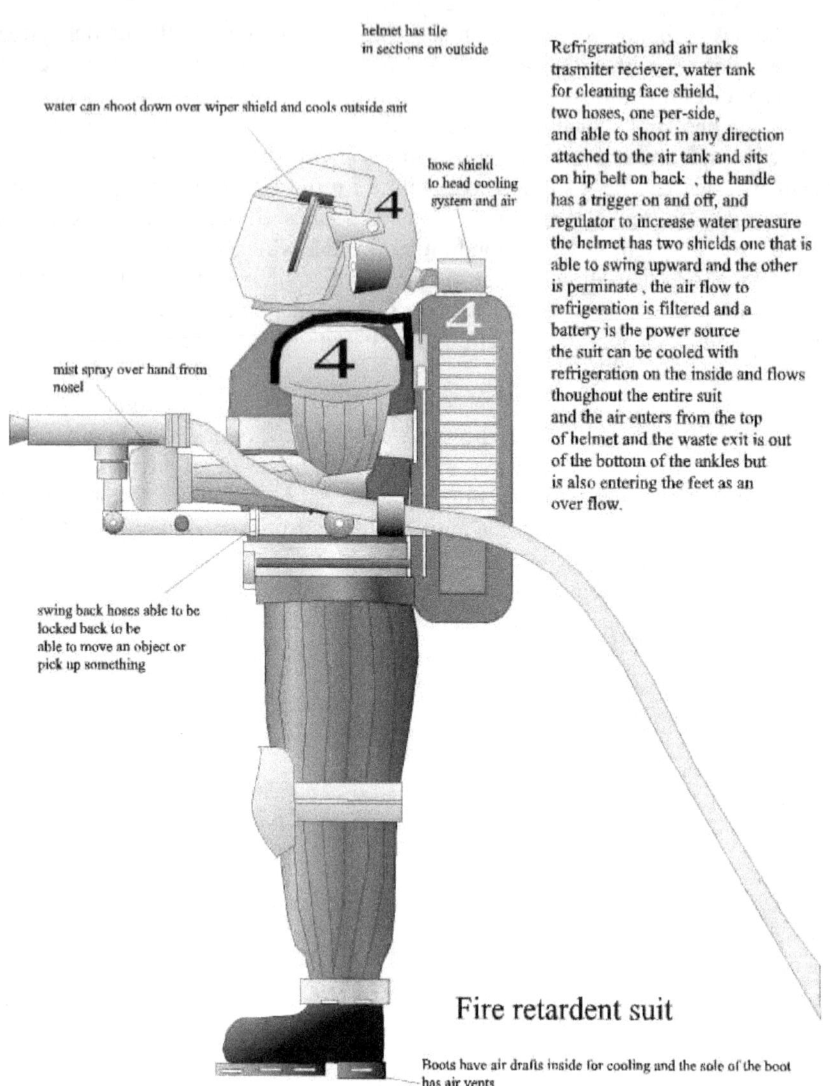

48. SUBJECT: TELEPHONE HARASSMENT

Many of us today receive a great amount of phone calls from robo calls or people who are irresponsible when they call us. So, are there ways to stop these people and robo calls, and how are they responsible for disruption? Because they are calling you, it is as if they have walked up to the door of your house and asked for you in person. But our laws have reflected that this is not the case, but have instead refused to act upon complaints made by the public in regards to invasion of their privacy. This writer has had many experiences where reports were made, and no action was ever taken, and there is no law on the books that says that you can be fined for calling someone and doing the following. Calling up and hanging up is an irritation and has been done quite frequently by many people. Another example is people who call you and do not tell you what they want or who they are. In the future laws should be passed that state that all conversations can be recorded upon your request that show the source of the caller to the phone company and you. People have the right to block any phone number, not only phone numbers that are listed but also phone numbers that are unlisted but have a series of numbers to attach to this phone, which the phone company recognizes because if it were not recognized, then the phone company would not be able to collect money in billing to that phone number. So, for the phone company to state that they cannot recognize a phone number coming into your home is a false and is a false statement because the facts of the numbers can be granted to the police if you make a big enough stink and are visible in most phones. Yet, the numbers may be from signals bouncing from one server to another. The law should be clear that people should be able to have control over who they speak with. If someone is infringing upon your privacy, no matter if they think so or not, it can be grounds to issue the proper legal action to go after the caller, you have the right to not be harassed, because this is an indirect invasion of your home, and once again you have the right to refuse to speak to whomever you choose. The law should be clear that any person calling a person, whether it is home or business, should first announce who they are and what they want. So a system of identification as to who is calling should be with an exact Identification, not just "spam caller" But the actual name of business, or the persons actual name no matter if you have them installed on your caller list or not. Bill collecting is a national issue because most of us have acquired much credit debt. This fact is something I'm sure most of us regret acquiring. But we have had an invasion from bill collectors asking repeatedly for our information without confirming who is calling because they want money. It is not that we would not like to pay our bills, but most of us live from paycheck to paycheck. To create harassment over debt only causes our people as a whole to be something where we now are apprehensive and become easily irritated because we are suspicious of whoever communicates with us. So these people continually harass the public, and that is the issue that they do this on a large day-to-day basis. Harassment is their goal, even though they do not accept this. The stress that they create overflows into society, and our days are made irritated by knowing people are listening to us and watching us, so we become even easier to be irritated, just consider it like someone scratching their fingernails on a black board repeatedly. This alone is why the law is now needed to fine individuals according to the extent of the crime. And what is the crime? It's an invasion of privacy. The proof of these calls with proper legal help could be a hot-line that could be like "111" dialed that retrieves the call info as proof as what was stated, and now is automatically submitted for review by legal means as to the formal complaint being addressed. You will be contacted for further assistance if the info is found to be a pure act of harassment with a fine to the offender, and possible retribution made to the person being harassed.

49. SUBJECT: NEW WORLD VISION

The Eyes & Ears criminal detection device. You only use this device when needed. You are in full control. The activation's are made when "you" push the button. Think of it like wearing 911. Control is made only by the user, not the government. In the future, there will never be a crime that cannot be detected of whom has done the crime if it is done near the person. The reason that crimes of abduction or bodily harm cannot be undetected is because of a new tower system that is located every 10 miles apart that stands to transmit signals that are located on your wrist or belt or headband that send images and record your whereabouts and your conversation whenever you feel threatened, a person in this situation now can send a televised image of their surroundings and the people and the area is now noted. Police are sent directly into this area when this alarm system has been activated. They come from surrounding areas and do just that: surround any traffic areas. Camera activation is now turned on for that area to monitor who and where you are and who may be in the area. This system, when activated, turns on a five-mile radius as soon as the system is notified. In our day and age, we have cell phones that give us images. This system is similar to what is on the market regarding the ability to transmit sound and signal. The difference is that it is worn on children or adults, and it's an emergency signal not to be used unless there is a true emergency. If someone uses this without any real harm, they can be fined and be given work details for community repairs for at least one year. So, the miss use of this device is not taken lightly.

With this device you are able to be protected instantly because not only does the device alert the police and your GPS location, but it also sounds an alarm within a five mile radius that alerts others who have their device set to alert in the event a person may be with intent of hurting others but this is activated only by the police image reviewer, if the person who activated their device looked in physical harm was possible. Let's not forget this device is only activated when "YOU" want it, so you're not being monitored like devices today. Those also near the activated device would be notified by hearing from their device: "Danger present- turn on your device now". This action will now record the surrounding areas to detect any movements of potential crime. These devices remove the need of street cameras that invade people's privacy, and allow crime to be stopped many times before it has started because someone activated their system just before the person was to attack them and deterred that person because they know "Eyes and Ears" are at work.

50. SUBJECT: ILLEGAL DUMPING

You can now travel to many areas of the country and find that people have taken their garbage and dumped it in creeks and streams in forest areas and drainage areas. You can find everything from garbage bags to tires to refrigerators, old cars, and sometimes just old junk that people have discarded. In the future, all people will be accountable for whatever they purchase. This means that when a person goes into a store and buys anything, they must present either a credit card for their identification or a cash card. This cash card will be an identification that attaches the code that is embedded into the item that is purchased, and this code is registered for all items that are purchased embedded. Each person is getting a purchase code attached to a product, so the item is now attached to the person, and if this item is found where it has been discarded, the item can now be traced back to the owner. A device similar to scanning in a grocery store would be used to identify the item thrown away. Once an item has been given, a tracking I.D, the person who has violated the law with litter would now receive a fine in the mail, depending on the item's size and weight. All items would be the tracking Identification, including everything as far as things purchased are concerned. Cigarette butts, newspapers, magazines, a metal can, refrigerators, tires, doorknobs, it wouldn't matter what is purchased. Everything would be ID. It happens upon the cashier scanning your items because in the future, before you buy anything, there is no need for you to scan your driver's license or ID card into the card reader, because inside your credit card is also all your personal information in another chip that links you to the sale and is being sent to your purchase server in the company computers that are also linked to the police who have access to garbage detection system. This credit purchase card brands all the items inside a chip of the product, in the same way, information is retrieved from your debit or credit card the information is placed in the chip on the product of when it was purchased and who purchased the item. This now makes a new department of waste management called the "Identification Waste Environment Officials" or "I.W.E.O, Which now scans found objects discarded outside waste management dump areas as illegal dumping. Depending on the size of the item, once again, the fines can increase, and the fines can also bring three things: money from any income automatically reduced from the person's bank to pay the fine, or fine plus community duty to pick up garbage and ID it, and also paint or clean any structure. This would prevent all illegal dumping. Offenders making these offenses more than once would then be given a ticket and community duty to could be sent to go out, find garbage, and retrieve and ID it. The fine can be as little as $10. For a cigarette butt, or as much as $5,000. For car tires or car. However, for even larger items such as boats, trucks, or heavy equipment, the cost can be even much greater depending on the damage. This program would be also with a home product tag which is with the I.W.E.O. has personnel that go to homes to I.D. large items such as old cars, trucks and any item that has no I.D. The difference is these items use a heat bar I.D. that brands plastic or leather, with a number and when it is done the number is recorded with ownership and location of stamp. Now, if someone tried to scratch this off and tried and dump the vehicle to someone else or just dump it in the woods. The make and model where also recorded, color etc. by an added photo of the unit which was sent by WIFI and is linked to the vehicle. So if this happened the model would be narrowed down to the potential owners, the owner who no longer has that vehicle would be the guilty party if the vehicle was not reported as to be discarded to waste management appropriately. Events such as hazardous waste being dumped result in even greater fines because if any containers are found, they will be marked as such. However, waste of that type would only be from large industrial areas, so those in that area and the surrounding area would be responsible for paying the fines. Moving Hazardous waste to authorized waste control: No longer will trucks be allowed to travel hazardous waste, as a new rail will be required for hazardous waste movement only, and the entire rail will have side rails that, in the event of a derailment, the cars will not tip over. The cars will only come to a stop. The waste is placed in barrels only as sealed, and the barrels are lowered down and stacked inside a rail box that is similar to an open-top box. Still, if any waste is leaked, it cannot pour out because the entire inside is a square bowl; each barrel has a metal chain hook that allows a hydraulic crane to pull the items out of the rail car with a large magnetic end that when activated can lift the barrel and release it after it is set down. These barrels are transported to deserts where nobody lives. The waste is loaded into space rockets called: BIO- TRANSPORTERS- where the items are launched into space to be collected at a spaceport and loaded on a massive transport ship that has no passengers and is run only by AI, which takes the contents to any destination and at a predetermined elevation will descend to a marked area and then have the entire contents slide out to be placed on the planet's surface and marked as section 1, and each trip makes the new next section- and the contents are checked for leaks that will now transport the waste from our planet into deep space to be released on planets that do not support life, like Mars. Also, worker digger robots can bury waste in the ground, making the waste nest before storage. This would prevent high winds from damaging the items contained. The items will also have an estimated date of compositions broken down. We do not do this to planets that have life; we place warnings of danger as marked for future travelers who may have crash-landed for repairs or had a mission that required them to be there. The markers for warnings are ten feet high and made of strong plastic. The result of the transport of hazardous waste means that nuclear waste and chemical waste can now be removed from our planet. Also, when these items are stored, they are not mixed together but are marked clearly as to the type of waste and must be placed in the same section as that waste. This will prevent possible mixing of the waste and gases that could form.

51. SUBJECT: WASTEWATER

People use clean water in the bathroom, in the toilet, and in the kitchen. Convenience is our only concern rather than whether or not it is acceptable to the environment. This water goes to a waste plant to be cleaned and brought back into society, where many times, the impurities are still in the water and are dumped without concern, eventually causing problems in our world, affecting either plant life or animal life. This problem has been an ongoing problem for centuries. The solution is for humans to develop new, effective means of cleaning themselves and the items in their home. We also need safer and cleaner methods to be considered better for our environment. I have designed a toilet that does not use water as its main element but is used for cleaning only. This toilet is designed to remove the waste from the home to an outside housing that is a dryer device, and evaporation is a large part of the process, with the help of solar panels and the final product is now a powder that falls into a garbage bag and can be removed periodically. These bags would be marked as waste for farmers and forest growth. This book contains a directory of how to regain desert land, and this waste could be used for that as the use of household cleaner products that have chemical base elements should no longer be used by the public and should be banned from public use. The reason these products are unhealthy is not only for people but also for the whole environment. There are other products that do not harm the environment that could be used and are not being promoted as environmentally friendly. Some soaps and detergents are environmentally friendly, which should be the standard. The water in a sink, dishwasher, or shower must go to a water-cleansing plant. This means a secondary pipeline is made for each home, the waste water is separated to keep chemicals from being involved with human waste. This plant is much different from what is now being used. Reflecting on my high school science days, I remember the process of water being separated from the contaminated elements The method would be considered costly, being that you must first heat the water to temperatures that cause the water to boil, where the water condenses upward into a pipe where now the water separates when cooling and creating clean water to be sent into another container. This process can be done very effectively by pipelines that allow waste water to travel to a local pump holding stations near all desert areas. Wastewater could travel to many existing waste purification plants even by rail if within a 100 mile radius to a desert location. These areas have temperatures that would evaporate the water without heating devices; instead, nature would provide all the necessary heat. The clean water would be then directed in a pipeline and sent back to large reservoirs, or would be sent on rail in large tankers that could be emptied into a town's water holding tank. Freshwater to needed areas. Since the cost of rail could be quite high, a pipeline would be the most effective way to send the water back to needed areas. The reservoirs to send the water could be strategically located to be elevated enough that the pipeline can reach the furthest areas. To conclude, the waste removed from the wastewater being if it is not toxic, could be placed in transparent barrels in a desert area to be exposed to sunlight, which has ultraviolet rays that break chemical compositions. Be assured that this process is not placing a hazard or a threatening chemical back into the earth; the barrels can be tested to no longer be harmful. Placing these elements inside a closed area like a greenhouse prevents any possible chemical not detected from going back into the environment, by wind or other natural elements, but allows the sunlight in to break down the chemicals.

52. SUBJECT: NATURAL WATERWAYS

Issue of streams, creeks, large lakes, and ocean waters: Inside our oceans are thousands of sailing and metal ships that have sunk for a long time. Many of these ships are made of iron, which has rusted in these waters and pollutes our waters daily. All ships and other components of metal at the bottom of these waters should now be removed. These ships have valued metals that could be melted back into use rather than be a destructive element to the ecosystem and fresh water lakes and sea life. These are not natural elements to these waters and are affecting the natural habitat of the sea. All the waters were originally developed to maintain life in its normal conditions approved not by man but nature. I have seen metal discarded into creeks and streams and in lakes and have seen all of these elements doing nothing but destroying what they're sitting in. This can be removed by teams of people who cooperate who own ships and who could be hired as salvage retrievers and to sell this metal in any condition that it may be in and if they cannot be sold, at least placed where he can be smoldering down into blocks large enough to be useful again. This operation would be a big lesson for humanity to clean up after their wars and mismanagement. In the future, ships that sink should be retrieved by large crane ships that pull the sunken ship first to the surface. Then large buoys would be attached to directly keep it not afloat but above the water line enough to drag it to a shoreline that also has large cranes that can be attached to it and pull it ashore to be cut into sections large enough to be transported to iron smoldering plants to be reused. Items not made of iron would be separated, and all materials would be reduced back to reusable means. Reclaiming this metal is worth it because the cost of iron in such large amounts would be profitable. Since some of the ships are immense, two ships would be needed, one for the front and one for the rear, to hold the ship while it is being transported to the cutting yard. The sunken ship would be held by long cables from both ends if this would include any materials that are not originally part of the ocean. I have heard that sometimes artificial reefs have been made by the sinking of these ships. This method is absurd and only adds to pollution. Instead, large concrete blocks that could be molded to copy a rock formation and even have holes in areas for sea life to survive inside could be placed in areas without adding rusting metals or causing pollution and wouldn't make ideal reef areas. So, to conclude, this project would be a considerable undertaking, but the reward would be our waters would be cleaner because metal decomposes and pollutes the water. Also, many chemicals and containers, including barrels of waste material, have been dumped into the ocean for a considerable time. Our mission is to retrieve, remove, and sanitize these areas so that the natural sea life can continue to flourish. As reported, this would include nuclear waste dumped by our governments and ships deliberately sunk.

Thanks Guys???

53. SUBJECT: WATER PURIFY SYSTEMS FOR DESERT DEVELOPMENT

The illustration of salt removal from ocean water is not new, but the difference is that the dry method is done inside deserts where water is unavailable. It will nurture the desert farming concepts included in this book. Also, using sunlight at higher temperatures will increase productivity or water purification. Countries that are in desert lands will now have their deserts brought back to life with a process of many applications that are required to make the desert go from a food source to a very massive forest, yes, even in sand areas, but before I get into the process, here is an illustration that shows that water can be shipped from the ocean as filtered for use for 1. Crops, 2. Forest growth, 3. Drinking water, and four. The salt can be used for consumption. But we can also add water to also be incorporated from my flood design of rivers that over use the over flow system, and what happens here is the process is from flooding preventive actions but the water from say the Mississippi river is now filtered at the source, but the pipe lines of clean water is now sent thousands of miles away , as from the pure filtered water which has had weeks to settle dirt is now pumped upward into a water line tower with a downhill water travel, this water is made to travel for thousands of miles to areas that are needing water. The priority is to send water to areas needing water most, All areas have underground water storage tanks, water tanks that are below the surface as to keep the water from evaporation , Water can also arrive by rail, trucked, a pre-test is done to make sure no harmful chemicals are in the water, and a clean test of the purity level as for anything harmful to people, and that the sediment is prevented at the highest level to maintain that the water is drinkable and pure to be placed inside storage tanks. There can be water that is not 100% pure and can be placed into natural water lakes if no chemicals are present but has a low concentration of sediment. This will help any areas that need water, such as a natural wildlife preserve. One more water development has potential for desert lands, and this device is not depicted herein. Still, the invention can make water for areas that need water and have far too many obstacles, such as mountain ranges. But water can be developed in the same way. This is not drilling wells, but delivers water to remote areas of desert lands. If there is an interest in this invention, please support word of mouth to others to buy this book, and the design of these inventions may become real applications, as I have worked out all the proper procedures to perform, and funding is needed, with thanks to your concern for fixing global warming and the future of our planet for our children to enjoy.

54. SUBJECT: HIGHWAYS

The elements are constantly at battle with our nation's highways, where cold and hot are constantly cracking and breaking the highways and causing millions of dollars of damage per year. I suggest incorporating rubber from tires used, discarded, and chopped into very fine pieces that would mix with the tar and now be used instead of stone where a stone base would be underneath. The upper layer would be flexible and in contact with rubber against rubber, which means our roads are much safer because the friction would increase where stopping faster would now be in effect. This idea could be worked on to its perfection, and plants would need to prepare it and then ship it to areas needing road material. The rubber would be flexible in the winter and summer, and small pieces of sand would be incorporated into the tar to give the road grit and improve traction. If other materials could be incorporated and tested to see their merit, one application I've seen is broken glass. The problem is that the glass would wear on the rubber tire to wear it out faster. Tire companies wouldn't mind if this happened, but I'm sure the consumer does mind. One application that should be used more often is banking a curb or road; this is a successful application used in road races but has never been applied in many areas of the world. One of the biggest problems we have is obstacles in the road. Many people have lost their lives because of the deer that may run out into the road and cause severe accidents. If we used some common sense a fence that prevents small animals from traveling through the bottom and increases the distance at the top, this would prevent animals from accessing our highways. It is true that deer have the ability to. When we reach this determination, we will set that standard for fencing on both sides of major highways. Who pays for this elaborate fencing, media advertising, that's right, media advertising? Saying goodbye to the large and ugly billboard advertising; instead, signs will be on the fencing only. The signs are located every 50 yards at the minimum. And no longer do we see signs that have nothing on them. Because the signs are removable and lock onto the fence with a locking mechanism. The signs on the fence also allow lighting above or behind the sign, but there are only two types of lighting L.E.D. white light. The reason is that this light will be from solar panels that help drivers be guided with tips of fence post showing the side of the road, also help visibility to the driver at night and during storms. The fencing along the road belongs to the state where it is located. Still, the funds acquired are to be used after the fencing is complete for lighting up signs and repairs of signs and fencing that may need to be repaired, but any extra funds at the end of each fiscal year would be considered public funds. These funds would be extensive because so many advertisers use the road signs. The extra funds could be used in the following manner} poor housing funds. This is where funding is applied to houses in poor neighborhoods needing external work. However, it is only used for painting, where the owner can receive paint with a voucher. The house would be checked to ensure it is painted within a seasonal time frame. If someone is found not using the paint, they would have to return the amount that the paint cost.

Roofing: This would be hiring a designated team of roofers, but there is no bidding at a flat rate per square foot. A state inspector must approve funding for internal repairs and only use it in homes that have never had internal work done before. The following can be applied: paint, tile, floor, wall repair, and insulation. The funding for home repairs is no less than 80 dollars and no more than $2,000 Per home. It can only be utilized every ten years and is on a first-come, first-served basis, with approval to designate the work needed, and a professional is hired once again on a flat rate. The homeowner can file an online application for housing improvements at any time. Once they have done that, the home is placed for an inspection date for possible work applied to home. Sixty percent of the funds would be available yearly from the excess fund. Forty percent would go to entrepreneurship for the development of new jobs. The effect these new roads could have on the lower and middle classes is to create an economic balance that supports the lower and middle classes to be raised to the upper-middle class level. The first step to provide this is to allow education to be not only for college but for trades, and evaluation of the student and their self-interest to direct the student in the proper direction rather than forcing education that does not appeal to a student (see- "Schools" for more details.) The intent is that all people still can acquire more than they once had, but to keep a realistic vision of the actual needs of the many in a society where no one is deprived of quality education or the opportunity to continue their education so that when they approach the job market, they are given options rather than single-minded education. This can be achieved through state funding through the highway fund, where families are no longer responsible for paying for children's college educations in their own dollars. Again, funds for these colleges should be state-based with the exact education or higher to accomplish what once cost the public.

Their private funds are now compensated 70% to the student as long as they have a steady grade average of 70%, so let's call it the "7/7 program". The main question about how this proposal directs us to" money "is its origin. The highway fund is the answer, which I will go into more detail about, but we have libraries throughout the United States, and these books are currently being given freely, except for late fees or books not returned. However, we have an actual product base made from educational books, and we all need information from time to time.

Not all libraries will be free in the future, a small fee of 10% added on to the cost to replace a book not returned, or damaged. This means anyone over 18 years of age must have a bank debit card to take out anything from the library. The funds are used to increase the cash flow to higher education to low income families. Book trucks are also to travel from town to town with this same 10% added. These two methods would generate, per state, enough funds for a potential 1,000 students or more to be given a school grant. However, the library also has potential meeting areas, a small fee of $20 to use a conference room, and late fees for books not returned are doubled in cost. The reason is that when a book is given, it has the date of return posted on the cover with a date return sticker, not on the inside, so it is easy to see, and there is less chance that a book would go unnoticed. But we have these standing structures of learning, and instead of making them profitable to help college tuition's, the libraries of today make only small gains. So, with the technology we have, and why are we not selling top-rated books directly from our libraries? The ability to sell books could be another means to raise funds to also support higher education, the investment to sell books would have profits for even up keep of our libraries. 1. Sales of books, 2. Credit card debit card payment is due from books not returned or taken from libraries. The funds combined with the high-way funds will achieve the success of our future students and enable them to have what they need: a better paycheck. Roads- The highways in the winter are covered every year with salt. This salt is extremely harmful to our fresh water and to the fish that live in fresh water. The salt is more than harmful to crops, and to the metal on cars, all in the namesake of driving safety, the land is being slowly saturated to the extent that we can say that someday soon, fresh water fish will not be able to live in fresh water. The solutions could be with: Rubber incorporated into roads. The reason this will work is because, first and foremost, rubber is what we use for our car tires to adhere to the pavement. Add rubber to the rubber, and you will make a surface that increases slipping by almost 90%. This means even in wet conditions without frost, the road will be able to keep your car from hydroplaning.

Hydroplaning is caused when a fairly large amount of water can collect or distribute in a puddle or thickness that causes the tire surface not to be able to make contact with the road. I have seen the center line of roads have indented pavement that tells a person they are getting close to crossing over into oncoming traffic or exiting off a road, if we added the same indentation as to keep those two as good protective measures to not go out of our lane of travel, but no a new set of added lines are going in the same direction as the vehicle, this means lines would not make a sound, but leave indentations to allow snow melting to travel to the groves made to become a drainage and this would slow down the process of water collected on the highway, and a simple horizontal line going across every 50 yards would allow the draining. With this design presented, the need for salt may be removed completely, or drastically reduced.

55. SUBJECT: SALT REMOVAL

Morton salt is a famous household name. We all love salt in our food, at least most of us, and we can thank Morton Salt for creating iodine placement in table salt to almost eliminate thyroid problems caused by a lack of iodine in the diet. But we also need to examine how Morton has decided to take salt from the planet. We are not saying that they need to stop making salt. We are saying they are disrupting the earth by sending water downward by pipelines to a salt cavern and saturation of the salt, causing water to turn to brine, a heavy salt concentration. It is piped back up to the surface to remove the salt in a heating manner that takes the salt from the water.

The heating process is fine, but the places they have chosen to gain salt are horrid. This process will also leave massive gaps in the earth's crust. Water is the most porous element on our planet, and it can move into dissolve some of the world's strongest materials just by being in contact-including metals and other components in the ground, and this brine could even harm certain types of plant life and animal life, and if the brine escapes from it's enclosed area of removal being the process is being done near fresh water, this could infect the entire eco system and destroy the lakes fish population who are fresh water fish.

The process of removing salt in this way cannot ensure that they know what is down there, and what they are dissolving may not be but a mere fraction of the distance that this process can cause an underground stream to be affected. At this moment, some crazy, money-hungry people have decided to mine for salt. UNDER one of the great lakes, the world's largest freshwater bodies, they honestly think this could not collapse and infect entire lake regions with salt water.

This concept is not only a fool's disaster waiting for the "oops." This can potentially destroy the Great Lakes as a fresh body of water, and many communities, even today, rely on its water. It would take hundreds of years for the water to return to normal state only after the cavern below has washed away all the salt.

To be more realistic, it would most likely never recover in ten lifetimes. But the removal of a hard substance with a water weight of God knows what is on top? So, the ground pressure is compromised. What happens then? Well, that much ground falling inward will be another disaster as the ground collapses beneath us and causes earthquake conditions that could be catastrophic. The chance of this happening can be increased with even the salt caverns being filled with water because one day, the water will find its way out. Then we can have massive plate movement of just an inside collapse in the fashion of sinkholes the size of Cleveland, bigger being they don't know how much area they are covering.

The solution: Morton already has the technology to take brine and remove salt, and in California, they use Ocean water with evaporation to make salt fields and then harvest the salt. The salt fields are not something bad, but they take a great amount of land mass, and as time marches on, the salt fields could affect local farmers or water runoff if there is ever an overflow. But the real answer is to use the method of heating the water that has been proven to work at Morton, but instead of Muhammad moving the mountain; Muhammad needs to go to the mountain, which in this case the mountain is the ocean. Our salt supply is essential here in our world, I guess until a man can harvest other worlds in outer space.

So, if a pipeline connects to the ocean and removes the salt by heating the water and considering that runoff could affect wells far beneath the surface. If we use heating the water directly at the oceans location, the water will evaporate faster and only leave the salt behind, and this eliminates the open evaporation and the possibility of salt being swept into the air, which in time could reach the lower levels of underground waterways. So again, the heating process could be done with solar power heat coils or with sunlight, and it does not affect the lower ground and stability. In this world, we sometimes see something and feel well.

It's there for the taking as long as we own the land above, never considering the after-effects. This situation needs legal action to stop this use 100%, and we love Morton Salt, but we don't enjoy the potential of the severe water damage that this has pointed toward. To those using these methods underworld effects can cause damage to others. In the future, I hope Morton and their CEO, will be bold enough to care about the future and not just their present wealth. But a word to the company: thank you for your years of salt enjoyment. I use it with great confidence that it is some of the finest and purest. I realize the method you use works, but it leaves us with possibilities that are real.

So, the time has come to use sea salt as it is available, but do it because you care not only about profit but also our world. Thank You, Morton. I know you'll do the right thing for all of us.

56. SUBJECT: NUCLEAR ENERGY PLANTS

The use of nuclear plants will no longer be needed. Still, the threat of nuclear war will continue because of man's unwillingness to let go of weapons of total destructive power. These devices are not a deterrent, as they say, but of the extinction of all humanity.

Humanity is unable to comprehend that this is a fact that when it does happen (Nuclear War) it will be used simply because someone will be willing to use it, and we see many leaders in the present day already expressing that intent as a real threat and consideration from Russia and North Korea and China with an increase in missiles, and so we only wage war not for peace to prevail because the end of this has no peace for anyone,- but humanities extinction because with one will come two, and so on, and we only have a total of estimated worldwide 13,246 missiles as reported by online at Wikipedia, so enough firepower to blow up the entire world about ten times over, and with enough radiation fallout to blanket the land and oceans, which removes all life from the oceans, all life as we know it will no longer exist.-?

Humanity can unite, and- we can dismantle and condemn the usage of any weapons of this nature. But could this be accomplished? This would be done only with an act of compassion unheard of in all time by humanity and unlikely to happen except by force. What force could cause the removal of all nuclear weapons?

The force ability to control the entire rim and outer area of our planet with a military weapon called Star Wars was introduced in former President Reagan's administration, but was not developed because, with this concept, no missile could be launched without instant destruction, because the ability system had the ability to fire lethal exact laser beams at targets on the planet that would devastate any enemy without the use of military ground troops.

But the technology has made it so that a warhead can be detonated utilizing a hammer. Sadly, we are not prepared for such an event, and the people who would do such a final act cannot see that they are a tool of evil.

Instead, they believe that by the destruction of humanity, which would be the final curtain for all mankind, not one of their loved ones would survive the massive retaliation of missiles now destroying everyone they ever cared about, or their so-called faith in the belief they are gaining will also be wiped from the world. And they will have one last request as they dissolve the life they had and those around the world: they will be welcomed by their creator to enter an eternity of pain and suffering that they caused humanity.

There will be no joy that they expect; it will be a real wake-up call when they are introduced to Hell, and the torment they will gain will never stop. This you can be promised. However, uneducated people will believe almost anything, so if you are someone who has a death wish, do it, but leave the rest of us alone to live because, unlike you, we like life.

So there you have it. So, is there any good that could come from our friend Einstein's nuclear destruction? Well, no, because everything that he created gives back a negative.

The real answer is to burn all that research before it burns us. But can it be used to protect our world? We could use nuclear weapons to destroy, say, a faraway asteroid, but then the radiation would follow the path of the asteroid that was coming our way. So we can X out that idea.

The good news is that there is no good news. Now, for those of you who call nuclear plants safe and clean- this is Chernobyl nuclear plant, a place of destruction, and you can't live there. But for people who like special effects- you may glow in the dark if that is your desire. But I wouldn't recommend it. Also, we can include the massive underground storage of nuclear waste, just waiting in the future for a groundwater catastrophe. SAFE ENERGY??? No, nothing about it is safe, but if people worldwide finally said- Enough is enough, we may actually have a future.

57. SUBJECT: NUCLEAR WASTE AND HAZARDOUS WASTE

The planet has collected a vast amount of extremely dangerous waste, and so we have made our mess that cannot be cleaned up. Instead, the powers that be take the time to place our lives in deeper pearl when suddenly their transportation of these massive dangers is given to the public as a gift in a disaster, as train wreck or truck transportation on highways. And now the mess has killed unsuspected people who are contaminated, and the hazardous mess never ends as legal battles pursue, and billions are taken or given in the long run. The waste that does arrive at its destination is a real threat as they believe that if the storage is not seen. That this is this the answer? Well, we do have a choice, and this choice is extreme, but it is the

146

best answer I can come up with. Let me clarify: we made the monster-(nuclear waste and hazardous waste), he has full control of everything, he is the threat, to keep him around costs billions, and to keep this monster also can turn on you. So they give the monster what they feel will keep it under control. They have made vast underground caverns, similar to ticking time bombs, that could erupt and enter into what is well known that lies beneath millions upon millions of acres of land, is "underground rivers," these rivers flow everywhere in directions they have zero idea how they travel. These rivers are connected to our wells that are dug directly or indirectly to our water sources in hundreds of thousands of areas throughout the world. So to place these waste storage compartments with the fact that earthquakes are an element that could happen, or let's say some upper explosive action disrupted this vault of destruction, maybe a plane crash, it could be a terrorist being dumb, but a flash flood that rusts the barrels, or floods the cavern, the ills of a containment breach is with vast possibilities. And can we, as healthy people, need this threat? The answer is a resounding no. THE ANSWER: We have talked of traveling to Mars, and for some strange reason, the only thing I have seen and you may have seen of the surface is "dirt," "stones," and more dirt. There is no life, just wind, dirt, and more dirt. This says to me- "perfect for waste." The ability to use this planet, Mars, without water and vegetation is a task or undertaking that I would say is next to "insane" as it would be just as foolish to place anything on our moon, and I will get to that next. But- Mars does have one real value. It has so far shown "zero life," and if this is exact, then I would say it is a perfect location to take our unwanted harmful materials and, guess what, bury them there! Now, a space station with water and food and even air containment and some vegetation plants to make air could be established for, say, a manned crew of 30 people, and the trips with the hazard waste could include supplies needed for crews and even replacement crew members. The removal would be in the same areas where the waste is made, and in the future, all factories that make waste- must be located in the same area so they do not allow the waste to "TRAVEL" around the population. It's called building the factories in the same area, folks, and being more exact-in areas that can support less energy to maintain the structures, such as underground structures in deserts, the ground temperature is maintained as cooler. So, the ability or the need to heat the structures decreases greatly. So allow me to include that all our nuclear missiles and bombs could be dismantled and also sent there. And good riddance to all of it. But how could that happen? A world family movement. (Who say: Enough is Enough.)

58. SUBJECT: PLACING STRUCTURES ON THE MOON

I have heard these claims of the United States and other countries like China's desire to make their field of ownership to be on the moon, but let's be fully aware that this is a well-balanced satellite that rotates in a manner that keeps us in a true line of travel around our sun, so we could say it is like hitchhiker who caught a ride and has to stay because he has a gun on you, so let's say for the short term, the passenger (The MOON) is essential to our survival, with one major factor that if the weight of this round orb if it suddenly had some massive amount of structures developed could cause what I call the wobble!

A displaced structure, and if some explosive device were to damage this, humanity has many devices that can cause explosions, so if this should occur, causing the moon to be, let's say, "not stable" as to fall off its orbit, well- the possibility of the largest rock in our immediate location called Planet Earth would have the gravity to pull this into our atmosphere. We also need to consider there would be nothing left of our planet but a mess of billions of lives gone (Extinction).

It's a word that keeps creeping up in my text, but a very real potential. But also that the moon is a counterbalance of our planet, as we in circle the sun, let's call it a device that works like an element of perfected balance that keeps our planet in orbit as we travel, as the perfect balance. So the conclusion is to stay away, even far away, not to disrupt its balance and never make anything there. If you desire to build in space, the ability to build space stations away from our planet has fewer chances of disruption of orbits, so- be bold in the future. Just be smart enough to know that when we mess with the natural order of things.

Things become unnatural. And to China, who believes they can make a claim to the moon, sorry, the American flag was planted there a long time ago. So, in a sense, we claimed it just like people used to claim land. We placed our marker on it. But once again, to be on this mass would be with the potential of a world-ending event potentially once the structures of materials brought to the moon were placed on locations. So, for everyone, including NASA, staying away from it would be the wiser. Let me put it to you more directly, - If we look closer at the moon during a full moon, you can see him say: "No trespassing." Good advice from the moon.

59. SUBJECT: OUR HEALTH and LAW

We have been forced to pay higher and higher costs to maintain our health. Any one of us can go to the ER and receive temporary care, but at a cost that costs our pocketbooks and wallets. The days of the doctor visit are now long gone, and a doctor treating you doesn't work on you. He prescribes pills! But they ask you to be there at 2:00 PM and see you at 4:30 PM or worse. The time has come when laws of standard procedures and costs are the same across the board, the days of inflated costs and doctors charging massive fees to include dentists who rape the public. Also, the law needs to be clear that all medical items be predetermined as for-profit margins not to exceed 20%, which means that all costs are maintained, and sorry, boys and girls of these professions, your hay day has come to an end. Shortly, all health care will be free, and they will no longer have control of their pills and items needed. The law will be passed that says 5% of all people earning income will pay this amount yearly to be placed in a national fund pool. So if you pay into this system at any time you are hurt on the job, or sick, and you get free care. The way it is done is simple: you see, the amount of people paying weekly from paycheck is far more than the number of people sick. So, there is always a balance that says funds are always available. But now doctors and dentists, and lawyers are paid a straight salary, just like the rest of us. The sum is straight across the amount. No more will these providers make big bucks for services be required by the public. Instead, they will now serve the public with a slight pay increase, but only to exceed 20% of the actual minimum wage, will be determined by monetary legal adjustments that are passed by law as to the exact amount a doctor can charge per visit. What they charge for treatment will all be under a fine microscope. If they are found making additional profits, they can lose their ability to be in their standing profession. Yes, they are well educated and should still be compensated with a wider margin than an average salary. Still, this salary should be based on the person's household and their profit margins not exceeding others. No profession should be so much that it costs the public to become harmed financially. Plus, the person with the lower income will be given compensation from the 5% across-the-board payment system. We need a person group per city to be established to ensure that hospital patients and persons seen have a place to go to for their complaints, and can give fines to doctors who have been found to have proscribed incorrectly rather than lawsuits that drive up the cost of medical insurance the fines would be established in categories, including loss of life, no doctor will no longer not be responsible for their failures, all of us have penalties for making the human error, and they in the higher professions should not be excluded. Still, no one person should be paid more than what the set price of the fine is established. Now, we have people die every day at the hands of doctors who have no clue what they are doing, so we need to make a record of practice "percentage." This means we must have the doctor's success and non-success ratings from people's families and pairs of every doctor, every dentist, and all medical-related conditions. No longer can a doctor be without a clear record of the mistakes or the successful treatments. All types of comments will rate the performance of every professional. This method will also be reported to every dentist, every lawyer, and every car mechanic. Why not? They take advantage of customers, so the price gougers need to be pulled in, and a website is needed called: "C- RATE" You can type in the location and name of the business and get free information that will tell you their status as good, bad, and track record of success, and you can add your rate as the rate marker is from 1 to 10 clicks. And a 500- word text to place any grievance regarding how they were treated or complemented a service. To be direct, the website is to be able to check the availability of services and see who has the highest ratings to accomplish the needs of the people. Now, to the points of services rendered, all these will be flat rates, and legal actions to regulate them must be implemented by law.

60. SUBJECT: THE NEW "ER"

E.R. most of us know stands for "EMERGENCY ROOM," not "Extended Resistance," so the ER that makes you sit in a "WAITING ROOM" even to have this is unimaginable to maintain as an excepted practice. To avoid a possible wrong communication during 911, the person sick can now call directly to medical assistance with the new 912 emergency calling. This sends the caller to the closest medical emergency call station, the call is recorded, and the ambulance is sent from that area without delay. this divides emergency calls of crime and fire are now separated from medical emergencies, so there is now a direct line to the party of concern, rather than a call to a call, a call as today, and the call is now directed to the closest emergency pick up as to give the address and zip code. When the emergency vehicle arrives, they take down the information for the hospital. But more importantly, the new medical truck has a call doctor. This means patients with heart failure, for example, could receive instant care such as nitro pills from a certified doctor, which, with the proper medical equipment, could help in preventive bleeding or other possible problems to a medical problem someone may have. The onboard doctor can now treat a patient with a simple authorized signature from anyone who knows the person, or the person can be recorded verbally to agree. If a person is not able to make the call as agreed, being under the influence of drugs or heart failure, the doctor then checks marks- coherent, and this clears the doctor of a suit. So, the patient is not questioned until they have been seen and given either medication or repair to the problem or have been admitted to the hospital. This reduces the time factor that may or may not save a person's life. We must understand that not all bodily injuries are visual, and some people have a higher tolerance to pain than others. All people must be given free medical help and have their fingerprints taken. These prints are placed on file to be accessed in the same manner as police now have fingerprint files, and the print is matched to the person. If prints are not readable, an eye scan is also required. The need for these procedures is if a person has no one present and the I.D. of the person is unavailable. By the medical fingerprint file, the system can locate a person who may not be able to speak so that the fingerprint would give the medical team any information affecting that individual's medical complications. When a person enters the ER either from their own effort or that of an ambulance, the person is fingerprint scanned to get all medical records and information. All records established are to be entered into a national archive available to any hospital at the touch of a search match on the fingerprint. This scan is also linked to police fingerprinting, which could reveal a possible criminal. And with today's technology why don't emergency rooms when attaining a local patient just have picture I.D. cards made that have a scan just like a debit card that attains the person's full recorded visits, and reasons for visits, and a reference file number? The V.A. hospital uses this method, so why not all medical institutions and emergency rooms? This is also for private doctors, where each patent record is kept in the same manner, so this file has the time of entry, condition, and prescribed physical repair or medication prescribed. The new E.R. is equipped with the following equipment directly in each room, which means faster responses could save more lives. As stated, each room has the following X-ray machine, blood supply, blood testing equipment, and lab tech to test and give instant results, which is also less than feet away. These lab technicians are located at the ER but can test blood from the hospital patients admitted between the tests being done at the ER. However, persons who have been established have priority because they need their results first. Now to address the new contact system for medical, let's take this even further. [911 for crimes] [912 for medical] and [913 for fire]. The person picking up at each is not an operator but someone who can respond to the call and instantly has the location to send to the personnel needed. A GPS gives the operator the exact location instantly, so the need for info is not to be delayed in asking questions. The call is also transferred with GPS location, so the info is in text format, and the team of people answering the emergency is now in transit. So, calls made inbound already have the area of the call. The only question the operator asks is this: Is this the area of the problem, and then repeats the address location? So let's say the Police are called- It is directed to an actual policeman in the area closest to their cell. The call is recorded and sent on the server cloud as to the actual conversation, which cannot be altered and is linked by the time of call on the cell phone as to ITEM # case. The Medical is now directed to an actual medical vehicle operator on the cell phone, and the same conversation is recorded. The operators who receive the call can even play back the info sent to ensure they have the correct address. As this was being sent to the cell phone, the signal was also being sent to the GPS locator in their vehicle to give exact directions instantly, including how far and the time that you would arrive. The fire department is similar, but there is only one difference: the need for load fire alarms to get people to go to fires is no longer needed because the system that sends the call searches the closest GPS location instantly, and now the call goes to all firefighters that are located the closest in that fire department. When they arrive at the fire station, the truck already has the location displayed and gives the firefighters info on the type of fire described in the call: house fire, chemical fire, train fire, etc.

So they can take potential equipment needed for larger problems. This system allows faster response time and more accurate information exchange.

61. SUBJECT: THE WORLD GOVERNMENT

The mere mention of this seems to give an instant negative reaction from the majority of people, and the thought seems to be of this massive loss of freedoms, and we have lost touch of the global effects now facing our world that we are no longer separated by distance and the issues such as over fishing of international sea boundaries, and pollution mismanagement of waterways that are destroying our oceans with plastics, and chemicals and hazardous waste dumped directly, and please know the U.S. has more damage done than many nations so we are just as much at fault. Known fact- That nuclear waste in barrels was once dumped into our oceans! In addition, N.Y. The city was dumping mass amounts of its garbage into the ocean only to have a great outcry from New Jersey as the items from baby diapers to hospital needles washed up on shore. I saw this personally. So the time to see our world as it should be is this fact- "we are all the same". I said it before, and I will say it again: I have all the same parts as any human male, and my language and outside appearance will be different in skin color, but I still have two eyes, two hands, two feet, and all the same equipment inside and out! People have been fighting over property for centuries, so where did it get us? Bigger bombs with a way to destroy us all in a few minutes! Is that the future you want for yourself and those you care about? So, people – get out of bed and make a wake-up call! Get your life to be more than breathing and dying! The world needs unity to be a family. Yes, a family considers the rest of the people around them too, and cares enough not to kill them. Okay, sometimes we feel like we could without really doing it! (Lol) But we make an effort to try, yes, try to see their side, and we work out our differences as people who have brains. And for those of you who do not understand what I am saying, don't worry.

You may get one later (REALLY!). PLEASE SEE EXAMPLE (B) BELOW as a reference later: Okay, back to my point of why we need a world government. The process would be similar to the makeup of the United States, except on a global scale. Each country is now a state by principle but not but the designated title of the country stays the same for easy identification, Instead of being a separate entity they are called the United Countries of the World The case with all countries currently. Or we could call it "The United Nations," which only makes sense because they are all nations forming a unified governing and already exist. But unlike the United Nations of today, all countries elect their leader like the president of the U.S, each country submits a prospect for a "world leader" who is also voted by the country's people. Also, they're a member of the "world counsel." Still, the difference is there is a world leader elected; this person is to have the votes from all countries as the majority vote, and must represent the people of the world as a whole; there is no room for error here by the "world elected leader" because the world leader can be voted out by the majority of the nations also, this can happen if there are more than three nations to dispute the standing leadership, so we have someone who can be replaced if they are not doing what is correct by the world community. If removed, a new president can be elected within a two-week window from the standing group of the world council. And who ever has been removed from world leadership , that country now holds a new election for their representative to replace the former world leader, but this individual is now only a member of the world council. If someone is doing well, their term can last as long as 30 years. Still, once again, they can be taken out of office at any time the national leaders vote them out if they are found not to be doing their job for all nations; once they are voted out, they are to be replaced within four weeks, the leader leaving stays with the new president for four weeks in a briefing once a day if requested by the new leadership, and must be on call 24 hours a day to give any advice on situations that where prior dealt with, no elected official would be replaced and not stay to show the new person the ropes, instead all people replaced will be leaving in three months, this gives the person time who replaced them by vote to be able to be given all the proper information and procedures. The reason a world leader is needed is that people of different nations have different agendas at this time that conflict with what are proper actions that do not affect the world community, whereas, in a one-world nation, we gain the ability to help every country as to give it the proper attention without the loss of human rights. A leader leaving stays with the new leadership president for four weeks in a briefing once a day if requested by the new leader and must be on call 24 hours a day to give any advice on situations that were previously dealt with. No elected official would be replaced and not stay to show the new person the ropes; instead, all people replaced would be leaving in three months; this gives the person who replaced them by vote time to be given all the proper information and procedures. The reason a world leader is needed is because people of different nations have different agendas that do not always address the most important factor our planet, and it's ability to sustain life. As a united world problems of the world would be resolved smoothly without long delays. The urgent needs of our planet are required that we do unite.

62. SUBJECT: TO MUCH "BLANK"

People are being controlled by cell phones, this mean our world has an attraction to wasting their lives with playing games and texting rather than investing time in quality-building activities. They are being absorbed by game devices that take their time as productive people and are now spending hours per day in so-called entertainment devices that only suck dry their existence. There is zero, nada! Nothing is gained by spending hours upon hours in front of a screen of any size to produce a victory of meaningless value. Our enemies in the world, aware of our weakness in these devices, are making every effort to bring us more of the same, as our upcoming youth have been recorded to have much lower education scores than many other nations. Why is this happening? We have far too much access to entertainment devices that do not help a person advance. Television and even radio have been strong contributors to time-absorbing devices, but we have gone further and further away from education. Our homes are filled with devices that take our time and our minds to places empty and void of learning, but I have games that do destruction!, or a fantasy world; I can even kill people with real-looking blood everywhere! , but the fact is these devices have made us--- stupid. No other word can be stated that says it all: "ignorance." Ignorant because that same person who spends hours making la-la-land (drug) as their escape to get away from their daily humdrum, is that possible or potential person who could be the next great artist or inventor, or maybe could have been someone if they had spent their time correctly, could have become a doctor, and the person who discovers the cure for some horrid problem. The possibilities are endless. But there is one thing I know: people who waste time waste their abilities, and abilities that they may not be aware they have because they have no confidence in themselves and most likely had no other people in their lives who cared enough to tell them they had worth. But maybe you had a dream; you wanted to do something. It seemed impossible, so you asked others how they could help- (maybe? But, instead of gaining help, you were torn down or laughed at? Yeah! This happens to anyone with goals or an idea that could be good. But you're lost inside, feel alone, and wonder- why bother? Nobody else cares! You say- "I can't do it by myself". You are so right. It is hard, and others will say this to discourage you because they don't like that your ambition is to better your life, but hear this- that seed is inside you, every one of you alive. The weaker can grow stronger, and the cruel words of others do affect us, but we must learn to move away from those who say to us- "NO!" instead, Live for your hopes, live for your vision; it is not in a game, a game has no value for your success. I don't know you, you don't know me, but for 37 years, people said "NO!" to me. They do not know what was possible. People will discourage you, people, even those who claim to be friend's people who do that are not friends. If they do not support your hopes and desires to do good, avoid them like a disease; they are a heavy stone on your neck that, when you swim in deep waters of life- will allow you to drown and say, "Oh poor you, as you sank!", So escape from their grasp as soonest, and seek others who support you only, find likeminded people who say to you words of encouragement. Those will be your friends.

To all of you, -The person who has nothing but F words to say, or covered their entire body in hell tattoos, just way to much! , it's all with anger, pain, or hate. Let it go; let your joy of life shine. I am not saying that people are a bunch of roses; well, maybe they are- roses who have thorns! Lol. But stop looking back with anger, drop the luggage, and move forward into a better day. I know it is not easy: the car just broke, your home was found with termites, you had your dear friend die, and the list is long and horrid of all the trials we face, and is it easy? No- not at all. But try what I have said, and I truly believe that if you do, you can become someone who helps others through their day rather than burden them. Here is my real story for you to just know that I could be bitter, I could hate, but you decide: You think I have never hated people? At age three, I was tossed away by my mother in the middle of the night, I awoke from my sleeping to find my mother was gone, I remember everything,-I put on my slippers and my bathrobe, and walked outside and up wooden stairs to the upstairs apartment and I knocked on the door. A woman in pink curlers answered the door with a pink bathrobe, I asked her: "Is my mommy here? ". She replied : "No, little boy she is not here" And she must have called Social Services, because I awoke the next day in a new place with four other children, I was tossed from home to home for two years, and I was adopted at age seven. I knew everything that was going on: had people treat me like dirt everyday walking home from school, being yelled at by ignorant children, hitting me, and saying, "He has no real Mom and Dad." I was treated very badly all through my youth, and I had no friends accept one who even that person when I needed him most ran away when I was being hurt; I was even later rejected by my adoptive parents, and was and told to stay away because I was sleeping in the backyard in my van without a job. I was tossed away like a bad shoe. My only friend later was Sandra Allen, she let me stay in a lawn mower shed I slept on a wooden floor, yes, about twenty feet from railroad tracks with trains constantly going by, and three times in my life-, I became homeless, once in Orlando Florida, living in my van, and twice in New York, for two and half years both times! I was not alone, Sandra was there, as a firm friend who worked for me, she never left, we parked our cars in the woods in our cars, she stuck by me through everything bad, we joined forces to start jobs and inventions in 1990, and I had ideas stolen which made millions for many other companies, as I stated before, items like -the toy bubble gun, lighted keyboard, for computers are just a few taken. Sandy never gave up, and when I wanted to quit she would push me right back to work, she always helped fight our way back after one financial loss after another. Cops constantly bothering us just for sleeping in the cars, having people look down on us. But we never stopped trying; we stayed clean, and I landed a $38,000. Deal with an investor while I was homeless, and that project is still being done. So hate, yeah-I could hate, easy, but to forgive is what I did. I know that was the right thing, even though I still have not forgotten the people who mocked and ridiculed, lied, and took; yes, I remember well, but I am now better in my life because I demanded better from myself, "not from others". We can be broken in man's world, and many of you reading this may be behind bars; maybe they are real bars, or they feel like bars; you may be feeling lost, feeling as if everyone hates you, inside your fear, but you replaced it with anger, but deep inside you, you want to be free of the past, to start a new, then do yourself the best gift you could ever give yourself, give yourself a chance to be better, don't condemn yourself, but be determined to be different as to be a good person, no matter how others treat you, I'm not saying it's easy, no, it is very hard! But this has impact, worth, value, and self-respect, so don't toss the red flag for penalties; Jesus says: "I will be your strength." But hear this loud and clear, when you or anyone pretends, please hear that word: "PRETENDS," when you pretend you are better than another, you are pretending because no one is better or less than another. You're not better because you're taller or pretty, or smarter, or have more money, but you can be better if you help others and work for what you gain legal, and follow God because he will help over time, so am I saying never take the easy road? Yes, always choose the long road, to achieve, one must take the hard road, which has a reward at the end and a happy life. I am not saying a happy life will be for me, but this way, one has a better path to being a better person, a person you can look in the mirror and say: "You're doing okay, not great, but okay." There is a saying, and it is very true, "The coal that gets the pressure becomes the diamond."

64. SUBJECT: NEW BABY EASY

I just had a thought this thought could be researched, and yes, it's being invented right now, yes, right at this moment, maybe we have been performing deliveries of babies wrong all these years. Why? Well, we have all heard of the C-section to cut open the woman and remove the baby; the problem I see with this is the woman is at high risk of bleeding to death; who wants that? Now, instead of removing the baby in such a, please excuse me, dangerous method, what I feel could be possible, once again, with some research, of course, is to build a device that can be made to help childbirth as to be without loss of life to the woman, and allows the process to be faster, and easier, sound impossible?, well we can look at some applications that have proven to keep pain away and cause a condition that relaxes the muscles and this done by shot of Novocain used in a dentist office, could be applied to the woman's birth channel. Similar to what is used by dentists into a tooth gum. The injection used by dentists cause the mouth to droop, and yes, people have a hard time talking. But, what if we could make a device that is like tool, that once the muscles completely around the woman's stomach and area of what we will call the exit of the baby "the cervical" which has reached a full dilation, but- now the areas are numbed with a group of shots prior to full dilation. Our new process actually mirrors an old way, you may have heard that Indian women who would go out into the woods and have their baby by squatting and actually used "gravity" to deliver the baby. So the device we may be using is to use gravity and make the delivery faster and safer and with less pain, in fact, no pain. The device which helps the cervix is like an oval circle that now is inserted with gel lubricant to hold the cervices with an expandable holder that now opens the area and actually below the birth canal , we have a very smooth slide covered with Vaseline this slide is a device to catch the baby in a foam cradle at delivery, but before this happens, a very thin and flat "soft" rubber tube is inserted into the womb into the sides and top and bottom, and with sonic detection to see placement of the very soft tube, as soon as this has been inserted, then a gel pure aloe lubricant is now pumped into the womb, the areas are completely numb and the muscles now completely numb and because the use of, a well-known drug called Novocaine, this could allow the birth canal to be wider without pain!, and we have one more device, it surrounds the upper area of the belly, it is very large circle that is like a half donuts similar to a upside down horse shoe but adjustable , and made of a semi soft foam around three inches thick and it has a small hydraulic electric motor that now press's gently downward on top of the belly area only, and at the top of the belly gently and slowly is ever so slowly pushing the baby downward and out of the womb as soon as the child's head is crowned, this device is only used when the babies crown has been seen to not cause harm to the baby, we will say this a very slight pressure is in a southern direction. The because of the lubricant injected, and the wider expansion of the cervices, the baby just merely slides out! Woo-hoo! And why did this happen? Because this system uses more gravity. Now let's look closer at this wild contraption, Today we have a women sitting in a position with her legs spread in front of a doctor, the device I have imagined here has a similar start as the women is viewed in the same sitting position as her sitting position is her starting mode, as to allow doctors to see progress of expanding , and see the size of her cervix as needed for proper dilation for delivery of a baby, but before she sat down, our pregnant woman is fitted with a loop around each leg similar to a horse harness but on each leg, and is flat with comfort adjustable straps as so she will be actually be sitting on this later, and an arm harness that straps around her upper body are connected also, these two parts are together and make a way for the women to be lifted comfortably above a saddle with an opening with a large hole in center , she is placed on this saddle when she is ready to deliver as state before, her baby is now being pulled by gravity, and with the applications of liquid lubrication and the device allows the legs to relax above the saddle seat similar to a horse saddle which is adjustable for any size woman, and with again a wide opening, the expansion device is attached to the bottom of the saddle , and sits just below the woman, a curved and lubricated slide which is the shape similar to a large banana slide with an upward curve at the end so the baby will not slide out when in motion, but is comfortable landing which the baby will enjoy, so similar to a children's slide but curved at the end to stop baby progress, when the woman has reached her peak of cervices expanding, she is now lifted by her arms and her legs above the baby slide which is now moved under the new mother to be, so as soon as the gel has been applied inside the womb, and she has been placed in an almost straight up and down position with her legs are spread very wide over the saddle, the expanding clamps will be activated slowly by hand movement of the knobs to expand the cervices further, the woman will feel no pain because of the shots given, but allows the expanding to be ever so slightly more than normal births done today, after the baby has slid out onto the baby slide, as soon as the baby slides out, the clamp device is released from her open womb, and we have a new born child! , no grip marks, no damage to the woman or the child. I hope that the medical community will consider this proposal. It could become standard, and no woman would possibly ever die again.

To improve this device, the woman being straddled could have the device shake her ever so slowly as in a vibration up and down, and use a very small dropping down of the straddle to help gravity along to allow the baby to release. And could cause the baby to drop even faster. If the womb has complications. If all fails, then a doctor would provide a C-section. But this method has never been considered, and I feel it would work not only effective, but be the safest way to have children. And I hope you women bring some nice people into this world. Now, to close this subject, please forgive me. I cannot give life, and you may claim you have a right to do with your body what you want, but that body inside is not you; it would be someone else. But you have a right, a right to be afraid, a right to pray to do the right thing, and a right to do what is right. You may not believe in God, but he believes in you because you are here; you're the living proof. Never allow others to push you into doing something bad unless they say, "Push Now!" "May you who are burdened be of good heart? May you who doubt be of good mind, May you be happy to hear your child sing for the first time".May good follow you both.

65. SUBJECT: PEOPLE RATE

Imagine an online entry that you only add your email address to get on this website and that is to every person listed in the world through computer entry in a cloud server, and you type in the person you were seeking. The website now gives you a rate chart of 1 through 10, with 1 being the lowest amount in quality of the person and 10 the highest quality reported, but you can now share your experience with this person, as one is considered a very bad person and 10 is considered a great trustworthy and kind person.

To join people rate means you can even attain a name badge that gives you a rate color code. "People Rate" is for anyone you can now know to let others know them, all for free and people can add their comments about someone they met, worked with, or did business with, and ratings from people who are shown as real people who dealt with the person, this can be fact or fiction. Still, fiction can be sued, so you best be with a witness that is not related or false, so proof would be needed, such as photos or many witness's.

Having an online record of every human being on the planet should be accessible to anyone at any time, and people can place their names. This record is statements about where someone has worked, how they have worked, their job performance, if they were honest or dishonest, and how they treat others. There are two columns: the good and the bad, and now anyone can see your track record as a human being. A rebuttal is located in the middle column, and this column is where the person who has been given this information on their ID list can now say what they feel about the rate issue. On the far right is a witness column for each of these columns, and the witness column is strongest if it is done legally. Each column has a witness column placed next to it. A rating is at the top of the ID list and tells the person reviewing the person's personal information which column has the most. In other words, if I have more complaints than praise, there is a good chance I am incompatible with others. This would also help employers to see job applications from someone with details that they may have been fired. As we know, there are so-called pay forums online as people search, but they don't allow you to enter any information showing whether a person is good, bad, or indifferent.

The good news is that people will always know someone well, even before becoming friends. Second, a person can improve their ratings by doing Community service, helping older people, joining a help group, such as March of Dimes, The Rod Cross, etc, and giving something of themselves that helps others.

Let's face it: many times we meet people we have no idea who they are, so I have come up with a new way to show exactly what a person is, with a legal device similar to a driver's license, as police can scan this device. This device helps police see automatically if someone is a potential danger, has mental problems, or is just a good person. Well, in the future, people have their driver's license inside an I.D bracelet, a color-coded I.D. bracelet/watch/phone that is worn. The device can give any police officer 100 percent identification of any person instantly, but the color bracelet is a colored silicone which is placed on the person as on the wrist, and is able to even be even shower friendly, the device has a GPS tracker, and if taken off sends a signal to last location, but the main reason this is important is because the device also shows others your rate which changes from year to year, yes led lights can change your rating as a good individual who has no criminal record, or very dark, this means , people may not want to be around someone who killed someone, each color has a value as to the quality of the person and the moral reports come from collected online , similar to face book but with any business or service can be expressed with a rate of 1 to 10, the collected good and bad will determine the color of the bracelet each year, the score is also determined by arrests, and each of these scores change the wrist band color, or disturbances, or threats, these all are collected and then reported to Police Department of I.D's, and D.M.V. could handle the distribution. The future of being bad is now easy to spot and better to avoid, but this also gives police a strong visual of what to expect from a potential color someone is wearing.

To wear the I.D. is stickily up to the public, but it must be on the person as to carry it. The future could be very colorful; I see clothing that reflects this same rate standard, as the shops give massive styles of clothing but place the clothing combos in a color style to give I.D. to others as to their quality as a person, either a good or bad person.

66. SUBJECT: TREES, MAN'S BEST FRIEND

I know, you thought a dog was man's best friend, but the real truth is trees are man's best friend but he treats them very badly. Many people consider trees very bountiful; most of you today in the United States and many parts of the world have computers. So I would like to give you a wake-up call that you can do. So go online and see Map Quest to those who are not familiar with it is a way for you to find a destination and track your destination to find and track a destination on a map. Still, I'm sure many of you have never noticed that the aerial view on Map Quest shows a distinction between forest lands, and you can view used lands for farming that once were forests.

When you select the aerial view in Map Quest, you can look at the future of our world and the direction the trees are taking. It's not a pretty sight. Please notice that the trees look like clumps of dark green or white green the thicker areas are distinguished as trees on the aerial view. As you decrease the size of the aerial view, you can get a clear picture of what is happening to all the forests we once had. They are being chopped down. People are selling trees for money. What is obvious to me is that farmland is reduced forests down to mere patches in our world, and the demand for food has outweighed the practical application of keeping trees; why do I say practical? It is because, without trees, life ends in this world; that is not a joke. I would say that the practical solution is to keep the more trees then there are, the more the merrier.

We all know that trees create oxygen and remove CO_2 gases. But that is not all that trees do; trees also cool the ground. Anyone can make a small observation while standing in the sunlight under a tree; the temperatures will increase greatly from a person standing directly in the sun. Now imagine millions upon millions of trees removed from our world, and we wonder why the temperatures have increased. Many of you know the trees have been removed over time. But I feel it is everyone's responsibility as intelligent living beings to determine what is right.

At the rate of forest removal we see today, and as the population increases, we demand trees. With more people, we need furniture, housing, and all the products that require wood, including winter heat stoves in the world's snowy areas. So, it is like a domino effect; as they fall, the opposite happens with our global warming; as temperatures increase, the more temperatures increase, more trees will die, and more vegetation will die. They continue to grow in the deserts as they have been doing for many years. Well, gloom and doom is a wonderful thing that scares little children in feeble fairy tales, but this is one writer who believes that we have a chance, a chance to make a difference, but it is up to us to make it. You see, we need those trees, and they need our help.

So my first suggestion is" When you remove a tree," "plant two." If you do not want two trees on your property, find an area to plant two. That's all I can suggest. It's just a matter of math adding and subtracting. So I have heard recently that there is a mission in the UAE to plant hundreds of thousands of trees; as great as this may sound if the areas are not prepared properly, the trees will only die. Another observation of temperatures rising is of highways we have made all over our planet; these roads are made of black tar and stone—a serious heat absorbent.

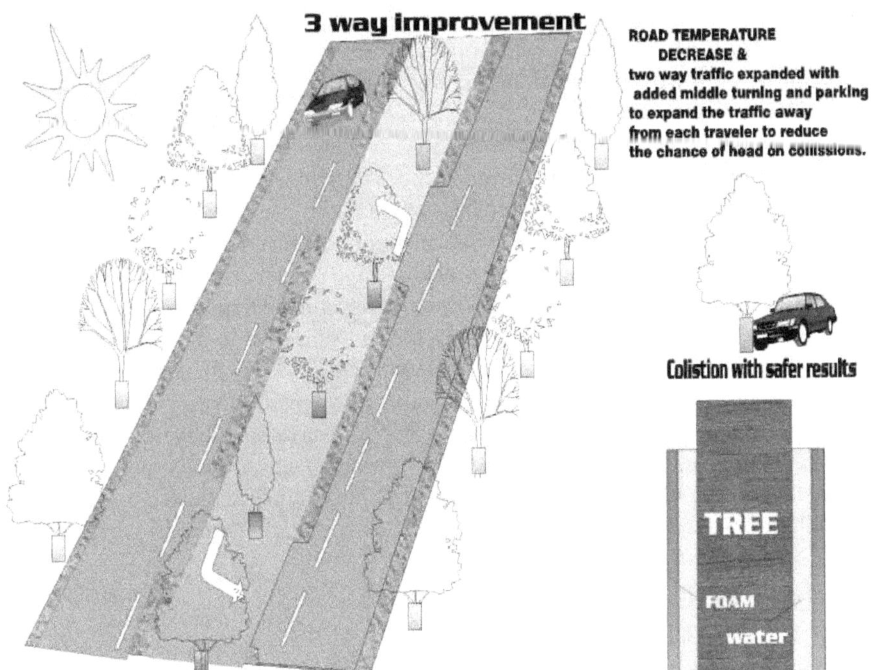

Shading all roads on the sides and the middle of passing traffic makes a vehicle stoppage in the center for turning lane,and the
roads now have the burms on all two way traffic made so the lanes are further appart and this can prevent many head on collisions.
If travel causes a vehicle to come off the road in any area center or sides ,the trees can be rapped with a expandable foam rap
that works like a crash barrel, and the foam actually has a open area to collect rainwater as to
allow crash foam to have more resistance. The trees in the middle are not at turning lanes, or traffic circles, but on open highway they are
reducing black top heat to collect heat and this reduses the temperature of the planet if used global.

On a hot summer day, if you place your hand on the ground next to a highway and then on the actual road, you will see that the temperature is extremely higher on the road. The obvious reason is one, black absorbs light more than any other color in the spectrum, and two, stones also absorb heat and hold it longer than any other element; we all know what it feels like to be sitting in traffic on a hot day, and our air conditioner broke. It's not fun. Now, consider how these temperatures are affecting our environment on a global basis with these roads all over our planet. This alone is a great reason for us to plant more trees, so we could be shading our roads as to place trees on the side of the road that best stops the sun light, so to have the trees along the sun's path would be ideal for shading roads. My suggestion and solution to highways with hotter temperatures is to create an umbrella effect from trees because trees are like umbrellas that shade our ground if we use the proper tree, and why not make the famous pine tree the real shade tree? This means the shade is thicker, and the roads would have a blanket of shade. If we plant our trees close enough to our highways, interstates, and even in our communities, we will shade the highways more. The people who could handle this task could be our forest departments, they could plant the trees on the side of where the sun is traveling from east to west. We can shade most highways up to 70 or even 80%. It is all about positioning the trees, which would be like a sundial; the shade moves similarly. The solution isn't cutting down trees so that we don't run into them; the solution is not to place trees at the end of roads so you can see past the other trees that are there and not block the view of oncoming cars or trucks. Now, with some concern in each community, you can bring back our downtown's with wonderful trees and save a valuable commodity: our planet. Trees make a thing called dead leaves, which is also needed. Why? We could help our deserts become fertile lands again if we dumped the leaves into those areas yearly. So I propose that all the leaves in the future be used, not discarded. They are essential to the revitalization of our deserts; they will be picked up by every town community and transferred to the local trucking station where trucks will travel to drop this mulch or compost to the outskirts and edges of all our deserts in this material will include any fallen branches or any other compost material including vegetation from our home that is discarded and buried in our garbage dumps up today. People have not realized that these materials, like banana peels, leftover lettuce rotted, and anything else, such as roughing vegetation, are very valuable; if it is used properly, it is something that actually could save our planet by being a good fertilizer. If we took every household that discarded this material and sent it to a desert area, we could revitalize the ground and bring nutrients to the plants that can flourish in the desert and a new ground layer that could allow trees to grow again one day. (Sounds like a plan.)

67. SUBJECT: 1992

The subject as follows below was documented from an earlier date of 1992, and even has reference that seems similar, but I feel it still has merit to be known again, for one very important reason. It speaks of " US " as people needing responsibility, and using this life, we have been granted in a better manner, rather than pollution of mind and body and the ignorant need to allow destruction, but seeking inside a pure and exact behavior filled with sharing and caring. And from where I sit, that observation looks dim and fuzzy. So these are my older thoughts-made in [1992- As you have gathered, this book aims to share inventions and solutions to human behavior in the hope that my efforts will be a help. Because we can look across the globe and see that many people have no hope for tomorrow, I say that with a clear vision of the future. You are most likely someone who lives to have a comfortable life. You want good food and a safe home and deserve real answers to our problems. But you may ask this question: "Why does this man believe he is the only person with solutions? The answer to that question is, have you seen the mess out there? I always say, if you're someone who has a better idea than me, then I'll do that idea. So I'm aware that inventions can come even as I reference mine presented, but they will not be the same as my ideas in the exactness of their applications. My ideas may hopefully be the cheaper way to achieve better results. Still, as the old saying goes, "Great Minds do think alike. "Well, you may also be someone who says, I don't care about the world, the future, I don't have time, or I am too busy, and if it doesn't affect me, why should I give it the time of day? Well, let me tell you why all those things stated are important; I am talking about "urgency" and the concerns for humankind's future to survive. To be as direct as possible- it's simply you, my dear friend, live in a time never seen prior just before the 1950s, the event that not only changed a war but changed the world to be introduced to its end of all humanity, I am speaking of the first atomic bomb that was dropped, the world changed from survive, to any second gone. And later I go into the subject much harder, but, you my friends are in the middle of a poo. And what is insane, -you are ignoring it every day, every day, right now, the only time in history humanity lives in actual.

68. SUBJECT: "END TIMES"

As a nuclear threat could annihilate all humanity and all life on this planet; unfortunate that the majority of us who live on this planet don't even consider this to be a real threat. In contrast, this writer became extremely aware of this fact at age eight, and your ignoring it every day, and I am still aware of it, but I wonder why people do not resolve this issue of dismantling its terror and concentrating on it so that humanity can live rather than expanding the threat in many countries that may lead us into nonexistence. But this reminds me of the old science fiction movie "The Time Machine," in which the people who lived above never questioned the people around them when there was danger from what was directly beneath them daily, but instead, when those creatures in the movie: "Time machine" ate some people up top, the others went about their weak-minded ways of acceptance that Bob was no longer at the dinner table, and Janet was also gone. However, those taken away were never considered the problem as important, and this is the adage – "If it doesn't affect me, why should I care!" Well, luck has it the owners of these dooms day devices have not been used as of yet, but that is like saying-people are rational and never get angry? So the first thing to make a better world is to have a better attitude. We've had many great men and women throughout time trying to tell us this message, but the facts of personal greed and self-awareness are that I come first and you come second, and this has no nationality, creed, or color; it is all mankind and women kind. In the past, many people who were great men and women who declared this message oddly never gave the real reason why it is necessary, and without their message of hope, love, peace, charity, compassion, awareness of others, consideration, awareness of your actions, and the result of your actions, truth, lying, the killer of truth, all these can cause the fall of mankind because without truth then you can only have lies. With lies, we do not know who or what to trust, so when it comes to decisions or to decide, the word no sometimes will rise without knowing if we did the right thing or the wrong thing. The ability of humankind to function as a whole is what we need. All of this sounds easy, wonderful, peaceful, and tranquil. It's a wonderful concept but has flaws because certain people consider it good and profit from destruction. Strange Demonic faiths, anti-God, or war profiteers. They think of ways to control others and how they think and do; groups that are fanatical religious ideas that go against our creator to say that harming another is what our creator would want? When the same religion will say that our creator wants us to love. These entities that refuse to acknowledge that God if he is real, which it is obvious to me that someone made the start of all this, that if it is science that made all creation then who made science? Because if we look at anything, we can see that it has a foundation of being made always has just that – a starter. So this creator with all of his DNA wisdom and ability to form matter is deemed to be a creator who created us in his image and I speak of a spiritual image. Because he did so, he did it out of love of creating. And we have fables that he has spoken to many people, who then told us of his truth. So Love must be the real reason that we are here; some people made love, some made lust, but they loved what they were doing would be my guess, but this life we have is a two-way street; you have to give something to get something, maybe it's your time, maybe it's a kind word, but whatever you give will return just bigger, some call it karma, I call it the enlarged image of doing. If I do well for someone; later in life, I get good in return, but in larger amounts. If I do badly, the same happens again; this time a bigger bad thing will rise for the return action. And the return is never always the same type of thing, but the return will either be a help, or a hurtful action that was previously made to someone else. So we can say that God – our if you prefer our creator, is real because we are real, and you're on and off button is not attached. So if love is our purpose, what are we doing, it is not love on any grand scale, miss-trust that is the replacement for love, it is the Kryptonite to love. So, what does this have to do with nuclear war? Well, a lot, yes, you see, we have those horrid devices because we fear each other; we fear each other because we don't "trust" the many leaders of who have countries who seeking ways to get what someone else has; they refuse to be satisfied with what they already have, and their greed for power is their downfall, they are blinded by being considered important by their people who do nothing to say to them, "why are you demanding of others to live as you do? If a group of people did not like how they lived, wouldn't they make the change themselves? I guess they can't observe that people tend to live in different style homes all over the world. Is it because they want to be the same as others? No, it's called being unique, and this principle carries over into even our way of doing our lifes choices, so to be different can be really good. So we as the living need to unite globally to put an end to this military mess that threatens our existence. And if you don't I have built an invention that is now at every persons grocery store in the world and it is called the "Silly Egg" It'sa raw egg that sits in the sun for weeks- When it explodes every human will be breathing in laughing gas! (Well it may stink, it's a prototype in development), so now finally the jokes I make in this book will be finally be funny! O-kay, if you believe that I have a bridge I can sell you, it's only $2.98 on sale! it was $9.95 but not one showed interest, darn!, But, you can put a toll booth on it ! Money! Maker! Yah! But I think I'll return from this moment of silliness back the subject. I am going to talk about "that serious issue" more later, why? Maybe because it deals with Planet Earth and you, and just maybe, you and people you know, might get the point of becoming more than a waiting extinction.

69. Subject: MAKE THINGS LAST LONGER

Why not make things better? We can reply that things that break now need new items made, so companies or people keep a security by making things that don't last, but what we gain is massive garbage. We have many structures that we make in the world. I am reminded of the many buildings and cars affected by water erosion and rust that happens when salt is placed on our highways. So we destroy what we build almost as fast as it is made. All in the name of profit. But the garbage dumps rise like mountains, I know my business was across the street from a major dump. It grew far more than it was supposed to. This could be decreased immensely if we took the time that the Egyptians took in the creation of what they made- the pyramids, but as usual, today man did not consider time, or if he did, he does not look at weathering and corrosion from water. Water and wind are two of the world's greatest enemies in terms of structures to include bridges. The solution: Instead of a man making buildings from elements that rust and decay, and instead of making bridges that have a lifetime of only fifty years in many cases, causing great financial decay as well as bridge decay, why don't we invest in long term bridges that cost more to develop in the front end, but save billions because once they are built the last 10,000 years. At least until man no longer cares to travel on the ground. I see the future roads like transporters of goods only and underground so that upper land is used for food production or just a wonder landscape of plants and trees to enjoy, and have nothing to do with travel for people except in the inner cities. Now since we are talking about things that can last, we can even consider all plastic cars so rust is no factor, and we could build vehicles that last 50 years. Unfortunately, this next comment has no design presented here in this book, and it has no longevity as to last thousands of years, but allow me to change gears for a moment, and it is a fun concept and maybe you remember skyway transportation like the old television cartoon The Jetsons? Okay, let's consider this next content like a T.V. commercial. And I'll use my narrator voice. "Are you someone tired of being stuck in traffic? Hate the long lines, and the people who cut you off? Well great news for investors this could be your lucky day! And be the answer for the individual driver. Why you ask? And yes, because I have a design for such a vehicle an AIR TRAVEL CAR – this vehicle runs on no fuel, and has the ability to even travel around the world none stop. Coming in the near future to invest in, the cost is extreme.

I'm back - If you can imagine structures made of stainless steel and bridges made of the same material, the future could be shinny. Even our tallest buildings gleam in the sun, and last for hundreds upon hundreds of years. A legacy not a wasted investment.

70. SUBJECT: SO MUCH TO CONSIDER

Here are some more added thoughts made in 1992. Made before time was important, lol. : Let us look back at how thoughts designed people , thoughts of change made from needed change , such as leadership that caused people to drift from truth, to embrace those who projected great authority and power because they themselves where wanting greater safe living from outside oppression, or even oppression pushed down from their leaders, the need for self-improvement is always a high bar , and many people in the past and present live under leaders who have their own agenda which is not the hope of dreams of their people, so the need to change to make a better world for themselves rises and is called out like a trumpet.

To make a difference. We with pure thought can believe we have the right answer for all, but time has shown that even the best plans can have a wrong turn, the fact that doctors prior to the year 1787 used bleeding to repair an ill patient. Or how about the use of birth control pills which today are still used, but could be directly connected to birth defects, and loss of a baby born premature. Trust, these are just a few elements we can say trust is essential, trust that a medication will not kill us, - Matthew Perry former T.V. star of show called "Friends" who trusted his doctor, Michael Jackson, well known singer pop star trusted his doctor. The action of trust is one of the most important factors of our lives. Yet we place such little concern for outcome, because 99% of the failed trust is never fixed. We now trust Japan, a former enemy, and we see them re-building a massive military, which was previously not allowed by the United States, yet the buildup of China making its military might be like muscle beach of California with men flexing their muscles. The action from China has every nation around them building up, and so trust is making an exit. I believe the time for these people worldwide need to have their people be the power, I have known great people from all these nations in my travels, wonderful people with poor leaders, so how important is trust to return? It is essential if we are to remain alive. So trust has been the element that mankind is missing, this is because many times men have made many agreements that they toss away, and then attack the same person who trusted them. This action killing trust has no value of honor and has caused war after war, as millions of people die all because leaders cannot live as working friends. We all need to look at the priorities as to what is most important that face us all at present and in our future, as well as what is being enacted by leaders of humanity. These leaders often prefer the power stance instead of common sense of right and wrong, and I think we can agree that it is not always the real benefit for life for tomorrow to continue to be. Again, trust is essential, so when it comes to decisions the word trust is the main element we need as the thing most needed. I guess we could design an electronic device that when leaders agree insert a trust implant, which is placed into their brain, this device could not kill the person, but make all communication monitored, and locate the person 100%. So the action of breaking the trust could be punished with an activation of an electrical shock, (laughing). I'm not saying that leaders would agree to have these devices, but could form a stronger trust. Putting all kidding aside, I would hope that mankind could come back to the table, to sit as adults with brains that see that cooperation is the way to trust. So again, without trust progress stops. What kind of progress? The ability of humanity to function as a whole. Some leadership think that they can control the way others should think and do; groups, like militant governments who desire to conquer, and we have those that use God as their reason to kill? These that are fanatical religious entities that refuse to acknowledge that God our creator made us all here in one place, so maybe it was for us to be together? You see if we don't get along as people we will never survive in the end. Let me rephrase that for anyone who can't agree. If we, as "WE PEOPLE" don't get our act together- we die from WWIII. Is that plain enough? I have said it before and I confirm it again, the majority of us do have a nose, ears, teeth, feet, and eyes, and all go to the bathroom in the same way, blunt but true, with this recognition is how can we say to another person I'm better than you? And you're certainly not better if you have more status? More money? More influence? Or if someone put a badge on your shirt and pumped your head up with I am better than others? The answer is a resounding one: none of us, including this writer, is better than anyone else because we all have something to give the world and we should look for the good in another person. Still, we penalize those people in our world who do not have a higher education, or who are without a job and many without hope will go out and steal because they need to eat or they are looking for a better life, and instead of helping them, we put them behind bars like animals in the zoo. When they are released they do not feel happy that they have been treated like animals. They come out angry and willing to hurt others. My point here is this- life should be with love, not just lust, not me first, to be with consideration that you are made just like everyone else, and from the maker who placed you in your mother's womb, so you ask, - so how do you know if there is a God? Well science claims that the universe was formed by a big bang, so who made the bang? We need not to be religious freaks to grasp this like a gold ring to hold life dearly. But pure common sense says everything made was made by someone or something. So let's look harder at ourselves and to be more willing to give others a chance to prove they are good. And don't get me confused with that I believe that all people have the ability to be good, that would be a false statement, because some people have turned bad inside from the abuse they had been given, and many are sadly non-reversible no matter what you do to try to help them. That is why my next subject here is:

71. SUBJECT: PRISON- CHANGES

Before I say anything about prisons, please understand that I am against this method of treating people like animals, and I feel they could be punished for their crimes against our laws in a more effective way that can turn them into productive people who have a chance to leave with some trade or job waiting, and actually gain money for what they do as incentive to be better. But to be given guidance to see they need to change, but to change from crime, and be a responsible human. And when a person has done time that they do not have this on their record as to be presented to anyone, and are given the same opportunity as someone who has not been in jail, and we may find that respect can go a long way. Not saying that a murder should be walking free, or a rapist, but those who did lesser crimes could be placed by into mainstream life. Now there are different types of prisons mentioned, because as we know, some people are what is called a hardened criminal, and this type of individual does not change, but still deserves to be considered as possibly a productive person who may be able to be somewhat normal. This said; as bad as prison life is when they place the bad with the bad, do you expect good to come from that? Prisons have become a free ride for those who prefer no responsibility and get free food, free rent, and free electric, some have free dental, and free medical, a lot of free stuff, and even their clothes. Yes, they still have worries because of the inmates around them. Still, I hope the days of inmates mingling to learn to be better criminals will be replaced with what I suggest here. The real issue that guards are in danger is not imagined, especially when surrounded by inmates who may have bad intent; the way to prevent all this mingling in the future is to keep the prisoners separated. With complete separation from others, the ability for someone to harm someone is impossible, and so each inmate is separated and given separated meals delivered to each room on carts that slide the paper tray designed only for food into the cell, and food is always finger food, so it does not require utensils, hamburger, hot dog, I refect on this subject again later, but what the prisoner can attain to make their visit more productive could be inside of the cell could have a small belt conveyor that when not used a metal wall slides downward, but when used the door is open, and the opening is just big enough to allow piece work assembly inside the prisoners cell, and rewards are given for doing job skills taught, such as assembly of plastic parts, welding, soldering, or electronic parts assembly, but when they have completed the item, it is now sent to the next cell in line; the list of possible jobs is endless. They can even earn pay if they do as they are told and make items correctly; prisoners can even get items like videos, books, movies played, and even movie night, with popcorn, food, and drinks, as long as they have been willing to do what is needed with proper correctness in the application. Or they can get "NOTHING" GAINED: This is not just a prison; it is explained in more detail later. And "NOTHING" is a place not to be wanted.

But for what we can call prison lifers or hardened criminals who have no parole, they are not paid but will attain benefits as stated. This type of prison is for those who refuse to work, who have a life sentence and are never released. Sadly, the days of people who kill people being let out are over. Instead, the prisoner has lost their rights because they have committed a crime so horrid that there is no forgiveness; they can't bring back the person they killed, so why are we to care how they are treated? The only thing I will say on their behalf is God, who will judge them for their misdeeds and who will judge them, but they should not be treated like animal either. The fact is this prisoner will only be removed when the person dies. We have people who have done these actions placed back into our streets; this is purely unacceptable. They cannot be placed back into life with normal people because they have experienced a full decline of apathy or concern for others and can easily return to that state of mind; this has been proven repeatedly. But they can live a somewhat normal life as even the worst person can have a slight bit of good in them, and yes, some have great abilities that need to be expressed. So, as condemned as they are, they can help our world if they choose a work environment. Now, when I speak of "NOTHING which I will explain this even in full detail later in this book, "but this prison is a new method of placing hardened crimes into a place that they can never leave, but this system I am speaking of is not a working system of making products. No, because some people won't cooperate no matter how good you treat them. So, for people who decide to be stubborn, we have a new place that will be monitored from the outside only, but I will talk about that more as we go on. Today, our prison system has improved immensely, but to the point that now some institutions treat the prisoners better than some people receive in the outside society, and these people cost taxpayers money.

For example, they receive three meals daily and have a warm place to sleep, not paying for heat in winter. Consider a person who is homeless, has no food, and sleeps in a cardboard box in the middle of winter and many live in cars, or under bridges. And if you say that does not happen?, please hear this: I was in New York City in the middle of winter; it was extremely cold that night, and the snow was about five inches deep, even on the sidewalks. I was walking toward where my sister's home who lived in Brooklyn, and when I turned the corner, I saw this large cardboard box on the sidewalk. I then noticed two legs sticking out of one end. A man was sleeping inside this box; no blanket, a worn-out coat, and layers of clothing could be seen, and I had nothing to help him except my sadness. So yes, prisoners do have a better life than some, much better than the homeless person who can't afford a bath and lives eating rotten food out of a garbage dumpster; yes, I have seen this happen. Our homeless didn't ask for trouble or try to make trouble, society just closed the door on them in many cases. The fact remains that a person in jail is treated slightly better than a caged animal. I dare say that not one of the prison guards would change places with their captives, "Man the animal in a cage" but we can give these hopeless a new direction, one that says better is good, and I am not saying they deserve it for some of the horrid actions that they have done; what I am saying is they can be a benefit to the needs of business growth desperately needing a financial break, now please know that I am not saying these prisoners should stay working if they have done their time , as we have heard of a woman prosecutor from California forced many black prisoners to be withheld from leaving , but I am speaking of prisoners who have not completed the sentence, and if the prisoner decides to cooperate, they can have a better stay. Now, to change the direction slightly, we can find references to desert farming, and people could be used from prisons to work these farms, working outside can be very healthy. What I suggest is that we create an area that includes strict fencing that is a perimeter of our sentenced criminals, and this prison has no walls, no mess hall, no basketball court, and no weight exercise; this is strictly a labor camp and every prisoner has to wear a restriction electric collar, a collar that if a prisoner shows the slightest aggression, will be delivered a shock that is direct to this collar as presented earlier. Now this may sound harsh, but if you do not attempt to leave, you're fine and good. But if you're willing to buck the system, you'll pay the price with a solid deterrent that only happens if you leave a designated area. So, the use of these collars can maintain a workforce anywhere. The prison camp is only equipped with tents, an upper shade over tents, an upper water delivery system, and an water tank pool that is stainless and covered with an overflow lid that keeps water from evaporating It and has two spots to attain water.

The prisoners have sandals and wear clothing that is cotton and loose. The cost to maintain prisoners today has very little gain to the States who house these men and women, so the mission if to have prisons become productive and with substantial gains. . This prison could be designed inside of desert land and does not give hardened crimes any real comfort. The mission to be in these deserts is to rebuild them and make productive greenhouses for food for the outside and inside population Many areas in our world have desert land that is not inhabitable but could be used no differently than nomads in many deserts in the East. This camp has access to a free way of life, but a life away from everyone outside, a life that if they do not work to grow food at food harvesting plants as pickers and package makers for retail selling. These prisoners in "NOTHING" have a choice to either be willing to work together or they will be placed in areas that they don't work, but gain very little benefits, as the food quality is less, and they are not given any perks or rewards. But when they cooperate, the day to day life improves. When they work, they will walk through a cooking style similar to the old truck wagons of cattle heading days, beans, bread, and water, and they can fill their plastic canteen for later use during work. The work is they grow food as instructed, they will be given not only a portion of the massive harvests done to maintain, and even given ways to store food, but they will be given a payment of food dropped into the camp that is all water-based for use. The prisoners will build structures as they will be taught a trade and instructed to build warehousing and greenhouses during the day; at night they return to the tent camp. And enter inside the only structure there is which is a gate that enters into a gate that enters another large space with an added gate. This is where prisoners are unloaded from buses used. The entire facility is monitored with six towers at the main entry; they do carry firearms, but something pointing down is more powerful than guns; what they use is a technology microwave directional targeting equipment to push back any disturbance. This is device when activated causes the sensation of a person's skin burning. Not friendly, once the bus is unloaded, all prisoners are now to go single file into an opened long gate and are to walk into this slanted wall extremely high with razor wire at the top, a very large metal gate is at the far end, where the last two towers are located with six other towers, these guards are not your normal guards, they have three options a long- range rifle as to shoot over 1000 yards, and another rifle an automatic gun that is a mini gun, last a microwave heat system, these towers are with air-conditioning, and the only way to get into the tower is a cable car located at the back which goes to the main entrance office areas. This camp has no walls, but a fence with motion sensors turns on lighting if something is near that area. The overhead water distribution pipe is on a swing gate principle, and below some 30 feet down is a water tank/pool, which, once a week, prisoners can take turns to wash by a sprinkler that they can bring their soup and bath outdoors. There is a concrete pad with a drain that only sends the water further away into the desert concrete open drain, so future prisoners have three choices. 1. Hard labor tent camps, 2. Standard prisons like today without prisoners can mingle, and an assembly prison, three. Desert food growers at a tent city outside a "Stack framing facility." But if they have done their job well without problems, they can achieve another method that could give them their freedom in a way that could help them free from the past mistakes. Now this means a tough exit but one with greater rewards:

They can join the Army and understand that they could be taken to a front line war or disturbance; if they survive their time given stretch as in action, they are released to have a military pension enough to live in a good apartment. So, joining the military is a possible for early release for anyone has a life in prison sentence, can serve in this branch or military group. I served in "Charley Company" along with other soldiers in-Bravo, Delta, Alpha and But these prison soldiers are called "Bad" soldiers, they will be regarded as a tough group that is similar to the Army Green Beret but slightly different. The only bunk with former prisoners not to be in barracks with regular soldiers; they are in their own barracks area and are called the "Bad Company." And they are given a firm understanding that if they break the rules of the military and kill another soldier without a defense measure, they are given no quarter as to ever being released, and are taken back to prison. Bad Company, will have the well-known song "Bad Company" as their theme song made by rock group "Foreigner ", they will be known as the toughest men able to survive like Rambo in real life, and trained with every known device. They are given a 15 years before release, and they are in constant training twice as hard, and to be with an extensive team coordinated systems of fighting. Quality programs are available for these people for self-improvement, which include schooling, self-awareness class, that improves a person's self-awareness and the right, and wrong way in training classes how to treat people and how to except criticism and use it for self-betterment. Those who gain this program are prisoners willing to work and who do not cause disruptions. Now, the offer to become a military soldier is not handed down instantly; each prisoner who has done three months of hard labor is offered the opportunity to join, but they are told what to expect, and they have 16 weeks of training. They are shown the "tree hugger's," a military prison camp called the "stockade" which houses on towers 50 cal. Machine guns with barbed wire and 50 call machine guns in towers; the prisoners there are forced to march in a massive circle and carry a log on their shoulders as they walk around and around until they drop. But this is shown on their orientation entry day, so they understand that making a mistake in the military is not a picnic, it means they own you. But if you pass the program, your now on your way to possible freedom and a better future, with pride knowing you helped your country.

72. SUBJECT: SOMETHING or "NOTHING"

This next presentation is still about the prison system but this is in more detail as to the profit system where prisoners do make money to be given upon release. Still, as stated, they can attain rewards for good efforts. However, can we consider that the actual people are caged, and they feel as if they are treated fairly? I ask you if you were placed into a cage, and given only orders and lined up as cattle, how would you feel. This is far from how people should be treated, but those in charge expect them to consider changing. That is a good laugh; you place me behind bars, allow me no rest from others making noise throughout the night, treat me like cattle, and I am supposed to like it? I am not saying that not all prisoners should acquire a luxurious life. Also, please know that I am not a bleeding heart who says they should have steak and breakfast in bed, but what I am saying is all people, no matter who they are or what they have done, should be treated with some regard as fellow humans, and even the lowest minded who have no regard for their actions, should have a slight view of hope. A chance to make things better even inside. In other words, some purpose can be utilized to give back to the world they live in. So, we can say that separation inside a prison is productive and the doors should be thick plexiglass over the outside of the bars so they cannot reach out of their cell, but make it easier to monitor each room from outside, but as stated earlier, each cell has a back side wall that rises up in the center and is very small but has a powered conveyor belt to do assembly. They also have two cameras that watch their every move from up above in two angels; if someone keeps any part, that is easy to I.D. because all parts are numbered, and that number has to match the same other part numbers, so each part is inspected, if a part number is wrong, (Because the part was stolen) the line moves past the prior assembly person because a metal wall now closes the area of assembly. They are placed on a 3-month penalty without pay. And could face being moved to a harder camp which all workers in this program get to visit the harder outdoor camp before they start manufacturing prison for one week.

They have an hourly wage of $1. Per hour that is saved daily, they all get eight dollars a day. This sounds a bit extreme as low, but let's consider seven days a week and 128 days a year. If they have worked the entire year, they have collected $1,024 annually. Now, most crimes have a minimum of five-year sentence, so when this person leaves the system, they leave with manufacturing parts experience and, depending on the elevation of projects, could land a job for a top electronics firm, but if they have been inside for five years, could be leaving with over $5,120. A very good start for someone to afford a living space, but let's look even further: these people are not given training that says, "We have a job waiting for you." That is what is needed: These training classes are done by top companies who bring their abilities to give this apprenticeship so that when the person is two years from release and with good behavior, they will be given a job to support their life, and a good salary is offered that is to allow that person the ability to maintain for themselves, and even support a family if need be. Therefore, by working for a small wage, they are gaining release money and can be trained for a new start, filling a need of a major company, and we can now add bonus applications to their stay in jail. These are- If they have had a good work ethic and did what was needed as productive and correctly done. Then they can have a TV which is inside their room and a metal shield that can be retracted remotely and slid over the thick Plexiglas that are over the T.V. screen. This can be activated and controlled with voice recognition of the person who is inside the cell. This device is not on the market but could replace the standard remote control with AI tech. The second reward for good work could be a list of books available from the prison library, and an even better reward is the standard meal, which is for any inmate is now given larger portions and a desert; these are incentives to the individual to do good work, another incentive could be the person can attend art classes and painting classes, or be introduced to car mechanics as a trade. The program is geared toward building products to help industries develop these items at lower levels to compete in the global markets. Now let us look back again at the prison called: "NOTHING," a strange wording that leaves us to zero gains. Let's take a closer look. The term is what the prisoner without the desire to be involved or cooperate gets as their reward. We can say that they are even given a bit of "Nothing" before they are sent simply a video of what "Nothing" really is: a place where all criminals go when they refuse to be unlawful or non-productive. And a place that you never leave. This area is fenced with top to bottom barb wire and fortified with large towers that are automatic in response to any movement within 50 yards to 100 yards towards the fence, and there are large signs posted high to say: Warning shoot to kill area- you are entering a closed area, return back or be you may be shot. There is an actual red line made on the entire perimeter that designates the cross line areas that can activate there neck collars, the item depicted in art. These towers are operated by monitoring systems that work on motion detection and have silent alarms that are set off so that the guards are notified of a problem; when these motion detectors are activated, they alert a district that is 10 miles in a perimeter and can respond to a section of the wire or the fencing and are permitted to use deadly force if needed to repel any persons who have made any attempt to reach to the other side of this fence. Nothing is located in the most desolate areas of our planet, food is something that must be launched into the area, and catapults are used to send the food in, and are retrieved by the people who are there. All food is water-based, so a fire is not allowed because there is no wood to build a fire, but they stay in tents and thermal blankets are provided for cold desert nights. Medical help is limited to triple antibiotics delivered and Band-Aids, penicillin, and aspirin; nothing as sharp devices are allowed. They are watched by long-range binoculars similar to built in devices with scope ability on a rifle, and to have any device such as a spear or knife will be taken as controlled drone can drop from very high altitudes to deliver a sleep bomb, a highly concentrated gas that, when breathed in places a person or other around into a deep sleep in seconds, this is used to put an area of people to sleep temporarily as to send a team in to retrieve any devices of possible built to escape, or that can harm others. Any teams that have been ordered into the complex have body suits and stun gun drones that are controlled and hover above them as protection and can shoot up to four separate power lines, but every line powered is set to knock out a person. The team may enter because a group has made devices for hurting others. The team has direct authority to shoot anyone who attacks them with stun guns and have rifles that are fingerprint only use, and if used by someone who does not match the hand and fingerprint, the gun barrel will explode. The guards enter with a metal cable drop line; they wear a light metal vest that covers 90% of the person to protect against any knife or spear attack; the drop cable is a hook-on clip line that can be extended over the fence and is used for their entering or exiting, it has a sliding round platform that has at the very top a cone shape that is upside down and slides on the same metal cable, so to use this device is simply the person enters a drop cage which can lock from the guards enter, and once they enter, they stand on a semi-round platform, and there is nothing to grab on to but the cable, the cable is lowered by a spool release and lowers the person to the ground , and this cage is on a massive extended platform that can be extended and retracted when needed. The person who has been dropped down can now enter the area, and when finished can now request an extraction, and the platform is now extended again the chain cable is lowered and they are brought up, but, let's say someone has somehow reached the cable and is hanging on to the lower platform stand, or maybe they pulled the guard but are making an attempt to escape, and are attempting to come up the chain cable, this would be stopped by the massive upside down spool that will now only be lowered when the person has achieved a good height, say 20 feet high, because a spool is now being lowered down on the cable which is also still coming up, if someone is holding on to the cable , the spool will soon cause them to release as it descends all the way down to the stand platform, the spool is extremely smooth and very high, and it is impossible to climb because it extends in a slope that goes wider and wider as you go upward, so a person who is standing on the platform stand which could hold a person has this massive spool coming downward closing the ability to hang on to the cable and they can only resort to grabbing on to the foot stand platform which if your holding on will crush your fingers. If you could hold on, well, you cannot go up, and you will hang there until your arms can no longer hang on, and you will fall. A person who attempts to escape and falls is not to be helped but only is removed from the concrete base; their injury is self-inflicted and will not be treated, as all others who may have considered the method of escape will see this as a failing attempt. But we have forgotten that each inmate has a numbered electrical monitor that is an electric collar that can activate no different than a dog electric collar used as a fence because that is the same principle: the collar can be activated two ways, one by a guard who points the directional laser to activate a single collar or hits a red button that activates all collars and has stun level trigger as to increase or decrease the electric signal. So if a confrontation should occur and the inmates rebel to overtake the guard's weapons, they will find their efforts worthless, as the equipment's fingerprint matches to operate , but the device will actually burn up from inside as to not operate, as the trigger and the components that make up this device have a liquid glue that is released if the trigger scanner does not match the designated user, and every movable part becomes frozen; this is released once again if a fingerprint is not matched. If somehow the match mechanism failed, a number match to that guard's weapon can be activated by remote control which will override the electronics and seal the gun so it cannot be used. .In short,-" Nothing" is a place where you go to live for the rest of your life. You never leave there if you have killed someone and refuse to work. Now I have stated that this place is called: "NOTHING" but we need to pay attention to "Something". So before I forget as I did speak of assembly jobs previously, I would like to inject some further thoughts and observations on that subject as my ideas for job development for prisoners are not limited to work efforts by Government control as the jobs offered have been in the past made for prisoners, such as Office furniture, Mattresses, license plates, dentures, glasses, traffic signs, athletic equipment, and uniforms and the prisoner helps in cultivation and harvest of crops, work as welders, and carpenters, and work in meat and poultry processing plants, all this was information was researched and verified, and almost all is directly profiting the U.S. government. And I am not against the use of prisoners to gain trades who could be released one day and attain a job of the same ability needed. But the Government is running a business for profit actions, whereas in the constitution this document says this cannot be done, but is being done. These are the same people I have reached out to over 38 years to start a business that was to form jobs in manufacturing, the so called SBIR , and other government grants who all did nothing , and many of the projects which I have presented here in this book. But every government branch has turned me down, why? Maybe because they prefer to build their own devices to profit? I hate to sound like a whiner but this is just wrong. The United States Government wholly owns UNICOR or FPI, created in 1934, and uses prison labor from the Federal Bureau of Prisons to produce goods and services, and they profit from these, not Americans directly, and they have really no rights to do so, but I say- if the system helps business's and can be used to help support the public functions such as social security, and roads and business development then my ideas are to be bringing these devices to a higher standard of products made within the United States, and that the use of this labor force will support our business's would be my hope, and the funds made could also help support our prisons, but most important is that we regain our manufacturing which has been taken from our shores by outside manufacturers, so the use of the labor force in prisons could benefit business with new products that are with patents applied. So rather than just the Government institutions to profit from these developments, and not Americans directly, they really have no right to do so, but I once again, do not condemn our country using these people as tools to build, but to allow them to be trained and excel that when the day they leave, they are in good shape to enter a new outside world market without the need to go back to jail. And use them in a way that gives them added benefits to motivate them if their work is found satisfactory. The States that do not pay for their labor should reconsider the view and see the overall benefits for the day of release of many people who did their time required. Let me say that our country has had many improvements in the methods of prison life; so to say it is a broken system is not completely correct. The people who run this system have made advanced conditions; I am saying the time has come to make these a profitable way for both the prisons, the factories, and the people in prison to attain a sum they can attain when they leave. I am not going to say that I am an expert on life in prison as I have never been in prison, but have met people who have, and watched my share of prison real life T.V. that shows how they live. My conclusion is that our approach to crime has no real opportunity for a person to change enough that when released, they will not go back to crime, which is the case today for the majority of released prisoners. So here we are, a great nation of so-called freedom, yet we have around 172,700 women behind bars reported in 2023; if we check other references online, we can see that they don't want us to know the real numbers, yet a total number of men and women is reported as 1,047,008, more than any other nation in the world. Fun facts right? Well, this just shows we are at peak input for the residents, as many are released back onto the streets with so called good behavior that sends crazed criminals to act out the same crimes they did, and even worse, many times, but they are not to be held 100% responsible, because our methods don't make these people have the desire to really have a chance , as employers who see their records posted and which can be seen by a background check, are not allowed to function in almost in every case. But to have a population of over a million people who do not give to our economy and drain us daily with room and board, and free meals daily, and yes, even drugs are handed out daily, drugs as to keep them in order what is required to gain control of these inmates. What we need is to have crime be with an: "Exact mandatory sentencing."- The crime standard is national, the rate follows the exact level of punishment set per crime, and by the level of crime as rated, this means zero bartering with Judge and lawyers, a means of lessening a crime or it's duration, and that criminals depend on to even get off many times without a real action of correction or jail time. The need for judges to decide is gone as when a sentence has been determined as final, the crime is only set as what they have done to match the minimum standard, and the maximum is for conditions of extreme ill behavior. Still, there are only two conditions set in stone. Any crime over that standard only gives a higher rate of sentence to be activated, once again, set; if someone goes over the secondary condition rated, they get automatic life in hard labor without parole. Parole dates are no longer considered as a possible release; these encourage the criminal mind to think that if they are released, they can return to a life of crime. So we can come to a conclusion that it is time for a real change of our prisons. I hope what I have suggested will be welcomed and if used will help our country develop new manufacturing and new jobs, and help prisoners who want a second chance, and I hope that our people who have made errors of judgment and committed a crime are treated more like people than caged animals. To all you in law enforcement and prison guards, with the greatest respect I say, you are faced with danger every time you put on that uniform and we who depend on you thank you for your service, and I feel the need to say that you need better benefits and should be welcomed into the V.A. for full medical and dental if you're not already. I consider your actions no different than what a soldier faces so my hope is this will be agreed by our leaders. And to those who are behind bars, I feel for all of you deeply, but to you all, know that many people live without bars, who are more in prison inside their hearts and minds than even you are. They just don't see the bars that keep them locked up are bars that stop them from caring, or to be happy, so I only can say that you may have been someone who had people treat you badly, and maybe some of you are in there who did nothing wrong but found themselves in the wrong place or accused by people who disliked you, no matter, all the pain and all we go through can be horrid, and I know that I don't have to be there with you to know your suffering, but if you did do what you were accused of, then use this time to be aware of your problems, be aware that you can change, but no one can do this for you, it can only be done by you, I pray that somehow someway you and others like you will find what I am saying to be real for you, not just words, my words may be laughed at, I know, people have laughed or said " oh that can't happen!" A man I knew laughed at me when I said I was starting a toy company, he laughed, oh he laughed so hard, it was 1990, in 2003, I was listed in Javits convention Center as a manufacturer and had two booths trying to sell my flying toy called "AirFling" a foam airplane wing in a circle, I came home with 300 buyers wanting it, now my success was that I was "there!" I was exactly what I claimed I would be, yes it took a long time, but anything worthwhile always takes hard work. And maybe you've noticed but I tend to care about all of you, and so in a way I love everyone, and this is because I had a family that pretty much pushed that concept of love , and that it was important to be happy. And this is the reason I am making this book, to free people from the walls they make between each other, and to free them from hate, envy, lusts, anger, greed, and open their eyes to care, be considerate, kind, speak well of others, find the good, make a difference that helps a child, or an elderly person, be open to love. Be friendly without reasons, seek and ye shall find, ask it shall be opened, get closer to the one who made you happen. And when they persecute you, know they have not gained, only they will see when they meet Heaven, the entry will be blocked. The only sign they will see is- "NO TRESPASSING". But to those who repent, see themselves directly, see the need of change, and do it! These, shall be welcomed into the paradise that waits. A short word to our protectors (police and our judges): We need honest men & women who remember the oath they made to protect and serve us, the public, to those of you who have made efforts to be good to people daily yet have also have placed your life in danger, we thank you, to those who have only one goal of hurting peoples pocketbooks, we are seeing you like the days of Robin Hood and his not so nice Sheriff Of Nottingham, who took the people's money. So please be aware that we need you to return to caring of our welfare, and we all need a warning sometimes. The day will come when you may not have given someone a ticket they deserved, but later that year, your car breaks down, or maybe you are off duty, and somehow some bad person should show up, someone who threatens you with death, let's say, a few not so nice guys that you once placed in jail, and that person you gave a warning to, suddenly out of the blue is there they are to help you.

You never know because the wheel of life is a circle, and the saying: "What goes around comes around," well, that is for everyone.

73. SUBJECT: The MIND & VIOLENCE

Science and all of its wisdom and all the people the world would claim to have such understanding of the human psyche have left out the obvious that mankind from its beginning is a sponge, a mind sponge that absorbs the feelings and touches of others in contact and absorbing the surroundings, even from our earliest words spoken of Ma-ma, or Da-da we are observing and the behaviors of others learning the language is also a physical and emotional and visual learning to be absorb into the mind. So all behavior is taught by others to the individual seeing what they are doing this includes the language the way they speak includes the pronunciation this includes body language includes all things that the human eye human ear can absorb into the brain to our senses absorb whatever we see. Even if we feel they are incorrect or wrong they will still be absorbed and will be brought out of that person in any situation of reference that may call upon what has been logged into the brain, and the subconscious part of the brain which can be very powerful and very dangerous if used in the wrong way. Every emotion that we experience is taught that similar to monkey see monkey do, as we repeat every emotion that we do or act out as taught to us from another. This means man is his worst enemy. He claims that he wants a peaceful and harmonious world, but in a child's infancy, we teach not peace, not harmony. We teach our children to be tough. We teach them habits they carry forth throughout their lives that cannot be changed because they are now absorbed into the mind. Once you have learned hate, once you have learned violence, once you have learned how to be cruel, how to steal, how to cheat, these habits follow you all of your life: habits like lying, habits like cheating, and even habits that are acquired like smoking and drinking and drugs these habits are encouraged by the physical addiction that follows. Still, these habits were shown and taught by others to those people who do these things or introduce them to others, which is only teaching, teaching them how to do something that they should not do. We see professional fighters and arenas, and they learn their skills; the keyword is learning. So my belief is if we take our children to a higher level and separate them from each other from their beginnings and teach them as they grow without influences of others with their bad habits, and only introduced to them what is pure what is correct or what is the right way how to treat each other, and never strike a person or hurt a person physically or emotionally to not teach anger to not teach hate, to not teach non-sharing to not teach compassion to show the children that the future is to share to love the kind to be considerate be cordial to be polite to be understanding strive to understand to not argue cannot be loud but a presentation as much as possible to be quiet and gentle and to always work for the best of mankind as a whole not as an individual to encourage bad but only things that are good to be allow, and for self-suppression of fear and not be with frustration or anger but never hate , but be aware of danger, but not embrace things that are ugly and have high standards that are used from the past that show what is beauty cannot be ashamed of the human body but not to degrade it and treat it as if it is a thing to be bought or sold like a book for sale, but no we are to give an incentive that gives this person all that it needs to survive in the world, but also be in harmony with each other to teach the dangers like a shark in the water or a bear in the woods but to provide the knowledge and understanding of what to do in those circumstances and how to handle them properly, to exclude our children from weapons and harmful things and to teach this not just to us but to all children to everyone in the world that weapon is no longer a need because fear is gone and hatred no longer exists and acceptance is standard in love and understanding, this may seem impossible because of the behavior of the recorded history of mankind has shown nothing but everything that is vulgar, cruel, hateful, possessive, envy, with greed, which all lead, to war and destruction. People today see the images of this hate for their fellow man being posted even to the youngest of our families. The journey throughout time shows us the many things in this world built by man, and today destroyed, such as the hundreds of thousands of standing structures lost in WWII, and will never be seen by the eyes of the future; wonderful things were destroyed because of humanity's foolishness, destructive ways and desire to rule others, the city of Troy, many parts of the Puritans empire, ancient Greece, and so much more has been lost in the destructiveness of humanity. And where did they get all these ideas? They were taught to be cruel; they were told they were soldiers to protect their homelands, as they pillaged and killed innocent people. Mothers wept for their children, their cries ignored by powerful governing that fell to their deaths for lost causes. History has taught this main thing: evil will advance and seem unstoppable, but it never prevails, because even with the most controlled people, the desire for the human spirit of being able to live as you chose, this inner spirit always comes forward, always makes its way out of darkness, and then grows like a giant tree that absorbs the evil until the process starts again, when evil forgets, yes forgets to see the desire of men's hearts to achieve good, as they also desire this same exact thing-to achieve- but reversed they want to achieve their goals of domination, but they do not do it with the understanding that by controlling others' desires, you now suppress the human nature to be better also, so they fight a constant battle even after they rule because men and women always will feel the desire to grow as with their heartfelt desires of improving their lives. This is why communism and socialism are failed structures, and today, we see our lives, even in so-called democratic nations, adopting regulation after regulation to suppress people, keeping them from their basic natural freedoms, an example- NO TRESPASSING, NO PARKING, NO STANDING, NO ADMITTANCE AFTER 10 PM, PARKS CLOSED AT 10 PM, BEACHES CLOSED, ENTRY to basic things once open. They sound so normal today, but in truth, humans need to walk in the woods, be a part of nature, and have the ability to sit by the shore or watch a sunset of the moon over a lake. At the same time, you cook fish you just caught and sleep on the beach with the sound of small waves hitting the shore; these things I did, and I will never forget how wonderful they made me feel. Today, those things are rare because only those who have purchased the lands that they own can enjoy these natural things. But know this also- as in my youth when we slept on the beach and later had our beer parties, we never left trash, we never made damage, we never tossed garbage in ditches, and when camping, never treated our campground like a trash bin, and we always, always made sure our campfire was out completely before we left. These things of basic respect for other people's property are why many of these once wonderful actions are now gone. So here is what I suggest to you who control our daily lives as governing or police- let people be left alone to do as they need as long as it does not offend someone directly, so some kids want to sow oats. Did you not do similar in your youth? But if they disturb some people, they are not being considerate. They should be removed, but- if they are having fun away from people, as we once did-up in the woods far away from everyone, and they respect the property by cleaning up trash and ensuring a fire is out, what is the harm? They will learn, just as we did, that getting drunk has a horrid next-day result of getting sick and making the error that is called learning the hard way and growing up. Well, Now please understand to you teen that may be reading my past, I admit I was with friends who dragged me into things I should never have done, believe me when I say – I paid the price the next day with headaches, and saying hello to the ceramic toilet bowl as I bowed before it. The good news is I grew up, and moved away from foolishness others said was going to be fun? And sadly some never see the difference, as we know, but over time, even they find the mistake of using these so-called good time applications. But, as I said, they will need to find their way until the day it is taken away again, which brings me to another subject. So what is the conclusion of this subject - "We are being trained to hate, and to harm others."

74. SUBJECT: LEGAL- BUT IS IT RIGHT?

In our lives, we have seen the process of substances that had been illegal and now are legal!

This means that the governing bodies of that time either gave in to the pressure of outside opinions or saw the profits of adding taxes on the substances and now controlling the substances to gain money for the standing government. But the outcomes– In Russia, they have allowed alcohol to be a big part of their daily lives, and for this reason, there are now millions of people who have alcoholism and are devastated by its use. And many in the United States also partake daily, and I have known people firsthand, who had severe drinking addiction, and yet they never stopped, and the result-- died. Liver damage and heart failure; the drug has many ways to kill and even take away brain cells, so it makes your intellect decrease, has been proven by science. BUT! It is allowed; it is now being replaced by other drugs of so-called escape from real life, and they run deeper and deeper away from the real world, don't they? And for a very high price, they use cannabis, cocaine, pain pills, and prescriptions of codeine, and the list goes on- and on- and on- but what do these people gain? They gain a false life, a dim look of what is real; many millions of people who do all these items stated, and any other that numbs the world, ar4e also controlled by the world; they refuse to see the truth, so they live with an attitude of this almost 99 % when asked about a political issue that could affect their lives, oh, they may have an opinion, sure. It may sound very sincere, but– they never take action, and they never partake, and most say this: Well, what can I do? Even more today, today's youth say right out when asked what they think about issues, and they reply: "I don't care!" The real facts are that we have not allowed this progress completely, but our leadership did. They allowed the items to enter this country saying they can stop the drugs, and yet the drugs are still everywhere. The real action to stop drugs is at the root of where it is made, not after it is made. Now, personally, let's be exact, you're doing all these things to get away, to find refuge from your pain, or your loss of love, or loved one, maybe just sitting around with your so-called friends that are also liking that you're as messed up as they are. However, maybe you say, well, I am not hooked on anything; the real worldview- is this: If you do it every week, you are hooked. So, little fish, try to release from this shiny metal hook that has captured your soul. That is my hope for you. I did it, yes, I smoked, I drank, and I did drugs; I saw friends die, and that wasn't going to happen to me, that is the lie was accept! But the truth be, I am lucky to still be here. But yes, one day after hearing of a guy I met in Pittsburgh, Pa, And a man who stood six-five and a giant of a body, his arms were bigger than my legs! He was reported to me by someone who knew him; that he died from a drug overdose.-I am sure all of you have heard how some people never stop, and they never using drugs and never saw the end coming. Today, I live healthily, play basketball like I am 35, run on occasion, I can do 40 jumping jacks, can lift over 150 pounds without a problem, recently lifted a full size box freezer in the middle of winter with snow on the ground, and made two sliding supports to push the freezer up, but at one point had to lift the entire thing to stand it up, I placed it into the back of my 53-foot storage trailer, now consider how high a back end of a semi-truck trailer is. I did it all by myself, and yes I'm bragging, and you would also, if you where my age-I just turned age 71. Now, do you think seriously that could have been done if I did any kind of the drugs people use today? If anyone doubts this, email me and I'll send you the picture of the freezer in the trailer. The conclusion- don't follow others, but let God lead.

75. SUBJECT: CHOICES, GUIDANCE- and GAMES

I grew up with great adoptive parents who did their best to keep me in line as they gave me all the good advice and fed and clothed me. I never once appreciated the gifts and the receiving of the things they provided because I believed this is just how things are!?; as a child, we sometimes can have behavior that can run amuk, but today, many parents believe that the public schools are their answer to not having to be the guide, and instead, let the school teacher give their child actual personal direction. This abandoned action has some very serious consequence for their children, and what I see is a lack of concern. Hence, they, as parents, seem to prefer that their children be taught the wrong ways to live and allow their children to be misguided by liberal people who refuse to see the path they lead as one of the destruction of the individual later. We see it growing daily as students act out the anger caused by improper teachers, yes many teachers are causing a student to be confused and filled with hate, because many of the students listened to their ranting of equal sex and that women should be men, and men should be women, and let's doubt our sex, as these irresponsible actions which were not approved by the parents show a pure neglect of the improper promotions, that the human body can be changed even if you're born a certain sex, sure, it can. Still, a tooth can be pulled; does that mean we all need to go around without teeth? These children being torn inside are not even of the age of consent, and are minors. Yet the schools who have started these belief actions feel that parents won't unite in court to sue their ill minded ways and that parents are not responsible for the child's metal status as to be in control of their own child, and to not allow this hurt and mental strain caused by these teachers and those in charge will also be held responsible to those actions as they approved them, and the teachers agreed by doing the actions which makes them as guilty as sin. The back lash made from their parents has not even begun, but if it was my child as to be disrupted even mentally or taught a belief that was not accordance to my family, and it caused the child to be harmed or harm others, I would be gathering other parents to flip the cost of court, and bring them down to their knees. $$$! Big time! So we can conclude to a small amount that to learn is to observe what others present, let's not forget a major action which is teaching our children and yes many adults today. Teaching daily to be with- (Violent television, and Violent Video games do the opposite of kindness, but, blowing up things and killing things) to see anger is to learn anger; to see killing is to become a potential killer; why? Because now a person knows how to do that behavior, and restrain remains. Still, the process of the mind starts on a course of withdrawal from others, and when disturbed in the slightest, they rebel outward in anger or withdraw once again, but these behaviors are all being given today. So we can see the results given to them by parents who approve because it keeps their child busy. Hence, they are free of that child as a disruption to their privacy and to their willingness to use the devices as baby sitters which now project the "how to," why, and where, and the truth is people will act out what they see when the acts are provoked enough and when pushed by others. These games are also the escape of millions who waste hour after hour doing an action they preserve as a reward or entertainment. When, in truth, as these devices absorb our nation, nations like China ban them from use, why? Because they know it distracts the user from being productive and allows them to be lazy, non-productive, and gain only – zip! Nada,-zero! The way life used to be was with families that we need to interact with each other, and others do affect us as how we are raised, but we also need to learn a behavior that is proper, or not a bad behavior, and this is a great attribute, and regrettably for some children abused- a terrible one, but the fact is we need inner action to coop in this world as our children grow and they now leave the comforts of home , and so here we see so many children who have not been able to fit in, or be able to communicate from the constant withdrawal they live through games, they now stay at home, and stay, and stay, and they are never mentally able to perform as a normal human should be, and they don't marry, they just have a so called friend, but even that is rare, and with this knowledge what do we do ? Or should I say – what should we do? We as responsible parents need to monitor our children's devices, remove the violence of the game, and if you don't know how to find someone who does know, weekly find out where they can access other violent games, and block these areas to gain your children back- you must remove the devices that control your child's behavior and re-establish better traits as parents that are projected by new laws that say- "I can now discipline my child, not beat my child like a raving idiot, no, but to give my child a spanking, in a proper method that will be maintained as such, and if abused-yes a fine can be given if witness outside the family has videoed the action as excessive. But a warning is made first as to the correct method, and this could even be classed as taught to parents, so the correction is properly done in a manner that does not hit the head, arms, or upper body, not damage a child's body, with a registered paddle made for children that is used on an area very much protected with a great deal of fat, so it seems to be the proper area God intended as our method of correction. I am not saying excess is the answer, NO! A one-time hit is more than effective in getting the point across unless the child ignores the lesson and repeats it, so the reward for bad behavior is not "Oh, we never have a correction!" Now, as I know, some children never need to have discipline made because of their acknowledgment that they understand what their parents just said, as it was shown to the child that is was not correct, and they actually listened, and they no longer did the same action! This is great if the child does not repeat the same mistake, but -this means the child could comprehend and follow instructions without needing outside enforcement. However, as parents, I always say this is our first option and one that is preferred! Because we are now giving our children the correctness of how to accomplish what we want them to "do," which we know is best for them. But, the use of punishment with a paddle is only required when, after the first measure failed, the only course of action is now a reminder of the fact that your child is not your boss, that they do not control you, and now allow me to show those of you why this is essential, and the result of no guidance or structured upbringing. In short the government has prevented the parents of today to be in control of their family, the parents need to say – these are my children, and parents need to become a union as: "United Parents", and have the legal means to do what is right for their lives. Here is the example: I have just planted two exact-sized trees; these trees are planted not far from each other, so the land is the same type of land; let's call the land our world. okay now we have given both trees the proper distance to grow, they are now both placed in the ground (The world) and they are now watered. But we have one difference: one planted tree has a guide stick and small pieces of twine rope. The stick is a long and straight rod. The rod is hammered down to maintain a deep, strong impression into the ground (The World). The twine is now tied in loops that hold the tree, not too firm but still attached in areas to allow it to grow. It is now made as straight as the rod because the loops hold the tree straight, keeping it straight. The loops are like a mother and father. But now we have the other tree, we do not keep a firm rod near it, we do not allow it to be guided by our keeping a secure hold, but allowing growth also, instead, we do not apply any guidance at all. But every day, we tell the tree to grow straight, but there is no guidance as the wind blows at the tree, as "Life will always make strong winds against our children," And it will bend our child like the tree not secured, and even when we said- "DON'T BEND"! It will. As time goes on, we have almost but given up as we see the tree which had no firmness, no guide has now bent and is a twisted mess, and then we see the other tree, which had been guided in its beginnings, it has grown tall, has a full strong foundation. It now bears fruit each season; the other tree is not guided, not secured,- not cared for- it has no fruit worth eating, its branches touch the ground, and the tree is dying because one day its bent foundation breaks. To give no guidance or discipline is not acceptable for trees or our children. But this measure does not remain forever for our children because their foundation is still there, so as parents, we now see our youth growing properly and well-behaved. They do not scream, demand, be rude, show disrespect, but give generosity, have wisdom, be mindful of their actions, and act as people you can be proud of and well educated. They are now in a framework of mental development that works because they had guidance that was with action not words. This guidance is essential for ages 3 to 13; as in ages 3, the guidance is not spanking; it is "NO," or we take the item and say "NO," we allow the child to complain to disrupt, cry, beg, but we do not bend, we are the rod. Only when we see the behavior has not stopped and it continues for days upon days, this is the time for discipline, and the rod must not bend to give any reward for bad behavior, but only after the child has seen that they cannot achieve a reward. Instead, they can only gain care or love after changing their way to be guided by the rod. The Bible says this clearly, as its words are not required to be religious but as correct wisdom- "Spare the ROD, spoil the child." As of today, we see over and over reports of actions where our youth lash out, kill innocent people, and damage lives and even themselves, so here we are at a crossroads; which will you choose- one of the horrors of seeing some mother wailing after her child was murdered by a child which had no guidance? These are the ages of improper behavior and live between each of us. No one sees the great injustice of our rights as parents taken away, so we see the results of their governing permissiveness- that they make no effort to correct this absurd law, which now shows the real truth, the unmistakable facts of how they have allowed the schools to be without discipline, and the parents to be held as terrible people if they discipline their child. This must change, and we as parents need to come together and make a movement for laws to change and make it proper, correct, and legal to act as a responsible parent who cares about their children. I am not saying a child should be beaten like a mean dog, or slapped in the face or head, or punched like a punching bag, but when I was young, my father would only have to take off his belt, and I knew he was prepared to use it, and It made me stop doing whatever I was doing which was bad behavior without him actually using it. It was only used one time in my life when I broke a neighbor's window hitting a baseball and had an entire back yard to hit that ball, but I chose to be next the neighbor's house, and my father took action. Needless to say I never hit any more baseballs near his house. The other promotion of our ill-minded children is the entertainment they watch- stop watching evil behavior that provokes our subconscious mind and fuels us to do what is incorrect. Now to the real answer of how we could govern and be with our child: This will sound impossible- It will sound as if I have lost my mind, as if I have no heart, or as if someone is demented and ill of thought; I assure you that the harshness that I am about to reveal is not what it seems, it is truly the only way to save our world from destruction. And I address issue in depth later in this book, but also some things are shared coming next. But before I give you this answer of humanity gaining back it's future, know these facts that our world has embraced images of killing, constant violence is projected on our televisions, shows of murder, drug sales by gangs, the list is long of things that should not be shown but if this was started by concerned parents we could save our youth which works like this, but first the bad news: Every nation of power today has been preparing for war, a war that will never be won, a war that will devour the entire population and remove humanity, the final war with the same result as the dinosaur. So why do I say this? Because our children are being groomed for this anger, they are being taught to be killers, yes, but are they taught to be important. No, instead they are taught to accept things as they are; that do what they the teachers only introduce, even though things are unacceptable; as we see, we hear the voices of demands, threats, and the constant banging of the war drums, images of new technology! Of war machines! How to kill better! You may be thinking this: Oh, here is an anti-gun person; no, that is not my position as of this time in the world because we do have threats that require defense, and that includes even our homes, but this is the reason I am now presenting the method which is the only true and complete answer to changing the direction, as we are now riding a cart into hell! And the cart is a train wreck with billions upon the earth now will be dying. Now, I will address the gun issue later and explain why it is important to know the facts about guns and how this should be or not be. But the subject here and now is the "present- and the future" and the only way we can achieve a planet that remains a planet with humans still living! This method as radical as it may seem is the only way we can turn the world around on a global scale. And I admit it sounds bizarre, but in every evil generation such as Adolph Hitler's creation of children trained to be war soldiers. The opposite can be accomplished, but as I stated is the only way to change the future and secure a future that mankind is not removed and we are all gone from history. So this is it: In the future, the only way for people to return to respect, kindness, a generous heart, consideration, love, and true peace,- is if our children born for 30 years are taken from us all and no longer exposed to anyone or anything that says: "I need to harm you" but instead all babies are raised from birth in the most precise environment these children are handpicked as the cream of the crop before they are allowed to be born, they are presented only as with perfect DNA, with all the top attributes of clean, precise bodies, so the future is clear of babies infected with drugs or birth defects, or birth control chemicals, these chemicals which women use which also can be the end of mankind because the transfer of the chemicals goes from one person born to another and can eventually stop all children from being born, as we see the actions one after another today of premature deaths. So to have perfect DNA from each parent is also to be without any chemicals present in children for this program, and the known parents must be of the highly educated and have an extremely high I.Q. are only selected, as the offspring will have only the best to be in this forum, a forum of every nation, and every child born is taken to this what I now call:

76. SUBJECT: PG CAMPUS

And they are our NEO people, the future without violence is one which seems impossible, yet, the fact is all actions and understanding comes from one very single principle, "Whatever you do must be taught!" Yes, we learn all the good and bad, and the bad has been a part of our growth from the day we saw another child mimic their parent. I remember it well the day I saw a little boy push down another child and make him cry. The act confused me. "Why did this boy do this?" So I asked him, and he replied," My father said I must be tough and he showed me how and told me I should do that!" Now, this is the reason for my next subject, and the whole idea may seem absurd, but, we are now in times when the norm, is hearing about some violent act, someone shot some people, has become the rally cry for the left, and their real intent will be to stop gun use, but with this plan, the need for violence becomes a thing of the past because each new generation is given the proper teachings that will remove the understanding of violence and stop it completely; why? Because it will not be taught in any form or manner in this new introduction, I call PG, which stands for "PROPER GROWTH," but it is not parental guidance! But in a way, it is a program that will further advance the method of education. Parents will see the results as their child becomes almost a superman or superwoman, or we could say an extremely educated top-quality student that has no outside influences to distract the child's growth and becoming on a higher level never seen in human history, the only comparison which will not be even as close to the skill levels of these children is a monastery of monks. If anyone ever remembers a very famous woman who in her youth was only allowed to be around adults who learned from top teachers privately and became a talented dancer and singer, and movie star, her name; "Shirley Temple", of the 1930's movie screen, her movies can be found even today. She became this wonderful talent because she was formed almost from birth, her parents formed her into the world's most talented young actress, this is why I know this system what I propose will work, because it will be the world's strongest guide, the world's best educators, the world's best teachers of every known fact, and these children will not be alone, they will have a bond that cannot be broken. Again, only parents who have a very high I.Q. and proper DNA strain with strong mental and physical attributes will be chosen; the start of this is actually at birth for the reason that the parents have signed off to have their child be in the program after they have been approved, they will know that their child will be trained, and taught like no other child in history, so unfortunately as hard as it may be for parents to part with a newborn, they go into a contract agreement before the child taken after birth, don't worry, they get to see their child again in many ways. Still, they cannot communicate with the child directly, only by text messages, which are monitored before presenting them to the child; that is because the child needs to be free of negative statements and words that may not be how they should talk to a child. They cannot say how they miss them, or wish they had them with them, the reason is to not discourage the child, or cause confusion of why they are there, but instead the messages need to encourage them to do as they are instructed and study hard every day. So to have words such as vulgar words or comments that are derogatory. These children are to be given only information and actions that are kind, considerate, and obedient to their teachers and encourage their health. The child's growth is shown weekly to the parents through live feeds that the child does not know is happening, and demonstrations of athletic achievements are also presented. And they can meet their parents face-to-face after graduation when they are fully mature and ready to meet the outside world. Parents are to consider that their child is away at a military-style school, but visitations are not allowed, and this is because of ill emotions, which could be given to the child that could harm their progress. And what is the main element we are seeking? To create a new direction of education that meets the needs of understanding their world, yet understanding correctness in behavior and how to deal with misbehavior. PG works to improve the minds and bodies of these children to an extent never seen: the finest food, the finest water, and the best housing; they are treated with care so great that their safety is never compromised; they are the elite. These children are not told that they have been separated from their parents; they consider the facility their home, and there are no outside teachers to come and go; the teachers are on the campus and will be the formation of a youth that respects, cares, helps each other, watches the ills of man, yet can stay away from it, as they are above the anger, above the hate and above the concept of difference is bad. They are taught daily to be clear in their minds, clear and precise about all objectives, and will be able to do things that other children can only dream about. The babies are sent to PG vacillates upon birth; mothers and fathers are permitted to make their children the best they can be. It is a 100% free education, of college level, and even the highest degree is achieved; the students are called "Neo's, "which stands for "New Educated Oneness"; this program is free to parents who do not spend a penny on this achievement. At age 24, the parents now can meet their grown child face to face for the first time. This after the student has been given free college education and has graduated, once again by the top educators in the world. The parents are welcomed to the home of the child, first supervised as to what is correct to the parents, and also for that child's private learning centers. Now after the NEO's are released into the world, they will be given jobs to support what they achieved in schooling. And the jobs will be with groups working together, because they do not mingle with others not like them, but they are told to be kind to others only. The reason is they are to be examples, not to be popular, but to be showing people how they could be. As the nature of these people is one of perfection, but should never be singled out as to try and disrupt one of them, because they are united and the outcome would be one of a Hugh disappointment to anyone who would try and harm them. Now back to the way they are taught. The people who manage these centers have great self-control and are observed with the greatest care, as well as the children who attend. The children are taught no violence but are given strength and fitness training that is far more involved than even a military boot camp; the difference is all physical training is presented as visual training to be observed as correct in the application, and only one after age 17, and from 17 to 24 now all students practice self-defense. Still, they do not ever use it unless attacked. Some may ask, why subject these people to any form of anger? This is not anger; this is the ability to maintain a balance of self-strength and the understanding that some forces in the world and universe are not always friends, so self-defense is never aggressive acts that are to harm but only acts that are defense and ways to protect one's person or life. The child from the youngest age will be filled not with games and fun time, as the fun time is replaced with inner action, while learning is now considered fun, and all acts of aggression are never shown. A person who shows anger or disappointment by being rude or having a loud voice is considered a bad response; the children are raised separately unlit age seven, as they are to not be in contact with anyone who is not their supervisor or teachers; they are given strict computer lessons of achievements rated, but never presented as they are not given information of failure ever, but achievement is always rewarded, these children learn to cooperate not being forced, but because they want to know. The constant reinforcement that knowledge is to be your best possible person is always presented daily. Each child is given exams to see their comprehension ability as they age. Still, once again, the results are kept confidential, as each child will be given a change of menu at age 14, where some who have shown a less ability to be strong-minded will be the trade's people and given an insight into technical construction, while others may be in the arts or the management. Some may find they will be teachers to the newer groups coming in; these chosen are from the best who showed the highest abilities in all areas taught. Now, these children are taught compassion for life, helping life nurture and grow, being aware of being clean and being vulgar has never been heard. They will learn of every aspect of the known world, but this has happened only in the last few years. They are shown all the ills that wait for them outside. They are trained to group always to deter any aggressive action, and this is the stand of the exams to show the learning ability and the ability to maintain knowledge body of once self will be never to be aggressive of and to remove a weapon but never how to use one. This group will know only at an older age of the thing called hate and harm, and they will be launched into the world, and I say launched to teach others and to stay only in groups of five; they will always travel together for greater protection, they will never provoke, but when provoked will protect themselves. They will have monitored devices that record their actions outside the classroom, which is their free time to explore nature and be in a clean, pure environment; they will be questioned if any behavior is seen as not appropriate welcome or a distance presented in their behavior, the mission is to present only positive actions, and never be a person who shows anything but good response to any situation, and yes they will be tested on reactions, these are presented in age 19 and 20. The teachers live in their areas, and each child has separate rooms cleaned daily. Maintaining the perfection of the students maintain a perfection made by the student, and is rewarded if they have made all things correctly, and they will teach others how to be like them once they have left the PG facilities, but, instead of forcing people to be like them, they are shown how wonderful these people are, and the cruel will soon learn they have no control over them, in fact they will be very perfect in demonstrations of why you will not mess with them , imagine a group of thugs approach our NPG, New Professional Growth, a, peaceful, and helpful, and courteous, polite, and never rude, they will know about the behavior of rudeness, and always give back intelligent responses ,or no response to the comments. But when together, they will dismiss the people and not express laughter in the event of any danger, but they will show confidence to others and each other. They will have many taught tactics for dealing with bullies, and code words are for these formations. Each group has no leader; they are all leaders, but each team has designated responsibilities in the event of an attack on any of them, which is an attack on all. The outsiders from their group, called" The others," are only making foolish attempts to gain upper control of confidence, which is false. But, the Neo teams are taught that being aggressive is only a last resort, but they are taught to watch signs of aggression and avoid it when possible. But let us present an example of a Neo protective stance. Pretend that an outsider group has now crossed the line; remember, these people are "never alone" They just look alone. They are always in groups that will spread out but be in a very close range of less than 30 feet or 100 feet, so they are always aware of each other. So our gang of misfits has entered an area, a dark area, and the Neo is alone? Well, they signal his location and now give a small demonstration of athletic ability that will stun anyone and everyone as they leap high into the air, spin, and land perfectly on their feet, and in an attack formation, which with a smile says: I welcome your action, and I warn you to move at me will be met with resistance, but, I will take you to the hospital." It is said calmly and is demonstrated again, this time; a group is now present which for the demonstration gave them time to arrive. Once the group shows up, they evaluate weapons or possible weapons held and take appropriate action only if needed or suspected as a possible threat. The difference between the taught public schools and even colleges of today will be no match for what these highly trained, massively motivated from sunrise to sunset daily learning are mentally stronger as mathematically stronger , and strategically stronger , but proper behavior is maintained which has no anger, or hate, it is only as reaction to protect themselves or others that may be harmed, in truth they are like a man made design of supermen and superwomen, with the greatest attribute, they will regard life, be a guide to others; and show the world we can be friends.- because these children will be better educated and have better teaching than any child in the history of this world, with methods that will give them greater confidence and be solid examples of well-behaved and strong-minded. As time passes, more NPGs are introduced as second generation and even third and fourth. Until only they who have proper ideas of behavior are now in our future everywhere. And now we have achieved world peace and common understanding and relieved of all aggression and its needs for these devices, we have retrained the human condition to care about everyone, to share with everyone, to be kind to everyone, and to live together with everyone. Now, many of you say this is impossible. We have seen this exact application done on small scales, never in completed ones that are for every nation or culture, but in Tibet and other areas of the world, monks and religiously taught people have been in this same way. The difference here is that it has no direct religious connection to a denomination, but all children are taught that the creator of the universe is the reason they exist and that this creator is to be acknowledged in any positive way; some may say, why is this essential? Because humankind has already shown the desire to know his reason for existence. It is no different than a child who has never met their parent wanting to know how they arrived, and yes, meeting that parent, if possible, fulfills the child's life, as the question of "who" is now answered. The other aspects of the Neos are that they have no violent movies and books these are removed from the world over time, and then people will learn better ways to live in harmony and give each other a better world. And after four generations of 80 years, the old ways will seem like a dinosaur. And the outcome is peace among all humanity. Now, if that sounds impossible, please remember the first time you saw anger, hate, revenge, or physical harm being done to you or someone else; you were taught that and went on to either accept it as normal or become violent or angry because now you had it in your mind. Now, science has proven that violence can be the nature of some without outside influence, but the fact is a close observation can see this trend and be weeded from the rest; in a sense, that is what we do with prisons: we weed out the undesirables, but we do it at the cost of sometimes Hugh consequence someone's life is taken from them. So today's pattern is teaching violence through TV, movies, fights, war, and many forms of physical protection, and then waiting until someone acts unacceptable, then taking action by removing them from society; my method takes this and reverses the application to a non-violent Environment. And removes the child from other children that show aggressive actions, to place them in an environment that does not have examples of hate, anger, or contact that is inappropriate, and no contact with the outside world until they have been given all the tools, and at age 25, yes 25 because by that time the outside world will not be able to affect them emotionally. Still, at 21, they will be introduced to what the world is like, and will be shocked is certain, but are given the bad in small amounts to prepare their lives in the event they are found away from their group and will be appalled that the outsiders do what they do. So, every generation will need to grow in these incubators away from the traditional family that passes on their bad habits and hate; the fact is the bad has become like an infection that causes other infections in the minds of us all; we must rid our world of hate, and violence to not enter into the only direction these actions can take humanity, and that is to its end. As extreme as this may seem to you reading this, consider today and how our children have been left to be on their own and how this mess has bankrupted the human condition as we see youth who have no idea how to communicate without being rude, or inconsiderate, the parents are in many cases worse than their offspring. The world is falling into a deep, dark place filled with evil hearts and evil projections; our internet, books, and television depict anger, hate, violent actions, and death as a normal thing, and the bloodier, the better the attitude. We need to clean our world, and these groups of every nationality will be the answer to fixing our future. Now as time goes forward, the Neo population will outgrow the outsider populations and the will bring the unity of all humankind. But with the teaching as stated herein, the new generations will be trained with the best job skills and have a job waiting for them. More importantly-they will have hope, the proper love and care for themselves and others, and willingness to build again, and not seek the ugly, the foolishness of wealth but will work hard for what they gain, and still respect others, and understand that the things are only temporal, but they will build a family of each other, and gain from their every day the desire to help and care for each other in a way that we cannot even imagine. Because it will be pure and clean thinking that does not place distrust in the center of every person, but the thoughts of unity, and to teach the outsiders the better way to treat each other by example, but the Neo will not to directly associate with outsiders, as to absorb their bad traits, so in the future, land zones are made for the Neo groups and new colonies are formed, over time the outsiders will not be allowed to have children and the new Neo will replace them. But let us not forget, these are people from every country, every nationality, so what is being replaced? Cruel ways, mean lives, imperfection of bad manners, and harmony is achieved as the final result. Last, as this system of development seems absurd, it truly is the only way to change the thought process of people coming into this world to give it a better concept of respect and consideration to others, with the intent of only doing what is right for others and themselves.

77. SUBJECT: LOOKING BACK 2014

At past thoughts to consider: The first improvement that I feel is the most important and outweighs all the rest is the inner thoughts of each person and child as being with or without discipline; Historians have stated that the foundation to civilizations is the family structure, and education, and these elements are now in danger of being dismantled. The actions of firm family development by man and woman are being replaced by families who either have no mother, or no father for full proper guidance, and children are given false directions from others who are allowed to dictate to their children as how to live correctly, which is the responsibility of the mother and father only but is being threatened, as with the non-ability to instruct their children to have their own children born one day. An action which needs legal unity by parents as these actions are negative family designed. But we can look at another group of strong family ethics which within the nation the concept of theft is not present, why? Because of strong moral values, and ethics of correctness. The Japanese have had a very strong family structure that if they lose face with what the family expects from them, in many cases, they have committed suicide, but this type of strong family structure has been in decline in many parts of the world. The reason is that the family was once established with the father as the hunter or breadwinner; in other words, the man was dedicated to leaving the home to attain money to provide for his wife and children, food, housing, and clothing. However, the family has changed drastically for political and economic reasons. First, the political structure of women becoming more involved with politics was considered good for women. So it was for women but not for the family because the mom was with the children before this while the father was out of the house getting the money. The mothers role was to nurture and continue the education of the child with proper behavior, which we call manners, and improper behavior was dealt with as soon as it happened with strict consequences to be dealt with either by the mother or she would warn the child of the fathers to return, and so with the establishment of women now in politics was the first move away from home, where it took time away from the children, and now time away from much need discipline in the behavior of the child being corrected when needed is void and the results are many times on the evening news. Second, the business arena realized that if they could convince women that there roll of child development was not important then they would have access to a work, and they could gain "money". Oh my, so now mom was not with the child and instead depending on someone else to be the part-time parent for reasons that now, because of the higher cost of inflation, she was forced to not in all cases but in many that she attends a job as the same as the father to be able to provide. However, an extremely important elimination now occurred, which was the ability to correct the behavior and give proper guidance as soon as it happened. The second was even much more important because it is a need that goes beyond material goods. It is the same as the animal kept in captivity, which dies for no apparent reason but from loneliness. Children and adults need the contact of others to feel needed and to need. Today, we only need to observe, and you can hear of many children who ended their lives because they thought they were not important. So, unless the structure returns that was once the standard that was most effective to be able to give the proper attention, which is most unlikely but could be adopted in families that can afford the correct standard, please understand there is no replacement for a mother or father unless they secure the position in its entirety, meaning- they resume the total role of correction and most important love. Some would say love is not a real need, but to define love is the time we spend with someone enjoying their company and caring about their welfare.

So, just by attending a child's basketball game, they consider that attention and show you that you love them. Now, at this point, you're probably wondering where I'm going with this; let me assure you that it's heading in the right direction. Consider a common observation of a young boy or girl, aged 6 or 7, in a backyard or playground area. Let's say they have a swing or are trying to balance on a rubber tire sandbox. But one of the most common phrases this child will say is: Watch me, mommy, watch me, daddy! This is inherited in the child from an extremely young age of childbirth. When the child cries, the child finds attention, warmth in being held, and the closeness that comforts and food, so as development continues, the attention becomes less and less, and because it is impossible to be in two places at the same time the loser of this battle is today's children. This answer is for men to become men again, and women to become what they only can do which is to nurture our children and teach them right from wrong, sadly the paycheck to Dad needs to be improved so Mom can do what is her job, to care, to allow her life to have worth watching what she created with her husband not to go out into the world and become successful parents like their parents who with them the proper care and understanding. It was in the 1950's a time when the people of our country had the right ideas, and our country was the strongest in the world, why? Because people took on the responsibilities of doing what was best for their family. Unlike today. We need to return to those old values. There is a saying: "If it isn't broken, you don't fix it". But in this case, "we need to fix it".

78. SUBJECT: FIRES & RESULTS

Here could be some added solutions for forest fires: The prevention of forest fires has been a long struggle that man almost always loses, especially when the wind is active, to encourage the flames even more. What is needed is for water tanks to collect water in a very unique way. These water towers work on a weight system that when it rains the water fills a small holding bucket that now opens the main intake that allow water to go inside.. When the tank is full, a float on the water now closes the intake and this prevents the water from evaporation. This invention will enable water to be saved in forest areas that can be used with water tower spray guns, the same that are used at sea for ships on fire, and can also have hoses unrolled. A group of six per tanker can be used. Tanks are to be placed on a hillside so that if a fire has overrun an area, the water main can be switched to drain into a tank lower on the hill. The tanks will be located every 50 yards and are permanent- and must be maintained yearly to check operational conditions. One of the most effective ways to prevent fires is to contain the field. In the event of a fire, in most cases, firefighters will be asked to go into an area and cut a pathway to contain the fire. Instead of doing this every time there's a fire, a 20 ft. to 50 ft. spacing between each forest could be maintained in large squares of five miles each. This way, instead of having an area that burns beyond this, you've only burnt one area and possibly saved thousands of acres of land by having plows turn the soil by the forest department. It could be done by someone mowing their lawn or plowing the ground, as in farming. By this application being used this would prevent the fire spreading from the ground area to another forest area being there the little vegetation or plants to burn.

79. SUBJECT VOTING-" ALL IN"

This should be done on the national level in the future in a democratic government so that the elected party shall be in control at all levels. The reason is clear, but let's compare communism and democratic rule; in communist applications, the leader is the main element, and what he dictates flows downward and has very little resistance because there are no people to say-no. However, in a democratic political structure, we can compare it to massive complications of people discussing every issue, and many issues get ignored because the level of issues has mounted so high that they are overwhelmed with newer issues daily. The groups have to pass the presented law and many issues can be in a presentation for a law to become effective or an action is dragged through the mud, and if it is agreed, we now have what I call the telephone pole bumps. The telephone pole has a line attached. Still, each pole has a different party attached to that pole, so a message is debated or even ignored at some point as the delay again is with years many times to get a bridge agreed upon by a state or a new project that may infringe on land or nature. In my method, "ALL IN" means just that- the party leadership that wins also has every other delegate who ran for office win. This allows the flow of ideas to move at light speed instead of "snail speed," as we have today. The exception to this set of applications is if the real need for a world-established government is ever so important, which I discuss as a formation that is truly the best way to guide an ever- changing world of nations and truly the survival of humanity is at stake as my ideas for Elected-Kingdom which is also presented. Still, according to today's standards, this formation would resolve election results. But when you get to that, try and keep an open mind. But here is some more information about voting "ALL-IN."

An election shall be done on the lower level first, but the two lower-level groups are not involved in a head-on election but are picked candidates that their party elected to stand if the "ALL-IN" is for their party. Example: Tom Smith is running for county clerk in the

Democratic Party and all Democrats-so Henry Williams and Patty Roberts, the three have a chance to be election nominated for County Clerk as Democrat-run, and only one will be chosen from their party. This a candidate running for presidential candidate. So they have three candidates who are also running for County Clerk. They now vote for the nominee they want, leaving one person they have elected. When the Presidential election is made and let's say that the Republican Party candidate for President won, now all the people that had been prior voted for the Republican Party would now be implemented into the jobs all the way down the telephone phone line to the State level. Keep the country running as smoothly as possible without interference from the non-elected party, which is always the case in today's system

80. SUBJECT: SCHOOLS & EDUCATION & PROTECTION

I saw how the schooling system has made a congested and elaborate way of educating, where a student is requested to move from classroom to high school levels, and for what? So the teacher doesn't have to move! So, it is better to move an entire school than to move a few teachers, isn't this similar? I'm not going to move to the mountains. The mountain is going to move to me. So there's one error I find in school systems: we must spend millions per year to move our children. From here to there, from here to there, and in this so-called consolidation, it's supposed to save the taxpayers money. All it does is because more gas to be used and cause greater danger to our children, who are now on busy highways being transported around, and bus accidents kill our children. And in the wintertime, this increases all over the North. So why not create a system that makes more sense, that allows our children be better educated, and allows our children to be better educated because they don't have to travel so far, and with my system? They will be educated, greater, and better than ever in history. We live in a technological world of information that can be transmitted in seconds. We live in a time where we can consolidate. Not our children, but our teaching can be done globally, and we have not considered teaching the world. Not just our country. So, imagine now that this is your first day of school. You hear a knock at the door of her home or your apartment. You go to the door, and you see a man who has a patch on his jacket or shirt, and it says school board educator, and he is caring a large box, and he goes into your home and speaks. He says: "I am here to give your child their learning station, I'll set this up for you". He now opens the box and reveals a brand-new computer system with a foldout desk that can be set up in seconds. He sets up the router, and he now connects the line, and now he tells you to sit down in the chair that he also has assembled and he adjusted to your height. He adjusts the camera and the sound with the sound test. He then turns on the system and says, say hello to your teacher. He hands you the mouse, says, click here, and points at a button that says teacher. When you click the button suddenly, you are face-to-face with your teacher, and above the teacher is the subject of what the teacher will teach you. Now, you are being home-schooled in a way that allows you to go as fast as possible, and other students no longer pressure you. There are no distractions; you must sign in on time to go to school daily. The education system is all-new because it is a program that each student is given from day one. The program instructs you and shows you documentation videos and everything about the subject at the end of each subject. You are given a small test. If you can answer. 90% of the questions you are to pass to the next level. But added to 110%. If you get all of the questions, you move on to the next level. The beauty in this. A child can grow at the ability that their mind can handle and can attain and is not being pressured to be in a group where one child may be more apt to understand and can grasp better than another. What schools have not understood is that a child. Just because it is of the same age does not mean does not mean that it has the same ability. Let us consider an example of growth here. We are looking out over a cornfield, and in this cornfield. You'll notice that some of the corn in the shade and moist areas has not grown very high, because these plants are in the shade, but the crop that is the tallest is because it has greater sunlight which will have the crop gain maturity and growth faster, higher, and fuller than the other stocks of corn. Then, there are some areas where the corn is not as tall, not as full, and has not matured. Nearly as fast, but once they are allowed, they will mature later. This shows you that in nature, all creatures and all living organisms do not grow at the same maturity rate. So why do we force our children to be herded together like cows and be formed as if they are all the same when a closer observation shows that no child is alike except they have the same features, but each even lives in different conditions? So, to this writer and inventor. It is absurd that we put our children, as I remember being in this herd, being in this mentality, and being told that I was not as good in many different times and ways. But if I had been allowed to be who I was in the time frame, it took me to be who I am. Then, I would achieve greater results without the push and shove and the deliberate conception that all children are created simultaneously. In the same way, as if we were robots able to absorb at the same rate. The data shows that this concept is not workable and has major flaws. So was my conclusion was that we need a system that a child can either Excel or achieve greater achievements. Not being held back by the herd by the children who cannot grasp as fast as other children. But now, many shows grow faster mentally and achieve their graduation at much younger ages. Others will take longer to mature and tube to obtain the information they are being given and achieve at their rate. That is to give them the success of learning and obtaining what is needed in the world they are about to enter. You may say that having a teacher who only gives programs is not a teacher. No, because these teachers see when they see struggling students. That's one day. Now, they ask for counseling and give the troubled student any time. They only need to type or say the word help aloud if they have questions. The teacher will now appear on the screen and ask the student by name. What seems to be the problem, Bobby? Then Bobby may ask a question, and the teacher will respond directly. If the student still has problems, then the teacher can bring in additional help to address the student. Let us consider when any subject is in question: having more than one person available to address the subject means you have more minds able to comprehend and help the situation, so yes, these teachers have backups they can call upon online or are available. The teachers used to have the highest grades in their category and had the highest of all teaching abilities over other teachers. Only the best is used, but a history teacher can teach millions of students in one session. The assignee's or backups are on a keyboard that they can now use with a simple press of the teacher's name. That teacher will now show up in a box on the same computer and now is able to chat by texting or directly communicate with the student and the student can hear the teacher with head phones, and the teacher will explain what the problem might be, and now can address other students. If, at the end of the program. The teacher can now address any student who has completed and passed, and questions can be asked, as the teacher will have a large group of teachers who can now address each concern, whereas in the past, a school composed of 20 teachers. This system allows a subject like history, English, math, or any subject with about 50 teachers to address a homeschooling of over 1000 students. At one time, or even one million students. At one time. Plus if some children have food problems in the home, they can have the parents apply for food lunch delivery. Just like meals on wheels and a frozen dinner is available or sandwiches, with a juice or milk and these can be given in a large medium size box once a week for the whole week for each child. Children can even pick the menu. But all are checked for quality.

The cost per student decreases the tax burden, and the teachers are better qualified, have better credentials, and can get instant advisers to be added to help the student; even researchers can be inside this program, who have access to the top information of libraries globally.

Now, the only thing that this has that I have considered a weakness is a student. It also requires a healthy environment, which includes many a lunch and the ability to exercise. So, to address this subject, we will have included that each child shall receive a monthly package that contains the necessary nutrients of the highest quality in food bars and instant microwave containment. This box is delivered as stated once a month and is a supply that will be for the total days of the month. So, the child's first need is to attain quality above all. The food shall be a well-packaged meal for each lunchtime. There are two methods to address the issues of activities that promote athletic and sports and health requirements to maintain a strong body.

One must be graded and include a daily activity monitored by the health instructor. That is the new teacher, who has a time slot that presents the exercises and sees the child, and the child sees the instructor to do the graded exercises; remember that all progress has to be passed. So, students who did not comply with these exercises did not graduate. The second application is called activity sports. Activity sports can be performed after school, and the family member will drive the child to the school grounds within the township that have already been established, where a coach of the particular sport can now meet the student. Sports are encouraged, and the activities are presented in ways that the student performs against other students in the same manner as in the past, where schools would compete; the funding these activities make for the student also supports the cost of coaching athletic directors. So the funding for athletic directors is paid directly from the participation of the community that supports the basketball game, the football game, the baseball game, and any other activity such as soccer, cross country, or track; all of these activities are supported by a half-and-half funding, half from the former tax structure, and half from the ticket sales of sports. And speaking of school sports.

81. SUBJECT: PAID TO PLAY

Why should a child growing up not have the financial protection needed if they can no longer live a full, productive life from a school activity that makes money for the high school? Each year, thousands of individuals who play non-professional sports have injuries, and some of these injuries have become permanent.

Also, some have found themselves in wheelchairs and even have died in their love of a sport; it is time we address this issue with compassion and, more than that, with our pocketbooks, I firmly believe a national fund should be maintained for persons injured with a complete health plan that pays for these injuries by a city by city basis, where doctors are presented with the "compassion law" that says all doctors are to give one week per year to the attendance of these injured without cost. (See "OUR HEALTH" for more information)

The use of High school and college players to exploit cash into a college or high school will no longer be valid, being these young people place their health at massive risk, just as much in danger as a professional player, and all students shall receive a free federal "College bond "of $200.

Per week in the activity, when held on to for over 20 years, will increase by 20%. These College bonds are for the person who was in activities but not paid a salary; these bonds are similar to" War bonds" but can be cashed within one year of receiving, but would increase as stated if held longer. "High bonds" are also given but are at a reduced rate of 100 dollars per week in the activity, but can be used within six months. These rates are fixed and can only be increased by the increase of wage scales, where if the minimum wage was $16.25 an hour and increased in the state the school was to $16.50, an added .25 cents would be added accordingly to compensate for the increase. Now, this money for student sports activities could also be used for- say--- a field trip or a trip to our capital, Washington D.C, but something educational could be done that helps our youth with the proper supervision included.

82. SUBJECT: WHY THE VIOLENCE INCREASE?

We wonder as we watch the evening news about the horrific nature of our people who attack and shoot people at random, and we never question the real seed of the action, the "reason to be" you can call it. The main educators will say things like _Well, he was a quiet lad or, he kept to himself, and yes, abuse has some very real effects on a person. But if we look at the so called stress level being reported, we can look back and see that our world has less stress than, say, the young men who went to Vietnam as they returned either dead or, if lucky enough to return, had very real mental problems of either having Agent Orange infections or post-traumatic stress syndrome from battles and these are horrid situations to deal with. But teens today do not even have to worry about a draft to take them off to war and possibly die. But suicide and murder increase, and so where is this stress coming from? I am very aware of the games that youth today indulge in; they will sit for hours upon hours fixed on a tube or screen, shooting a gun and attacking an array of different people who, when hit by their bullets, splatter as real as the graphics have become so real, it is hard to tell real from none real. So, where did these fighting video games come from? We don't get this advertised by the gaming companies, but they were designed for the U.S, the military, and the police. How do I know this as a fact? I was in the military when I first saw the concept. It was presented as crude room-to-room 3D images that you, as the holder of the controller, would shoot at the enemy and targets changed for you to make fast correct decisions of the friendly or the child, as not to shoot them, but, as all things, over time the item was seen as entertainment. And so, war video games became a profitable business. The thrill of killing over and over day after day, the human psychic will absorb the visions of these murderous acts, and soon a child or even a teen or, yes, an adult becomes not only numb to the constant killing with the great variety of illustrations to depict the death being accomplished, but you gain levels or points in some games as a reward factor for acting out this aggression. If we define the act- It is murdering or killing people they do not know or care about but are the "enemy." So, the teen is learning the skills of killing. The fact of this statement is that crime is taught as a whole; it is best seen as a teaching element inside our prisons. The people released, on average, a high rate return back to prison, and we would like to believe they have reformed and are corrected in their minds. Still, in truth, the closed areas with other criminals only inspire as they all talk openly about the mistakes or the acts they committed. Video games are only a tool to show the same ill-minded ways of people who kill. It is my conclusion that the main element is war and criminal video games, but also the increase in what people read, and what they watch on television. The television in the 1950s and '60s was a fun time as shows like "Sing Along with Mitch" were a program in which people watched the bouncing ball to words they knew or didn't know, but families gathered to sing with the T.V. show. Various hours of entertainment, such as Lawrence Welk or Ed Sullivan Show, gave viewers the finest entertainers and singers of the time. Shows of laughter without sex being a subject, but clean, wholesome family entertainment. Oh yes, there were shows about guns, like "Gun Smoke" or The Rifleman" or "Bonanza," even war movies and T.V, WWII depictions. But the actors never were seen with blood or gore; they died, yes, but the images had no serious effect, as children of those days played as if war was not real, and so you got a stick and pretended it was a sword or a machine gun, you fell dead as pretending, and then you got up, and now someone else did the pretending. But children learned about war or killing without; I repeat- without seeing it as a real act. But as a fantasy, I remember in my youth playing as a Roman soldier because my parents got me a plastic Roman helmet, a shield, and a sword. But I never took the play as real. I used my imagination to attack a tree limb, and I conquered the tree! Yes, silly, but I never wanted to get a real sword when I grew up. But today, we are seeing more and more the actions of teens and people who live in a very real view of death and the use of a dangerous weapons, and constantly day after day they become brainwashed to enter into a life of violence that what they see and learned is now brought into real life. When asked why they do violence, many would laugh and say- "why not?" or have no remorse; why you ask? Because they became desensitized to killing and now wanted the next level high to do the action. It is my strongest recommendation that these devices be removed, and only fun games that have a person in a death or destructive view should be in our lives-period. Now for those of you doubters who say these are normal actions and that our youth are not being programmed to harm others? As of today, and I am speaking of July 7th, 2024. A 17-year-old "boy" entered a dance class, killed two children, and injured nine other children on Monday, in the township of South Port, in Lancashire, now this may not be in the United States, but it is still a youth attacking others, and why? Because as crazy as it may sound, these people who have suddenly jumped into the deep end of the pool may actually be under a control or programmed to kill.

A new possible- The military has researched mind control to use for soldiers. If you have doubts about that statement, then research it yourself. But it has also been reported that the government has a new way to help people, or control them? I admit this sounds like a science fiction story, but what I am saying is very real. A technology called "Nano Technology" which has been in development since 1989 is a way to insert miniature robots into the human blood stream with a simple needle injection. The devices are so small that they cannot be seen by the human eye. The devices are claimed to be strictly for medical purposes, and can be a tool to find bad cells and even do repairs to the body, because they are miniature-controlled robots, which can be directed through signal transfer that tells the robot information which it can now perform or even send information to the persons brain. This means they could attach to the brain stem, and could possibly control people. If you are someone who says – "My government would never make devices that could be used to harm anyone"- Then tell that to the Native Americans. They have been estimated from the time of the first Pilgrims landing to be a population of almost 100 million, today there is estimated around 6.9 million. You do the math. Now do I hate my government? No, I just want the possible to be stopped. But- strangely a recent report that blood drawn from some people when viewed under a microscope actually found these nano-robots in their blood. Gee, I wonder how they got there. Science fiction just became science- possible?

83. SUBJECT: TELEVISION / MOVIES

We need a stronger value for ratings presented for family use, and children should not be allowed to watch violence. The statement resounds a negative reaction. We can't remember when people took time to inner act because we now have more and more devices that allow us to reach out as never seen in history, but more so, these devices close our ability to be close to others in the same room. An example of the way people are now treating each other with total disregard is I had the pleasure of attending a friend's home one day. There was a gathering of many people: four children, four adults, and myself. At the same time, I was there, at one point, while we were waiting for food to be presented, everyone was either watching T.V. or on their phones: the older three people were watching T.V, and the children were all on their phones, texting or watching vidcos online, and not one person where speaking to each other. The person I was visiting was also texting in his chair, and everyone ignored each other. I heard this; I was even told that I was not welcome at one point while visiting because I talked too much while I was there visiting. The fact is that people have lost touch with the real world, live on Facebook, and have so-called friends of people they don't ever meet! This may sound like a complaint, but I noticed something else: all the T.V. shows they were interested in were about some weird event of horror, bad, or violence, and they constantly watched the news channel almost hour after hour. Totally absorbed by the next word of some major terrible event that was just announced; this entire summer I spent was time with people absorbed by their phones or TV. They rarely did anything together except eat. The people around me did only that; I went to dinners and restaurants, and people had no real communication skills with others there! Zero, the only person who communicated with them was a waiter or a waitress, and even that was limited. But phones and T.V. sets were available, and the tools of separation. And let's face it; today, we live with the so-called "NEVER TALK ABOUT'S!" But on Television today we have-a free for all! Do you remember a thing called Censorship? Gone! But- religion is put down, or politics have gone bonkers, well one party! To think one of my mentors in my life as someone I have always wanted to be like was John F. Kenndy, if he was alive today, he would be screaming to his fellow party today, run! Run away! This is not democracy! It's Nazi Germany in sheep's clothing! He was a great man, and cared about freedom. These socialists are "freedom thieves", doesn't a con-artist pretend to be your friend? The Socialist Democrats will smile at you while they stick the knife in your back, and for anyone who says poppy-cock! Just know if you allow these people to continue, the day will be seen when Americans are placed in barbwire camps for anyone who talks of freedom of speech, or faith, or building something without there permission. In short, Hell will be on top of the ground. And then they take you out and shoot you. Well the half the population doesn't even know what socialists are, if you over 21, pick up an old book , around 1960, do some research and look at countries who have socialists in control, not a pretty picture. But go ahead, just welcome the smiles and the laughs as these idiots run our country into hell and over 50% have been brainwashed by their communist teachers who the Russians and China handpicked indirectly, you now have been controlled with ideas planted in your brain from the youngest age, and their path jigs us to lead you in the wrong direction and to accept their will of how to run your life. They do not care about you, instead they want to bring this nation which has always been from its beginning a place that followed the constitution drafted by our forefather to protect our freedoms. These people have been attacking and dismantling these principles time and time again. They are a cancer which has entered the body of freedom. The teaching of socialism is to remove God from the view, simply - they want full control of you. So we wonder why our children are making massive withdrawals from us as we toss God in the trash can of conversation. And seriously, we don't even have "conversations unless it's about the weather, or did you hear what happened to so and so? Of course, something stated on– You guessed it— "T.V.!" Now, the fact is many T.V. shows depictions of crime, murder, theft, and cruel people, and we wonder why our youth are using guns. And censorship has been removed, as vulgar people now project their filth and teach our youth the words F-_-_K and Sh-T as a norm. This is okay, though, but talking about God! Oh no!! That is terrible!

The values are backward, and the message of our behavior is we want to show our youth hate and how to be cruel with our blessing! Okay, junior, here is a TV show that shows people shooting each other and has blood and some serious gore. Learn WELL! And the movies are even worse; the newer movies project all types of ill-minded behavior as if this is how we should live. And guess who owns most of these major movie studios? You guessed it? Other countries have bought our studios and are now using these places for their propaganda, which is all to disrupt and take America down while they make money to build their nations.

THE ANSWER: The time to remove these devices and people in power is now. We need to have censorship revived, and we need the standards improved to improve moral ethics, which are correct teachings of democratic values, and of how to live and how to treat another person as not to harm or be in confrontation, but to give examples of proper actions and build foundations in people to be better and to do better. In the 1940s, the people of that time understood that movies had a big impact on the behavior of the youth as many cowboy movies gave our young men an image of someone who was a hero who would take down the bad guys.

Today, the bad guys are taking control, and the some of the so-called good guys are just as bad as they are, allowing drugs for profits, or wars with people killing everyone with bigger guns and bigger explosive devices. So to everyone everywhere, take back your lives, reach out to your friends face to face, and be alive, not a machine; take the time to see real life, be aware of the world around you, and in front of your NOSE.

Get involved and make time each week to do just that; take one hour to let those in power know that you are fed up, want to be given clean and decent entertainment, and inspire representatives to do things in the office to change the laws to stop the violence and the sexual content and I must say this- We need to remove those who follow a form of governing that does not reflect the standards of what is in our constitution, the left needs to be left behind, the left is what China calls there form of governing, is this really what you want?

Please use the brain God gave you and investigate not from CNN, Or NBC, or CBS, or ABC, all these stations have been bought to follow the agenda of China and Russia, you see Putin was met by Kamala Harris, she gave him the go ahead to invade Ukraine, because it was just a few weeks later when he did, and she knew that we would be dumping billions upon billions of U.S. tax payers money into the war so the United States would become weaker and weaker to maintain our nation's military spending, you see when you want a business to go out of business -you first have to make a way for the business to not gain money to support itself, and then your able to make an offer to buy it, and sadly that buying of our nation is happening in the United States of our political people who have lost their minds and their honor and allowed money to rule over them, being promised a pack of lies that "they will be okay" all this is happening as I write this. You need to be motivated to make a difference in the world you're in.

Don't count on others to do it; it is up to you, but if you do this, then others may follow if you spread the word; the change is around the corner, and you can say- "I did that, and it was good."

84. SUBJECT: GREENHOUSE GASES CARBON DIOXIDE REMOVAL

This subject is a continued subject, not to be too boring, but because our planet is in very poor condition, so I am making every effort to give you hard facts that we could do , actually something maybe you could do,- or share with others who could do this.

It's well-known that trees diminish carbon dioxide from the air; our planet has many areas where trees no longer grow. These areas are called deserts. It is also well known that these areas were not always deserts, so to reduce our carbon dioxide; we must first recondition them so they can hold plant life and decrease carbon dioxide. Now, this is not an easy task, but a gradual decrease in the desert is being created because of an increase in temperatures, so procedures must be made to increase water to the edge of the desert and have the plant life nurtured so that it may grow back in towards the desert. And a worldwide establishment of new laws and penalties for any nation that cuts trees without this next regulation. I call it the cutting a tree, plant two trees. With this implemented, trees can be throughout the entire planet. Places called seedling farms would be operated by each government and financed through shared organizations, nonprofit organizations in contributions from wildlife organizations, Green Peace, and nationally televised advertising that asks for donations to be sent to a committee that would distribute the funds equally to each nation that has implemented this program. One of the issues that has been very crucial for many cities that are located by seaside are endangered of being flooded in the future by the ice caps melting. This is a legitimate concern in solving the problem of global warming: the ice caps should be considered sacred areas. But-we have scientists and many people who travel around these areas and disrupt the ice. Ships have been cutting through the ice, and the ice is created to become smaller chunks. Let's consider a very large ice cube, about three feet by three feet; we're going to place it in a room of 40 degrees in temperature, and time how long it takes for this ice cube to melt. Now, let's take another ice cube with the same dimensions and place it in the same room next to the other ice cube. But this time, we will take a hammer, smash this ice cube into small pieces, and separate the ice from each other. You'll find that the ice cube where the ice is still intact has not melted hardly at all, but the cube that was smashed into small pieces is completely melted much faster. The reason is the compression equal to the mass holds the coldest for a longer time. Our ice elements are not in direct contact with air, meaning the temperatures only affect our edge, reducing the size and the amount of time it takes to melt. So here we understand this fact, yet we have ships traveling through the ice areas, cutting out the ice, making it into smaller pieces, and doing it when we do not even need to be using ships in these areas. There is a device invented called an airplane that can travel to these areas if needed, and they even have ways to land on ice. It's called a helicopter. So, to protect this area and to increase the ice, the areas should be off-limits to all countries, and everyone in these areas must move out of these areas to ensure their longevity; this should be done immediately because if these areas melt completely we will see some of the greatest cities of the world destroyed, and billions upon billions of dollars, trillions to relocate the millions and millions of people will devastate the world and cause a disaster beyond disasters for lack of food and housing the displacement of people would dismantle civilization and bring it to its knees. I think you got the idea. A well-known fact that shade is always cooler than direct sunlight. Many applications that could be implemented to reduce the temperatures of our planet. Shading roads and farming lands with planting trees along the sides that reduce sunlight could be done. Farmlands, when there is plowing, increase the land temperature because the land has just been exposed without any means of the water being held. Because of the exposure of the land topsoil, the water in the soil is drained out. Grass and small vegetation retain water and shade the ground, Bush's and other plant life above this area also shade the ground and reduce the water evaporation. Lastly, we have trees that protect these elements. Without these, the water base becomes non-existent, and life becomes non-existent. The result of this is the deserts that we have throughout the world. So, by creating shading in farmlands, we reduce the water being evaporated, nourish the plant life that grows there, and reduce the ground temperatures. Although minimal, the earth's temperatures have been increasing, which is the main reason we have our weather conditions as horrid as they are. So, making every effort is needed and if everyone did what they could, in time we could start to see a slight decrease in temperatures. Man creates roads barren of all vegetation, meaning they produce no water absorption and, to make matters worse, increase the earth's temperature tremendously.

The black tar is exposed to the sun in direct contact, absorbs the sun's heat and increases the air temperature? If you doubt this at any time, go outside the next time it is 90 degrees. In touch the payment and see how long he can hold your hand there. You will most likely not be able to hold her hand there for very long. If you saw the image of trees along roads previously showing need of trees used to build shade, then just imagine this heat wave this is happening all over the planet where these roads are exposed to sunlight. To prevent this from happening, a very important application must be done. And it would also be very enjoyable to see it when driving. The view is large shade trees on roads on the side where the sun comes up and goes down would create shading on our highways and reduce temperatures naturally and effectively. The time it would take to develop this would be a mere four years. I say four years because trees that are already mature can be planted, and this would reduce the time it takes for the trees to become effective. Trees mustn't be planted too close to each other and become even a problem for the highway if they are too close to the road. I would say that some roads would not be affected by the shading because of their positioning when directly overhead, but the truth is the sun is not in that position for most of the day, there would be shading on the road in mornings and late afternoon. People love to have yards that have grass and no trees. This is something that also increases the temperatures of the earth and greatly. It is not considered that when we do not allow water to be retained this will increase the land's temperature. I realize that most people know this fact, but just for fun, the next time it is a sunny day, stand in the sun, then stand under the shade of a tree; you'll find that shade is much cooler than in the direct sun light. So, if you are a person who owns a home and cares about your world, plant more trees, and you'll do a good deed for the world because you and millions of people just like you can make a difference. And if someone says to you, what are you doing to help our world? Tell them I create a shaded yard, or I created a shade garden; what are you doing? And if you can't stand having more trees in your yard, increase the length of your grass. This will make your grass greener and hold groundwater longer, and yes, it won't be as effective, but it will be something to keep the ground cooler. Every little thing we do affects our world; people who throw garbage away into creeks, rivers, or large water streams do not realize that it pollutes the water. Chemicals on the object can also pollute the water. People who do not realize that every time they throw something in a ditch or out in the woods these elements are going to affect their world. Garbage needs to be processed so be mindful the next time you're tempted to toss some junk out your car window, or that cigarette, that wasn't out and caused a forest fire that you didn't know you did.

85. SUBJECT: SKY-TREK

The creation of Co2 removal from our planet is not as far from fact as this method devised is the answer, but the concept has major applications to consider. First, the containment of the proper vegetation that produces the most O2 and puts off the most Co2 at night and the research to have the best performance would be required; the structures to house these Co2 makers would be considerable but should be placed nearest to large cities at forest areas already established for starters.

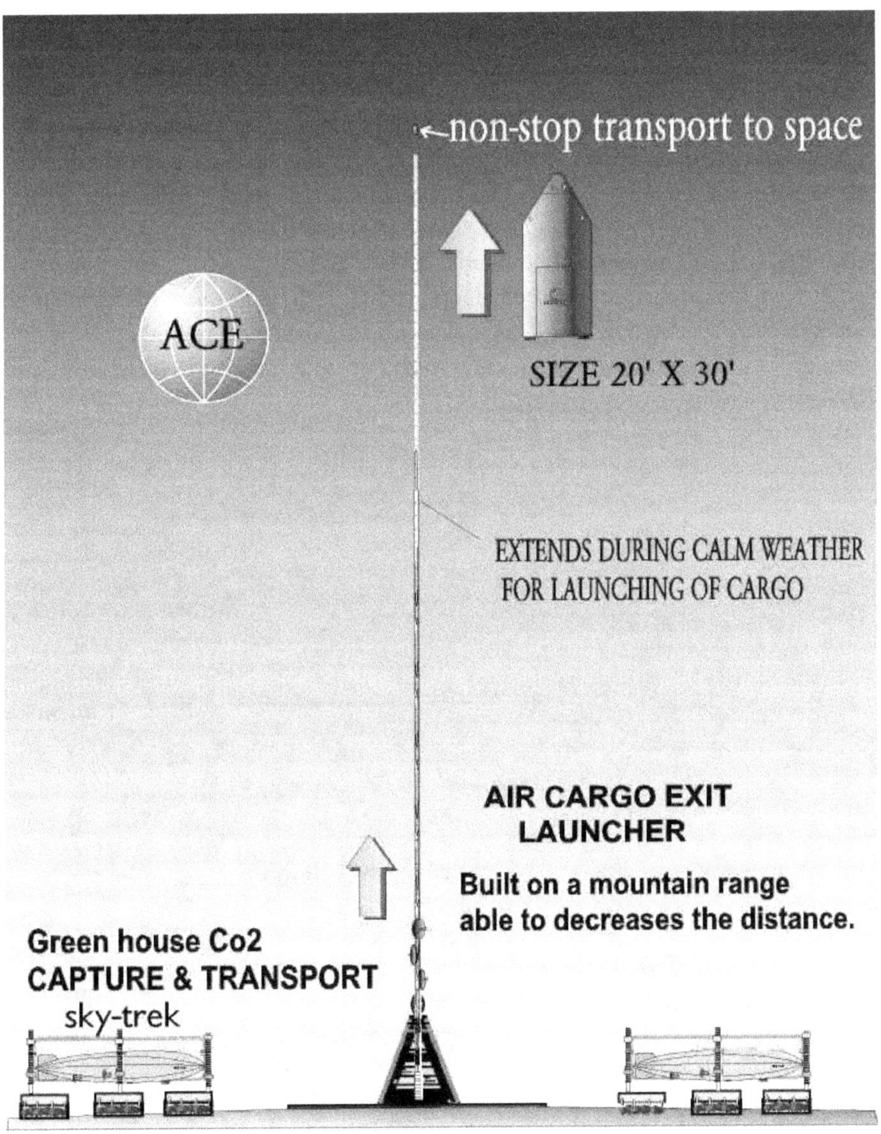

Trees would be inside massive greenhouses would surround them, and the sunlight would be in controlled lighting during the day; the windows would be open to allow air to flow out as O2 is made, but at dusk, one-half hour into the sunset, the windows would automatically close and contain the Co2 now being released. Two hours before sunrise, the vacuum system would remove the Co2 and place it inside the waiting Co2 housing balloons. Each of these containment housing units are the size of a football field with a height of a 12 story building , and hold the capacity of compressed Co2 that is massive in volume when filled.

The project will require a crew of men and women to maintain the propellant tubes that need to be attached and secured to each housing balloon to be connected to the main helium balloon transport air craft's, which will rise to a predetermined area that is done when the space station has been informed to release the space forklifts that are driven to attach to the containers upon the exit from the planets gravity, and those who are positioned directly in the path of the containers coming for the retrieval of the containment after launched from the aircraft helium balloon. From that point they are directed to the train connection waiting areas for the A.I. controlled journey were the containers upon reaching X-plant destination then releases the Co2 down to the planet's atmosphere. I have presented the gas be sent to Jupiter, but a more immediate need, and closer with the ability to be even stored could be on Mars.

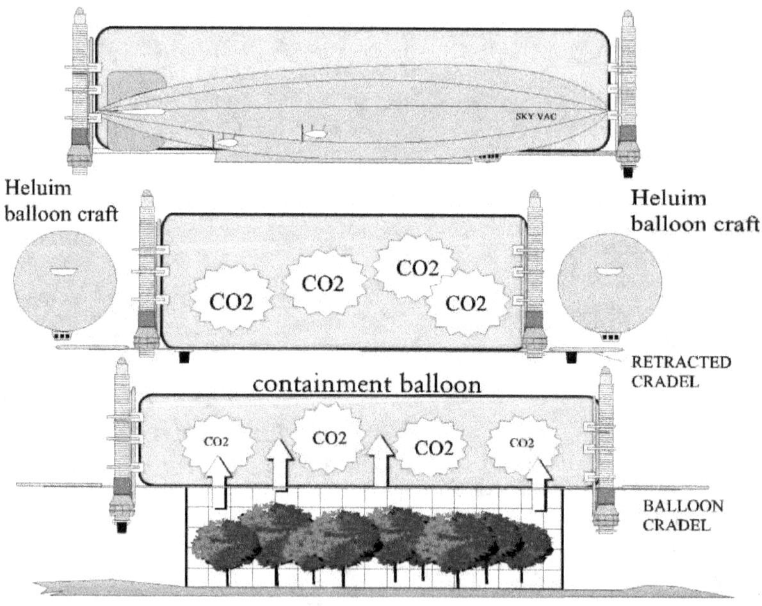

The objective is to store the containment far enough away from Earth's gravity to have these separate containment units be secured and attached in a train fashion to be pulled by a rocket ship that was developed in space as one of the largest and strongest long distant rocket ships, that will be making voyages to and from Earth to Jupiter to drop the cargo of gas Co2 to the gravity of Jupiter, and return for more units filled, and the process continues unless it is not needed. The delivery is made round trip by A.I. and returns to exact starting point empty.

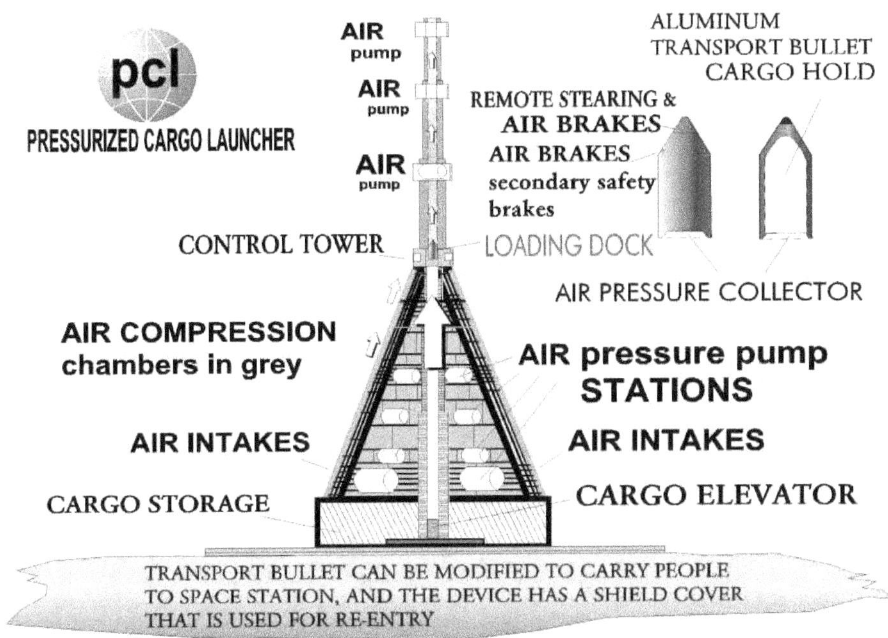

1. WHEN HELIUM BALLOON TRANSPORT HAS REACHED MAXIMUM HEIGHT, THE CONTAINMENT OF CO2 IS THEN LAUNCHED INTO THE WAITING SPACE TEAM THAT WILL STORE THE CONTAINMENT FURTHER AWAY FROM EARTH

2. CONTAINMENT UNITS ARE STORED OUTSIDE EARTH GRAVITY TO BE DELIVERED TO JUPITER AFTER 100 - Co2 CONSTRAINTS ARE READY FOR SHIPPING.

3. GAS IS RELEASED INSIDE JUPITER GRAVITY, this could also be Mars.

4. THE TRANSPORT ROCKET SHIP AND THE CONTAINMENT HOLDERS NOW EMPTY ARE RETURNED TO EARTH FOR REFILLING

SKY-TREK Co2 REMOVER

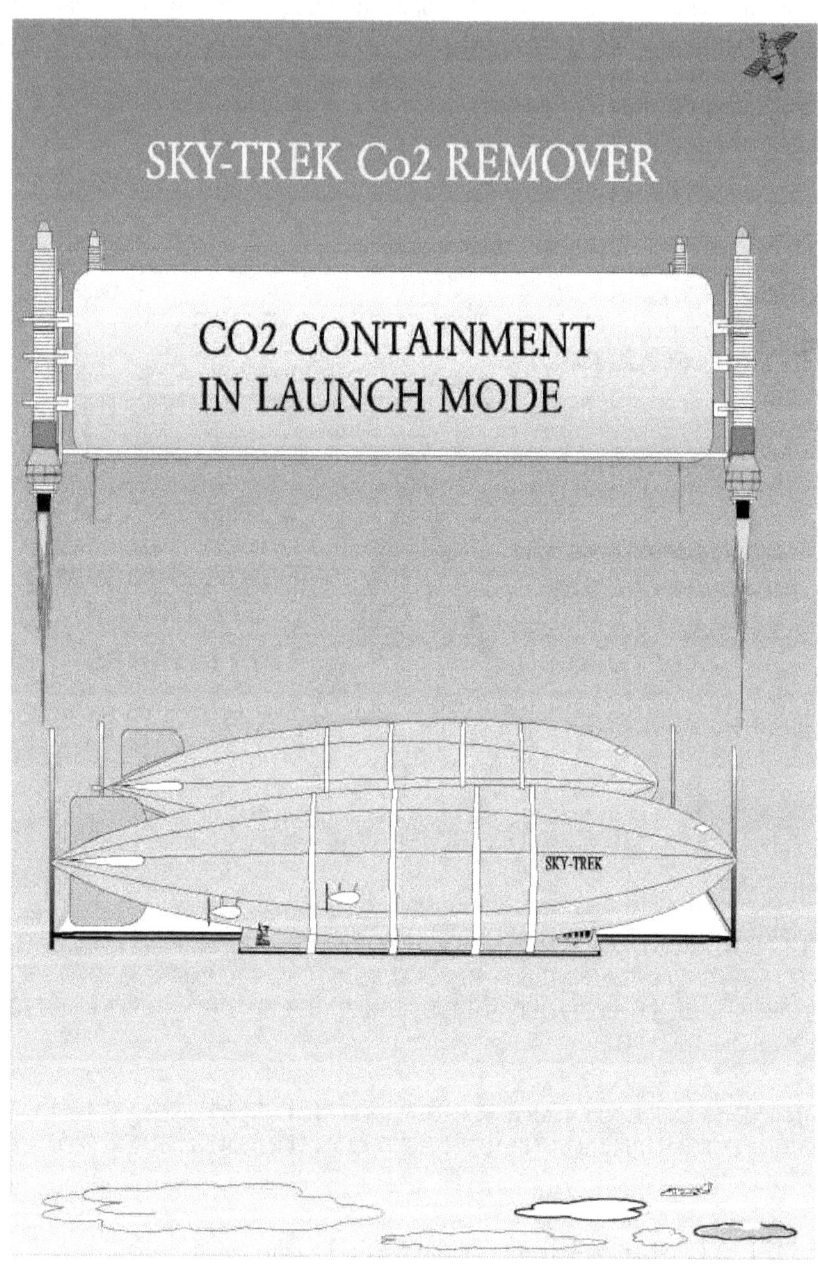

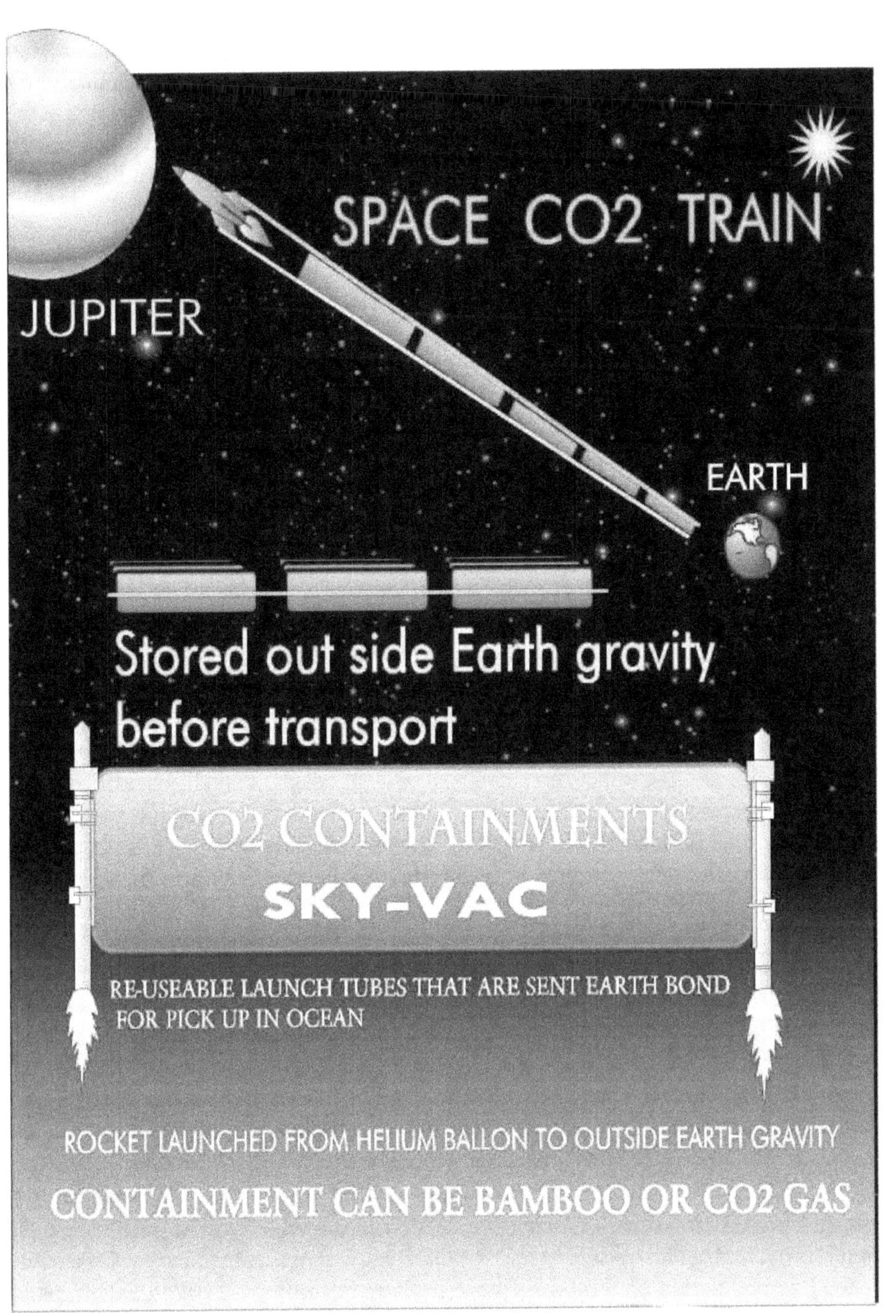
SPACE CO2 TRAIN
JUPITER
EARTH
Stored out side Earth gravity before transport
CO2 CONTAINMENTS
SKY-VAC
RE-USEABLE LAUNCH TUBES THAT ARE SENT EARTH BOND FOR PICK UP IN OCEAN
ROCKET LAUNCHED FROM HELIUM BALLON TO OUTSIDE EARTH GRAVITY
CONTAINMENT CAN BE BAMBOO OR CO2 GAS

86. SUBJECT: GLOBAL WARMING

We have all heard of this many times already in these pages, but as we go about our daily lives, we see more and more evidence that the claims are real, but little is done by the public, the constant warming and changing environment have flipped the norms of our world. The heated temperatures are directly caused by increased amounts of CO_2, with a combination of the increased ground temperatures, making wastelands of once fertile land, still, the heat is affecting the oceans, and the waters are now infected by the red algae. The evaporation is now bringing massive rainfall accumulations, and flooding is the result because our infrastructure was not made for an overabundance of water. So the water increases and the damage is in the billions, and will not stop until the issues of CO_2 have been removed and removed on a very large scale.

The temporary fix cannot repair the increase but can save many lives, give water where needed, and remove water from overflowed areas. But in no way is this solution I am to bear witness to a lasting fix. The reason is that the planet requires a massive effort from everyone, and many nations have zero regard for the problem; this is why we need a world united but under control, not of dictatorship and absolutes as governmental acts where people have no rights in prison. The true power is people being aware and making an effort every day to be less of a problem to our planet and stop raping our resources, "trees for the profits are the ruin of our world and its future. One of the most unnoticed developments that cause global warming is the depletion of space and air. This can be compared to a balloon that is blown up, and over time, the air escapes, decreasing the size of the balloon so the balloon, let's call the outer parameter of our atmosphere and the air has been depleting over time from many sources ,a few to mention, some billion or more of automobiles that burn air in the process of burning fuel, also the jets , the trucks, and let's not forget the forest fires, and gee, how about that cook out grill, or that bonfire at the school rally, or that wood burning stove, how about the relaxing outdoor fire in the summer? All items that burn the thing that is reducing our air and making our planet hotter and hotter , so oxygen is decreasing more as we kill trees at our homes, and make the forest a profit. We destroy tree areas for expansion of business or more structures that can make people money, all for cash without concern for the future. But we can look forward at our world like a fast forward even closer with another comparison or real fact is what we should call it- But our lives are no different than fish who live in the ocean; we are in a substance that keeps us alive, and we can jump out of the substance (air) but we can't survive without it. Like a fish, they jump out and die if they do not return to the water. So, the truth is our fish bowl has been losing that most important element we need most to stay alive; we are losing O_2, which is decreasing faster and faster. We have many devices that help this; we have a very large population of billions of people who are turning air into another gas we cannot live in, CO_2, and we are using devices that burn air and here are a few more-, fireplaces, stoves with gas, oh yes, the flame is there because it is using our most precious element and doing it very fast. A gas powered motorcycle, a lawn mower, burning air is everywhere, and a constant action done by billions daily, with never a thought that what if we have no air. So we now buy water, what's next – buying air? It is already being done for the elderly. So, the planet has been under attack for over 1,000 years, but only in the last 100 years have we seen the massive increase of all the air burning devices, and many make Co2. They also do something we fear: they almost all add heat that was delivered by man's mechanical ability, but the mechanical age also brought the fire with the locomotive and steam engines, and then the big change was when Ford made his destructive combustible engine, not to mention the worlds natural wildfires, forest fires and every back yard fool who has a weekend fire to burn the steaks and the hamburgers, I agree they taste great, but let's take the figure times around 1 million people plus per weekend doing the same thing, burning, oh you thing that isn't so bad? Ha, funny, let me demonstrate a small experiment that I want you to do; please humor me: find a match stick or a small candle , now get a large clear container like a peanut butter jar, and please invite the children to see this because it was shown to me as a child and it does have something to realize for all people and what I want you to do is to count to 100 when you place the jar on top, and make sure you count as fast as possible, but make sure the surface is completely flat to not allow air in, now light the candle place the clear jar over the candle,- ready- go, so how far did you count- three, four, eight- doubtful? Now, we just saw how fast oxygen burns with just a small flame; the bigger the flame, the more oxygen burns, and even faster. So now, if we placed that same backyard barbecue inside, say, any football dome, and then we sealed it so no air could go inside, it would become completely void of air in less than one hour. Possibility of two hours at most. Now, that is because I am considering the size of the flame be steady, and the fact that the coals will be burning after about twenty minutes also. So let's take these times one plus million fires from many sources, in truth we have around a billions of sources that burn air. How do you think an automobile engine runs? It runs on burning fuel. So how much more O2 is being burned per day; how much air is being lost? Oh, and there is a very good possibility that this calculation is off a bit, but consider that your fun time burning has lasted sometimes three hours, four hours, all day. I've seen people go to someone's house for a party and have a fire burning from 12 pm to 2 am. Gee, what a great gift they made for their children. What people have not realized is that as the trees are depleted, so is- THE OXYGEN! That means one day, not long from now, and I did say not long from now, the world will become void of air. That means you die. Now consider a forest fire that some fool started, just for a fun time when the same guy one day is told he has to buy oxygen, and one day that is now only for the extremely rich. Oh, that couldn't happen, you say, no, that's why you don't buy bottled water? Because there is a lot of water, but clean, well, there is a lot of air, but clean? No, and it will only get worse.

The solution- And I regret being a bit cheeky, but – if we used electric stoves this would help somewhat, and heating with electric would decrease air used slightly. But. I have seen so many people in the summertime sit around their innocent little campfire, thinking, gee, how nice, gee; this is just so good watching the air burn away.

What they should be thinking is; those flames should remind me of the hell I will make for my children one day. So the only word that comes to mind is -foolishness, get a fake fire and use an electric cord; for God's sake, you as a human already consume a great amount of O2 of air, and it is turned to gas; as you well know, most of you who listened in science class in school, so yes what comes out we can't use to breathe is Co2. So, wouldn't it be smart to place the barbecue with an electric coil? Yeah, but that would make it hard. Tell your kids one day when they fight for every breath. Asthma has increased by many thousands. Is it possible that our air quality has something to do with the deletion? Maybe, this is why we see older people wearing hoses to oxygen; the need for their bodies is more because the air has decreased it's amount of O2 naturally , is it really from the lungs break down?, or-? Could it be from years and years of breathing car fumes, factory fumes, truck fumes, jet plane fumes? Oh, there is the real pollution maker- see that wonderful trail in the sky, gee mom, how pretty the children say, oh yes little Becky, that's called Co2 at its best, and it's like 100 thousand cars leaving a trail in the air the size of New York State every time it travels from New York to La. (Gee, now that's progress.) But who cares as long as I get to where I must go fast?

87. SUBJECT: GLOBAL WARMING SOLUTIONS

Take the train. You've just helped the planet because you saved the air burnt and cleaner air quality. That is all well and good, but seriously, no, folks, that is not the real answer because we need to think on a bigger scale, MUCH BIGGER, and this is one of the main reasons we still have global warming happening, which is the devices they have developed are not able to remove the other devices that are still in motion, and! The large oil owners and, yes, oil developers don't want you to stop their trillion-dollar industry, but the answer "is"- to make a new form of energy, a new form of travel that no one knows about. Still, I invented this process as "continuum energy travel," which I have tagged as XCESS ENERGY. And this could be for cars, planes, and yes trains, all with electric engines, we can make the same travel time; in fact, we could be there possibly faster because electric turbines can move faster than combustible engines. I have the design, and it will work. Now, what does it take to make this- simple- a team of engineers, some electrical engineers, some good fabricators, and "money"? How much money? Well, more than most folks have, but by helping me by your verbal support, I may be able to bring these inventions to the world. That is my hope. And please except my apology for any subjects be repeated, the reason being is I felt the need is urgent and more important than spelling errors or repeating a subject, because I do care deeply as many of you do also of our future. And my inventions shown are with many additions required to function proper, but I am well aware of what is needed. So with your support, I will do my best to bring these ideas into reality, and make the costs lower of electric that you may pay for today.

Now, some of you may say:" Global Warming is a joke! Why kick a dead horse?" well, what if I told you something is happening that is related to global warming and very serious, but an entirely different problem, which is just not going to fade away. But before I tell you the issue, let's give our planet a proper name because in the real proportion of what planets can grow in size to, if we compare Jupiter, which is an estimated that it would take 1,300 or our planet earth, to fit inside Jupiter's diameter, our planet is truly still a "baby." Yet our baby has only been determined to be around 4.543 billion years old. Just an infant. So, within this content, we will refer to the planet as- "Our Baby." Unlike our human babies, this baby needs the formula to grow big and strong like Jupiter, and this food is something we call

"OIL, "made from coal. Many of humanity call it liquid gold because, for over 100 years, we have become increasingly dependent on this substance, but I will prove without a doubt that this substance is our baby's "FOOD" source. Fossil fuels take millions of years to develop from vegetation decay that is compressed in the ground. There is a great need for our planet's oil, but as I said, this planet is a very small child. So, we can say by observation that Jupiter has the potential for any planet to grow in size, possibly even larger. The main question would be: "What makes a planet grow?" Well, once again, if we look at how most living things grow, they take living matter and consume it for its ability to be sustained by the nutrients the particular living creature needs. And so, in observation, we have documented that the continents of our planet were separated over time from the fact that the planet's growth caused this separation, so we can conclude that planet Earth is a living thing or creature because, my friends, it is alive. Not only is it alive, but it also can die; such planets as Mars and others are no longer with life and also no longer grow. So you and I, you could say, are the caretakers of this planet, and we have been, sorry to say- not good stewards, guards, and not good helpers to most of us. But as a planet that has the potential to grow. That is no different than the plant that has the potential to grow from a seed to a larger living plant organism. So, our planet is truly in the millions of years old, no it is now reported a 4.5 billion years old, seems science gives it get a birthday every week. But still, as stated, our planet is still considered a child of the universe. And how does this planet sustain its life?

It sustains its life by feeding from the vegetation of the millions of years that has created layer after layer of different pressured vegetation, which was once abundant and covered the entire planet, unlike today- and the research has also shown that oil being made is from the vegetation that under greatest pressure is concentrated now becomes coal and coal is a byproduct, which in pressure now secretes and oil basc tar and his tar, or oil feeds the planet's core. Now think of yourself as a planet; you consume vegetation through your mouth well, for the mouth of the world is the vastness of the layers, and the under layers have crevices and spaces that flow toward the center of the burning hot lava. This lava causes the planet to continue its growth as it now performs volcanic eruptions on the surface, where the materials melted now make new layers on the planet's surface, and we, as people, are unaware that our planet is a living creature. We use one of the byproducts; this byproduct is the waste that is depleted, which is depleted waste. From this burning of the center core of the planet, it extrudes. This waste material is just as we do with people, but our planets waste is something we call tar. And his tar is the waste pushed back to the surface from the planet's burning of its food. And we use this waste material to cover our roads. So, we are like the ants on a giant rock floating through space on a living planet. Now, what is so wrong about using the oil? Well, it takes millions of years to make this oil. And we now are seeing what I call the grumbling stomach. Yes, the planet is hungry, and it has been disrupted by taking the food that allows it to maintain itself; the oil we use daily is a living substance for our planet. And now we see earthquakes, massive earthquakes. The grumbling of the stomach from our planet, we see storms that we don't understand. We see our planet dying, but the oil producers who make my money. Billions per day, they don't want things to change; they are older men with old ideals and ways. And they do not care about you, or your children, or even this planet. And that is the saddest part of this report, but here is even a much sadder report, as a planet Earth slowly starts to fade away, because it no longer has oil. Once this happens, it will be within an unknown time frame, which could be just a few years. The center core of the earth, which is no longer being fed, will. I repeat, it will begin to cool, like a crystallization. If anyone has ever seen a rock with crystals inside it and you see all these beautiful crystals. You'll also notice that the rock has an opening in the center. So we could compare our baby (Earth to a six-year-old; why do I say this? Well, we can look outward into the universe and see examples of how planets can grow to immense sizes and make our planet seem tiny; again, if we took Jupiter and placed it next to the Earth, it would be like comparing a basketball to a marble, now that's not exact, but it gives you a pretty good idea that the planet earths potential of how big it could get if we didn't interfere with its growth. And how did Jupiter get this big? In the same way, all planets did, they had life at one time, but an outside event happened which caused the life to stop growing from a event that killed vegetation, With that happening would eventually cause molten lava formed to be in a cooling process, As science has already determined, the continents are moving because the planet Earth is growing; this means the planet must be eating something to be able to grow; isn't that how you grew? You ate and went from baby to adult; our baby world is doing the same. Let's break it down to a simple way to understand it. Inside your body is a complex system that takes outside fuel- we call FOOD, and it breaks it down, but please notice that the temperature of the stomach is what has been determined as a hot zone of the body, and the hot zones are shown in photographic few many of you have possibly seen these with the different areas of the body in a different color , and the cool areas are in blue, where the hot areas of the body are in red and orange, well, now that we understand that the stomach , and heart are the warmest areas, this is how the FOOD is broken down and we could say it is the core or center of our intake of food before it is sent to the blood stream, and then is dissolved into the body. Well, our world is the same; it has a warm area. It's a bit warmer than our stomach. Temps can exceed any metal foundry, and volcanoes show extreme temperatures from lava flow. Science has determined that the planet's core- The "stomach "receives food,

FOOD? Yes, this food is essential, or the planet could starve. When this happens, the baby can be unhappy and shake up anything that it couldn't digest or burn in the baby's tummy, but I'm getting ahead of myself; when the core cools, it would be like hot glass cooling in a lump. It will make void areas inside and contract inward because when lava cools, it contracts and becomes smaller than the original mass, and the planet would then have a hole in the center. Eventually, once formed, this hole would make the earth like a ball, and if it cools too quickly as some planets have done, then- the planet could implode because the outside pressure- gravity is greater than the inside pressure that is now no longer there. This is how many meteors are formed, from planets in the universe that have imploded and now fly outward into space from the explosion or actual implosion. Still, the force is more powerful than any explosion because of the intense action of force against force, resulting in massive pieces of the planet being tossed in every direction. To be exact, when the planets have a center core that cools this can cause the planet to implode, because the outside food supply ceased to be available. So what is the source of food for the planet Earth? It is OIL, coal, that's right folks, you guessed it already, great, or maybe you don't believe it, well, let's consider the extreme temperatures, can you possibly think of anything else it could be eating? Well, what is oil? Let's see, its vegetation, vegetation, what is that? Trees, dead, that routed into the ground, pressed by layer after layer of dust and dirt being formed, and compressing the vegetation into coal, that after millions of years pressed the coal so hard it formed OIL, and this, my friends, is what she feeds on, now anyone who says this is not so, well- then show me what else she could eat that would burn, is it not so that you eat vegetables, what are they? Vegetation. People, we are stealing FOOD from our planet to make a means of travel. Now consider when the day comes, she gets hungry, really hungry! Oh my! Well, I hope I'm not here because it will be, oh, guess what, I am here, and she is making a fuss, earthquakes, what are they? What happens when your stomach is hungry? It rumbles. So does she (our planet), and we are taking every resource from her since 1901, that billions upon billions of years of vegetation taken that she requires. The sooner we realize this, the more I hate to say this, but she is almost hungry now. And that is why we have seen earthquakes everywhere, and she has Trying to tell us to stop. But greed for money outweighs what is good for the survival of humanity and the world. The problem is the older men who read this, who have made billions of dollars selling her food, don't care. The day the planet implodes, it will give a super display for about ten seconds, massive dried-up volcanoes will blow air upward first, and then- wham! It's all gone, and everyone is in an instant grave. No, it won't be fun. But most of you are saying, well, you have to go some time, yes, but what we leave behind is our footprint, and when this happens, there won't be a trace that we ever even lived here.

SOLUTION- Stop taking oil from our planet; we may have enough time to colonize another world before this one cools. My prediction of when this will happen is 2194 if we stop within the next year. If not, the end could be as soon as 2053, and I could be here. Great! But allow me to say to those who work for oil or coal as their way of financial support. The jobs can be achieved by the oil and coal companies investing in electricity, and jobs would be required to maintain equipment and run electric stations. Now, predictions like this do not know how much oil is left, but let's consider that we now have drained all of North America, Texas, Oklahoma, Pennsylvania, Alaska, etc., and we now are draining the Middle East, and now have to drill miles down into the Gulf coast even to find it. But soon, no more, and in 1979, we saw what that did: it was gas lines, people fighting and stealing gas to get to work, my friends that was a picnic compared to what will be, no more oil means no more world, because they have made us so dependent on their products that the world will become a total mess when it stops. Now, only fools believe it can't end; demonstration time: Get a bottle of water, drink from it, and consider that bottle,-The planet Earth's oil supply; now consider that every time you take a sip is like millions of gallons being used, and now consider that every time you take a sip the sip gets bigger, and make it into years, and your now consuming billions of gallons because the amount of automotive transportation increased over the last 100 years to such an extent that almost 5 out of ten people on the planet have an automobile, and this means billions of people are now consuming trillions of gallons of oil. Whew! If the oil companies told us how much oil is being used yearly, it would stagger the brain and make it want to shut down. But we as the public consider this normal, why because they have brainwashed you to believe it is the way we should live, why because they don't care about the future,- oh I'm sorry, they care, it's that what their commercials say / ha, ha, funny guys. They care only about their present and the future? What is beyond that present is no concern to them. They already knew everything I have told you. It's not something the average person considers; why? Because heck, I'm late for work, or my dentist appointment is today, or my girlfriend wants to go to Walmart, the point is we rap our lives inside our lives, and we only see what they show us. This, my friend, means the media blinds us, and the press now belongs to those who are the highest bidder and is controlled by dollars; I'm sorry to say that the media, on the whole, is controlled by China and Russia and both political parties have members or those who promote the interests of China and Russia as for their goals for world domination, so we must look at the full spectrum of who wants oil to be maintained, as Russia sells its reserves. China wants to be in the oil industry for their plastics for the hundreds of thousands of products they sell to control the world economy, and we can look at past presidential leaders like George Bush who had family involved in oil, so yes, even our presidents have been oil men. So, can you blame them? All they know- is it made dollars and cents, it's the same type of greedy brain cells that once tried to kill all the buffalo to be almost wiped from existence because someone saw money in buffalo hides. This is what our world is to some people; it is just a way to reap and not care. But" hope is alive ", and it is something that would work –It's a fast fix to increase our melting poles but if not done properly Could actually cause our planet great harm as to freeze and cause millions to perish, but if used properly can be our answer to start the reverse of global warming.

REMOVE GLOBAL WARMING BY INCREASE OF ICE IN THE NORTH POLE. THE QUICK FIX FOR GLOBAL WARMING IS THIS:

. SUBJECT: SUNBRELLA

GLOBAL WARMING FASTEST FIX to solve the problem, here is a "shortcut" to end global warming until all the other devices are place. Then, this device can be shut down, but it can be implemented when needed. I call SUNBRELLA PROJECT. A shading to a "TARGET POLE" This is a method that requires a sensitive small application to go very slow as to make the project a success is a game; it could be and excuse the pun," A world of trouble" if not applied properly, and the expansion of this device needs to be ever small once in place to make the climate change very slow, and- we need a team to be there in the event we have made an error. But this is a two-rocket application, and the rockets that are launched into space for the mission of placing a very large metallic giant umbrella to be opened, once again ever so slowly-and little by little, because we must not forget this device is to make a shadow the size of almost the entire U.S. but at a distance from our earth that blocks sunlight to the North Pole which allows winter snows and climate to increase the ice yearly back to its original state before the meltdown, and we need a ban on ships breaking the ice, This is one of the dumbest events I have ever witnessed almost as dumb as someone making a bomb that can blow up its own world, yes folks- gather around for the observation test is now available to you directly in your own home a science test as like a science fair.

You were told about the massive two ice blocks inside a 40-degree room, but I thought, why not give people a way to see it live. Okay,- Get two 'large' bowls of the same size and get a board or a hard surface and a hand cloth towel will be needed- from your water faucet, fill the bowl with lukewarm water exact temperature into both bowls half way- The water we will call the ocean/ now go to your refrigerator and your nearest ice maker, and retrieve eight ice cubes, as fast as possible place four ice cubes in the center of the hand towel, and holding the four edges of the towel together now swing these like a hammer against something solid and hard like a floor of the concrete sidewalk, now once the ice is broken into fine pieces, now take that and set it next to one of the bowls, now with the other bowl place the other four solid ice cubes inside the right bowl, but as you do this note the time on any clock of what time you placed the four solid not broken ice into that bowl?, now as fast as possible empty the other hand towel of broken ice into the other left side bowl, and please consider that the ice of the solid ice cubes entered the water "First" , with both bowls now filled with water and with ice, one with four solid ice cubes on right, and the other with broken ice cubes on left- , now sit back the wait,- the question here is- will the ice size matter for them to melt faster if smaller or will they melt at the same time frame? The results would be the same on a large scale, like a planet with a large bowl called the ocean, but- we have other elements that cause these meltdowns even faster. One is sunlight and heat made; Air temperature increases another, and water temperatures rise. None of these are in your test-, so let's take the bowls now and set them in a sun-lighted window because this will be a way to simulate sunlight and increase air and water temperatures. The results are why the poles are sacred territories not for subs or ice breakers but do not enter zones, and global activities such as so-called weather testing in the poles or samples done by science, which only cause more attacks on this fragile ice area, all need to be stopped, and a world agreement made to this fact.

The illustration shows a rocket ship that can be sent into the exact orbit an orbit direction between the Earth and the Sun, and when it is in place is a massive umbrella is opened from the inside and exterior of this manned craft; the mission is to use two separate space crafts, one is for the actual "Sunbrella" activation and positioning, and the other crew is there to assist, and secure the umbrella as locked in place, and for the first team to be in the space ship able to be returning to earth with a third craft that is attached to the second craft, but the main crew is to stay with the outpost as to build a space station observation and control function for "Sunbrella" as to make a power transfer to secure the crafts electrical power to solar cells and the secondary ship remains in orbit to adjust the "Sunbrella" when needed, new teams are sent up every year with half of the first team to stay to instruct and guide, and for

this space station is not as far away from the planet as you may think, it is because the mission is to shade a specific area over years to increase ice mass back to its former level, so this orbit is further away from our planet but is designed to stay in the actual speed to maintain the shaded area over our planet as we travel around the sun in the same exact locations to be controlled by A.I. navigation to keep the flight for a period of five years per team exchange. After five years, the system is closed as the ship continues orbit but has no refection devices open. But if and when the planet has a global warming issue again, the device can be activated when needed by a new team. The danger of the device is that the area of the orbit is not exactly done, and it causes the sun to be blocked completely. If this occurred, then the residents of planet Earth would freeze solid, a basic pop-sickle. So, the recommendation is to advance the distance of the craft ever so slowly as to position the shadow exactly over the North Pole and consider the rotation of the earth so the shadow should only be maintained in winter months, and this would bring back the pole as to its original state. And to those so-called smart scientists that use boats to break up ice in the poles to test their equipment, the fact is you're only causing the ice to melt faster; take a simple glass of water, place ice cubes in the glass, now take the same amount of ice cubes and place them in another same size glass, But, this time take that same amount of ice cubes that you placed in glass one, and now crush the ice into small pieces and place that inside glass 2. The result will be that glass two with the same amount of ice will melt much faster because the cold areas are now displaced and allow more air contact than the prior state as being solid. Be smart; if you wish to go into the Arctic or the ice-formed areas, take a plane, or a helicopter, boats, and subs, stay the "heck out," and grow a brain that uses common sense. Sorry, I called it as I see it.

90. SUBJECT: WORLD ELECTED LEADERSHIP

The real way for the world to be in a world government is similar to the structure of government in the United States, but with some very important changes that protect the public worldwide. What I would like to introduce to my list of bucket lists-to the American people and the world community is a better way to govern even better than democracy, remain a free and stable society, have all rights as people, and have a better functioning means of getting government projects needed done faster and more effective as the biggest problem in a democratic form of governing is- Hear this loud and clear- "Our form of Governing is wonderful if applied correctly with all parties working together, But !- the objections of different parties constantly disrupt progress like the speed of snails. Most needed issues never reach a fulfilled bill passed and are often abandoned for lack of support, even when an issue stands in a priority need for the people it represents. The debates and the constant back-and-forth objections only delay what is needed in the here and present needs of the people. So, my ladies and gentlemen, please allow me to introduce to you for the very first time in history what I call an "ELECTED World Leader, "a pure and exact need for all human needs to be performed with the best results, faster, without delay, and with a linked network of bodies of the single party working in full harmony. And a good title for this position could be called: a form of "Kingship", but this is very different than Kings of the past. The idea of a world government is frowned upon, but it is the only way to maintain world peace and a system of cooperation. But my concept is not a dictatorship of communism or democratic rule. Still, the world's people elect the global leader, or to be more exact, a world leader who can be a dictator with one very massive difference. He can be removed and done very fast without dispute or a means of stopping it if the number of world delegates also elected favor removal. It is a principle of the good kings of the past who had the authority to implement action immediately. And the defense is if you're not doing well and considered even the slightest as not to be doing what is needed, the people have the power to approach their representatives and file a complaint, and this complaint if matched with other countries up to three total. The acting leader is history. China does so well because they have no one to say NO to! When the subjects are presented, the ruler makes the action happen by saying yes or no. When it is yes, the wheels move without disruption, unlike the governing of the United States system with agreements holding down a bill, disputed for months, and some proposals have sat for years; Let's look at the system like telephone poles connected by wires. A comparison using telephone poles as party members connected with lines. An example: We have just fought for months, and the other side finally compromised after massive outside influenced groups who may have pushed for the passing of the new law, or the new project, but finally, it was passed. However, we have all these people who are still in conflict; we may have a Democrat as the person to start the process, but it goes to urban development as an example, and the program is for a new housing bill to be passed. But, the person at the "Urban Development office, well, they are a Republican, so the telephone line now has a bad connection, and in this case, the connection is delayed until suddenly someone realizes it, but this is going through so many different people, and so many of them just don't like each other, because they believe their party is better than the other party so the battle to get the process goes from one person to another, and sometimes gets lost in the mix. The project which should have been done is now back because those who needed the program which was decided never got done. The decision-makers never even see the result; they pat themselves on the back, never understanding that the proposal hit the dumps; why? Because of the many telephone poles of different parties, it took to get it done never accomplished anything. So, an elected world leader who considers the action requested after getting all the facts his action is- 1st. He hears the proposal and its content of what it should accomplish, and second 2nd. The person says yes or no, and if yes, there are no obstacles any further; now, to be clear about the type of person who is to be a leader of such power, well, the gender of the elected leader has no bearing, but do have the same ability as Kings of the past, but they are not deemed as blue bloods, or designated by God, they are just people who are good at seeing a problem and give a exact solution to the issue to fix it. Or is someone who has the ability to see further ahead of outcomes, and has plans to be ready for any event. Someone when addressed an issue, they make their actions clear and why, and when the action is stated the issue is resolved as law or by action being done without question. The process was placed to the proper departments, and the action was made directly, with no delay, and the process of repair or the issue was started as being fixed or built. The process: The means to vote are much clearer; candidates are placed on review charts regarding their qualifications and what they did in their country. The person's face is not shown, nor even their gender and their name is not revealed either on the application-instead, they are assigned an Identification description that they can design, such as- I care, or my ability, as your best representative -etc. Anything within three words can be in all languages, and all submitted applicants have their text in all languages. The text is placed in any language and can be played. We can consider it their promotional reasons of why they are best qualified for the job. Other candidates do the exact same thing, yet in this election, mudslinging is not allowed, you're being elected not from your past, but what you have accomplished is presented in your prior local political votes, so yes, one must come up the ladder to be qualify for such a high office.

Only after people are elected in all areas are their names, genders, and images revealed. This is essential because many people may be better qualified than others, but because someone has a visual attraction or a gender preference, these issues are no longer stopping the best applicant, but their facts, past actions, and negatives are also submitted as their hard lines but are not presented by the opponents, all entering this race have their good and bad presented to voters, and how they may have voted in other elections inside their country. However, having a record of being in an election is not mandatory; facts must be made for all candidates inside a country who are sending their elected representatives to be submitted as world leaders. So, an election is held first inside the country, requiring a representative to be sent, and then he is placed in the world election category for the world election to take place. After this, the person will present their ideas of how they will make life better, and these are all submitted on the same day so others running against them can use their ideas, or claim them as there idea. The world election is taking place, all men and women of any race and any form of education can vote for who they believe is the one person who will care about all people of every nation, creed, and language. Each country would become a territory of- The "United World Community", for example, the U.S.A, would remain the nation of the U.S.A. as all countries will remain the same names. A president would be elected to each of the countries, and from a popular vote total. To include that from one of these countries, once elected, this person would also be eligible to be the elected as world leader, so all nations do elect their president, but each now would have their candidate be shown to the world as why they would be the best to lead the world. If their president is chosen as a world leader by the votes globally, then the person who had run against that country's president is now the world leader and this action would require a new election to fill the empty position of their country's nominated leader, since that person is now elected as the world leader. The candidates are considered not for education but for wisdom, not for their wealth, but their voting record and ideas they contributed in the past, but a complete record is of all candidates of even, false statements, or changed policies made, everything this person has achieved of their political carrier is presented and displayed online with all other candidates having the same information presented,. Still, they would have to speak openly without speeches pre-drafted, or on teleprompter, but would have to speak from their minds, and then once elected, they would have to follow through with what they proposed, and here is the real reason this is needed. Because when they say- this will be, it is done, no question, no rebuttal; it is just put in motion faster than fast why because the action is taken as "DO", and no less, absolute power, but- if the leader has to prove to be someone with unselfish ways, and not only cares about his nation but all nations, If the world leader has shown biased, then they can be challenged to be removed, this can take place when any country with no less than three other countries state this challenge. If the issues they requested are being done in other countries, and not in their territory this is a main reason an Elect-king can be removed. Or a program request was given to the world leader that was never achieved. And once the Elect- president, or what I would like to call the "world leader" or (World leader is removed, which could happen even in a short time of three months, even less if found to be doing something that is to harm people, If this happens then the representatives elected from each country as their leaders, will now look at the situation to determine a withdrawal of the accused world leader. After the vote has been made of this persons guilt, and if the world leader is dismissed, then a new election is now done by the people, the world leader election is based only on all votes collected for each candidate and the candidate with the most votes wins. But each time there is a world leader removed or one that is nominated this is how the election candidates are chosen. The country candidates for world leadership are picked from a lottery drawing where only 20 candidates are picked from a rotational bin, all presidential names these are the country's leaders who wish to be a candidate as world leader, and is done in front of all the nation's leaders and is televised to the people, as one by one they place their name on the paper , then names are placed in snap lid balls and placed inside the rotational bin, and after all have placed there name inside, the bin is rotated, this bin is stopped after one minute of turning, With the view of cameras recording the balls are taken out and announced as who will be running and 20 are counted and as they are removed are placed a long row that holds each ball. The balls are opened one by one to disclose the candidates that can be running for world leader now, the world will vote for the world leader from these 20 people. All the candidates are placed online with all of their information which each had prepared prior to the lottery drawing. Each candidate with all information online in full detail, including what they are as a person and why they would be the best choice. Representative from each country chosen in the lottery. The world leader can be up to 30 years, but an election is done every ten years,; this can only happen three times after that, they must step down, and a new representative for that nation now applies with the qualifications for the country and act as representative, but is not able to run for re-election because that country already had a ten year world leader, but after ten years can be entered into the lottery again. The reason there is a drawing is because the world has way too many countries, and to hold an election would take far too much time to gain the results, and if every country only voted for their candidate the nation with the most people could be the winner the most. So when all votes are made, each country has no more power of numbers of people over another country, instead when someone has won the vote total say in India, and this person may be from India, but his total is only one vote added as the total, and let's say another country England tallied all their votes and the winner was the person from India, Then that total now is only one vote to be added to India's count, so the man now has (2) two votes. So in short all countries no matter how big only attain one vote from the person they have chosen who gained the most votes. The person who has the most votes collected from all countries is the world leader. There is no such thing as a political party. There is only one party because it represents all the people worldwide. Still, each country does have candidates for all elections of national positions and local positions, so Mr. Smith, running for the local office of, say- State treasurer, is now running against Bob Milder, whom the votes are made by the public, the same way, there are no tags of what group they are in because that is not needed. The days of dispute and division are gone, but people are placed in open forms to say why they are best for the jobs. The main reason this is being formed is that, in many political arenas, we constantly delay things accomplished that need immediate attention. However, because of parties who disagree, they drag on the process far too long, and many issues never get fixed. You see, in the past a Kingdom, was represented and controlled by a King with absolute power to look at a situation and determine the immediate result, the method I am proposing does this with a major exception, this King or dictator can be taken down and this person is elected not given power by force; in other nations prior that used dictatorship without representation, the leadership was able to be over the people with an iron fist and even kill their people who are out of line or disgruntled, that is not the world leader method. Instead, we have the people maintain control of whom they want to improve their lives, not for their lives to be worse. So being able to be removed will make the world leader more involved as to do what is correct, simply it takes once again only three to five nations that place a dispute to the representatives, that is reviewed by the representative committee which is made of leader of any country to sit on the world committee to review the situations presented to be done in their country. If it is found that the dispute after investigation is real, the world leader can be removed from power if the issues presented are not resolved. Now, the only thing that can prevent removal is "The nations who submitted their issues after the investigation was made, they had no serious claims, and the issue was resolved within six months, or is in development within three months, such as bridges needing funding for repair, or any issue needing to be fixed. But if these issues are ignored. The world leader is notified that he is in a red zone as a possible removal, and if the issues are not fixed, he has 30 days to start on that issue, or a new election will be in place; the fact is with our present voting system in the United States,, you cannot be removed unless impeached as that process, which is long and dragged out, and former president Bill Clinton was impeached but it had no bearing on his office because it wasn't accomplished until he was finished his term, If a world leader is not working on behalf of the people as a whole, this is grounds for removal. It can be swift and painless to the public, but the public is made aware of the offenses, which must be valid and approved as a big concern that affects the nations in that dispute. As stated previously, the representatives are the reviewers. As there are many countries, each with its representative, the only representatives that cannot review the problems addressed in the dispute those from that nation who would be biased. But the nomination is different to replace a world leader. Those picked in the lottery are the only candidates, but the person who came in second place in votes would be the new world leader. If this action of removal of a world leader continued within the ten year period of the world election, The people who had the most votes would be in line to become world leader, so let's say a leader was replaced again, then the third place winner would be designated, this is doubtful that this would happen 20 times in a 10 year period, but if it did happen, Then what was to be held on ten year stance as the next election held would be moved up to start the process as a new world election. But for 20 people to be replaced in 10 years would be somewhat impossible. The representatives have a vote of a five person committees —we can say is similar to the supreme court of the United States, but is made of five people each who will chose by all countries leaders as to who they are. These five men committees are groups of five that are constant in travel to review the problems stated at a nation who has delivered a grievance, this is their main job. There are representatives formed from each leader of each country to be placed in a group always of five individuals. Being there are 195 nations and some that are not recognized by the U.N. which is absurd, and rude. But in my system the U.N. is gone. Because they sit around making complaints and they do very little to stop conflicts, because they are all divided. So the need of disputes are gone because all nations are now unified to take care of each other. But since we have an estimated 195 countries, this means we have divided them by five for a total of (39) groups of five who will investigate all issues global. And will be sent out when needed, and also collect grievances to be reviewed. A submitted grievance is to be investigated by one of the five person teams as soon as possible which is three working days to attain more information and set up travel actions needed. They are to take clear videos and pictures and interview those involved directly with the problem, and they are to make a list of values of rating the situation; these are done over two weeks and as thorough as possible, and they are to meet not only with those who are making the claims, but they must see the problem directly. Once their report paper is completed and all evidence is finished, they will return to the other world representatives to submit the documents and proof to be given to the world leader and all world representatives. With these submitted, the world leader is advised to fix the issue within the delivered time frame that is determined by the five person team, only if the issue is real. If the issue is not resolved, a final warning is now sent to more urgent action, as notification (2) two is the last warning; with no action made, a vote can be made by the representatives to remove the leader for negligence or abuse. In the event of a leadership change. The information is sent globally so all people can see that this issue was not addressed, and the election date is now set if the vote is to remove the world leader; if the former leader is removed, he is also removed from the advisor position at that final action. Anyone over the age of 30 can register to be an elected leader or run for any office. The minimum age 30 years is required to be in any position that means local elections and national and world leadership. Anyone who presents these attributes and has the majority, and the fact is you can be a poor person in terms of wealth. Still, if you have maintained a solid foundation of these attributes, you can actually qualify each attribute is rated from 1- to 100 as the wisdom of understanding and giving solutions that are sound, caring people, but all people's ideas bring this is a serious attribute because to improve the conditions, the ideas are placed openly as to how they would solve but to ensure that no other candidate can apply the same ideas as taken from another candidate, all ideas are submitted at the same time on the same day, and cannot be changed or added to for each subject which is presented, these are to be seen by every person who votes that are not words of promise, but actual solutions that speak to everyone once they chose to run. This are recorded videos that are once again submitted all at the same day one month prior to election date which is held on January 1st, New Year's Day. And every ten years, and this starts on an even date in time, an example 2020, or 2030, etc. So every election is set as a decade. So in 2030, we will have our first world leader, but to make this event more exciting, we will start the process to establish everything with a world leader forum, and this will be held similar to the Olympics , the difference is the contestants are from every country and even countries not recognized by the U.N. The he oldest a person can be 80; they can rule as a world leader for no more than ten years, but after the rain of that individual, other nations leaders can return to run again if they choose. Still, they are marked as prior candidates so people know they tried before; this can be for them as someone determined or someone who was not strong, which is very important. The like of the world's leader must be a very strong percentage, and nations that voted against the actual leader who won will be the first of any country to gain a 5% increase of all needs. But once again, it can only be for ten years before any world leader is at the forefront, but a former world leader is now an advisor to the newly voted-in world leader, the program is designed so that all countries get a chance for elected leadership. The election popular vote: This is not like any past election we have in America; your name is displayed on television and at the same time online voting display, and everyone sees whom you voted for and why. It means nothing that people know whom you voted for when the voting is done legally and does matter- unlike today, where people vote but does not place the person in the office; in fact- they really should have stayed home because it was the Electoral College that decided not them. This method is a lie to people of freedom, and the people by the people is now false and should be stopped, but I know you are too busy to care about something so small: freedom, gee who needs one vote that counts?, not important; I know). Yeah, okay, so with my system we now have a vote that counts and the person is actually awarded the positions by the popular vote Which is displayed on a TV screen, and each name after voting, a check mark is made on-screen worldwide by satellite, and your button to vote is inside your home which is attached to your remote control, if you chose not to vote, no problem it's called- NO SHOW, and you see your vote canceled also, and if you vote it now changes the tally instantly, you are requested to be at the TV at an exact time you stated you can be there. Then the announcer will call your name and state, and a I.D. number which you received is also a verification that it is you, the screen on your TV will light up from red to green, and you will have one minute to push the up and down arrow and then the vote button; once the vote is in, the tally changes and the next person is requested, the election postings are first done on the local level, and each level is not placed in a pool, but is kept separate so every level can see the results. They are posted online, and the final count is projected on the screen for 24 hours so people can check it with the online postings. In other words, no one can cheat; why? Because all the tallies are shown. All new laws are displayed on an open website that needs no password but displays all new actions and results.

91. SUBJECT: REMOVAL OF ELECTED LEADER

This action takes only three to five nations to make it move into action, but the determinations are finalized by the five person teams that evaluate the situations; this does not mean it is automatic because it is like a court hearing; as we have the Supreme Court today, we will have the World Court, but unlike the Supreme Court now where the government makes it. The World Court is made by- The people of all nations, in the same manner as we have Parliament or House of Representatives. The people are chosen seats and elected seats that are replaced every 10years, or if someone dies, the difference is someone can hold the seat for up to 30 years as in re-election. I have made the terms longer because many political seats last long enough for someone to get used to the job, and they finally get a good feel for the job, and suddenly, they are taken from it. Then the next person comes in cold, blind, and unknowing, so my idea to add to this is called change of station; this means whenever any seat of any

political seat is being replaced, and this is at all levels of governing, there is a six-month change of chair, which allows the older adult that was in the seat to show the new person what is needed to be done and places them in a training position for six months after that- they are on their own. Okay, sorry, back to the removal of the Elected World Leader. Once any of the nations have challenged the Elected Leader, the evidence is presented to the world court; the elected leader must not be there but must send any rebuttal of anything within four days of receiving the rebuttal documents that show the claims. The court only reviews the documents, not the people; they are not required to be present because they, (The World Court) only need statements of the nation's grievances and the five-person team sent to review the people of that nation, any documents that show disregard or neglect, or misconduct statements. The actual testimonies are not required- "YET" is only the evidence; the next step is if the court returns with a COURT date, then all parties will be requested to a hearing. This hearing now becomes world-televised, and the world court has the final vote from a five-person court, if the Elected Leader is found to be guilty THE POSITION IS OPEN. THE NEW SEAT COMMENCES to the second in line.

Now all leaders of countries who have issues are to present these issues of great value to the world leader for approval before the action is applied, so a world leader listens to all nations daily on a massive screen that allows each leader to have an open forum which can last for seven hours, but once a nation has finished speaking they now are able to leave the forum once given permission by the world leader who only repeats the items as to the actions needed or approved. So, no world leader is safe from exit if they do not fulfill their obligations to the people equally and fairly to all cultures.

92. SUBJECT: WORLD HOLIDAY ELECTION CELEBRATION

I don't know about you, but I enjoy a good holiday celebration once in a while, and this will be something to be a fun event every time a new world leader is elected. An event held every decade. Now, the following is not mandatory, but as a creative thinker, I feel that if and when a new leader is appointed worldwide, the new leader should be celebrated by everyone worldwide. So here is my version of the action of celebration- Which happens two weeks after the election is final. When a newly World Elected Leader is placed into power, a weekend single-day event is called: "UNITY DAY. "It is where the people of every community share song, dance, and food. All cities have large festival's, and people in the countryside have barn dances or meet at their local fire halls, but no matter where they may be, they are gathered this special night to feast and dance as never before, something that is made to share what you have, or talent, and or poems, or words of praise and a parade of all the latest graduated students of the last ten years, and the banners from what school they attended. Wherever the newly elected leader is from, upon this day, the country where the person resides puts on a massive presentation for the elected leader and the entire world to watch in the evening. And it goes on until sunset; the sun rises to represent that a new dawn has arrived and a new day begins with better days. The actual event is for the new leader and their family to attend one-half hour before sunset and one-half hour just before sunrise. Opening a ribbon with large scissors starts a world parade that presents floats from every nation. Also- the top school's athletic achievements are displayed by those who made contributions march in the parade, and then the college students who acquired new jobs are introduced, and they are all brought on floats as to the highest honors. Unlike today's parades, the achievements of the best in their field are recognized. First, individuals who have made contributions of invention, found new medical improvements, or cured an illness are given up-close images of the people, and their names are under their image, and what they achieved is televised; the parade is also one of recognition of achievements presented of the goals of the leader who may have been elected again. So, showing progress is essential. This festival also acknowledges people in the community who have started something that helped others, and awards for the best creative worker are presented from a factory or- hardest all-around worker, most improved worker, things that show others what dedication looks like and has some monetary value also, a small gift, or food rewards at a restaurant are given to a man's family. These awards go back and forth from white-collar to blue-collar until all awards are given. The JOY -award is one given inside each community that says- we thank you for your help, dedication, and hard work, which can be awarded to any age and person. This is the highest award that can be given to anyone. If someone attains three of these in a lifetime, they are to be called "friend" Because they have received this honor from being a friend to everyone. And when introduced- "This is friend" and their first name, and last name can also be given, but is not needed. And this title name is of greatest respect. The person who meets this person is to hold the persons hand with both hands when they shake hands as to show respect. And how do you attain this award, it is asked by everyone who you feel has really shown respect to them the person or even child who made the list is now tallied and the award is presented at yearly festival.

93. SUBJECT: HOMELESS, & DRUG PROBLEM TO STAND

The world has some very cold elements, and one that comes to mind is the way we as people can become insensitive to the needs of others and forget that everyone has ups and downs; even the richest people have had times when they reached rock bottom, so this next segment is to address this situation as so many of our people fail from the depression of life and dreams are crushed, hearts are broken, disappointments that rained over and over, and these souls either gave up or never had a chance from the start, because it is possible they were discouraged by others even from their beginnings, so to address this task of fixing people who need our help, these are just a few of my ideas and reasoning of why this system "must," and I do say "must" change. And the next time you see this person, be willing to give a warm blanket or hot chocolate; never give them money, but food and things like a toothbrush and toothpaste, a bar of soap, and a washcloth; tell them where a gas station that has a bathroom is. So they can at least clean up, but be the bigger person, and if fear is your reason not to help, go with two other people you know well. And watch how the kindness spreads like warm butter. Let us start with the Homeless; they are the main group who have either mental illness or drug-related addiction or have lost a job without a backup income. And never forget- this person you're looking at could be someone from your family having a hard time or even you. Now I spoke earlier about how I had some rough days being homeless, but I never looked homeless. I stayed clean, used laundry, and worked my way back into the mainstream, but both times took two and half years to bounce back. Prior to these events, but some people just are not made like me, they lost something that somehow just took the desire away, drugs, alcohol, the list is long. But I once saw a homeless man as I was dressed in a five hundred dollar suit attending a toy fair in New York City, and in a rain storm with my umbrella, I was walking from Times Square when suddenly I saw a shopping cart filled with stuff, right in the middle of the sidewalk, and the stuff was getting soaked. I stopped and examined the contents: old shoes and clothes. Suddenly, a man came out of the shadows in a doorway; he said:" That's my stuff. "I said oh, I'm sorry, and then I said- Are you all, right? He replied to my surprise. "Oh sure, man, I'm fine"! "No problem here!" I was very sad; I didn't know what to say; it shocked my entire foundation to see a black man be so happy in such poor conditions. I never knew that one day, my financial life would crumble and my contracts would be ignored. I still remember that man, and his perseverance kept my hopes alive. So, as I said- never say- that could not be me. It can. And it can happen fast. Now, what I always did whenever I had extra when I went to the cities, I would see a homeless person and invite them for food and drink, no cash, no money for drugs, and I would talk to them, ask them how it happened, and let them spill their guts. Why? Because then I would tell them I also had been in their exact spot. It was always a shock to them, but it gave them hope; I would tell them the secrets of how to get back on their feet, but some are too brokenhearted and don't care about working, and they would rather be takers than givers. But we all know that being non-productive leads to laziness, and self-worth goes down the drain. As I said- some don't want to work and prefer the life of a beggar, taker, thief, or dependent on a food pantry or soup kitchen, all free, and if they have a crime record for theft, they can be placed behind bars to attain free housing and free food. Many prefer this to the outside, where responsibility awaits them like a cold dish; they prefer to ignore it or consider it not something desired. So, to be direct-, most of these people either have a reason not to be involved with work or cannot function normally as a productive adult with a purpose. To be or have personal goals. So that said- let's start with the mentally ill; these are people who have today zero help with a condition that can be considered, in many cases, a fine line to be either harmful to others or to themselves. So, what can we do to change this- Okay, we all know that helping people is never free, and the real issue is many mental institutions have closed. The local state governments have just thrown their hands up in the air and let the people be a problem they refuse to fix. Here is how the people can be funded and how the problem can be gone, but before I get to that-let's look at some of the facts- Fact 1. Well, first, we don't ignore them; we need to have our medical community be informed to make the mentally ill or handy cap individual be rated as coherent as to what level, and second, the physical level rated, can they do minor jobs? We make work placement programs that they work with others who they are close to in these level scores, and these scores are determined by tests made that are of an eight-year-old level. This is because they must have simple tasks and positive-minded individuals who can be trained to take care of these people. The training schooling should be a two-year course on communication skills, handling a threat, and things that will help these people enjoy what they do. The important factor here is to give these people purpose and a sense of pride in what they achieve. For some, just being able to hammer a plastic part together with another part can be a fulfilling action and reward to that person's self-worth. So, even the Drug addict could be in a program to be watched and taken from the environment of being able to get drugs. And we need to stop giving food away free not that they don't eat, but the action of free only supports a non-productive mind set, instead they can attain food, and a shower daily, and even wash cloths if they attend a work program. These work programs could be a simple task of picking up trash with a retractable handle that grabs paper, or a paper stabber with a metal tip that can grab loose paper on the ground. If they have more cognitive skills, maybe over painting graffiti, or using a weed whacker to clean sidewalk areas. The important thing is they are now being somewhat productive. And lastly, they need a safe place to sleep. So an old warehouse can be set up to house these people only at dusk, and the person has a locker at the end of the bed to place all clothes they wear inside, to include shoes. And they sleep on a fold-able bed, which can be washed weekly by the staff. And how many people are there, we have three cooks, one security guard who also points to entry areas if someone is new they are sent to orientation to be evaluated and assigned a bed which will be there bed for as long as they are homeless. Each bed is inside a framed enclosed room that is only big enough for one person. So we need our cities to open the doors to homeless, but have people there to give them help to get back on their feet, and training classes can be on the job training which companies need to present in the area as potential work for hire. The homeless hired need to stay at the safe home, until they get there pay check, and the money is placed in a two signature account, this is so a person who has say a drug problem or drinking problem does not have access until they have been fully designated by staff as to be recovered by certain tests that are given to see the result, if they fail, they stay at the safe home until they can be able to leave and maintain their life without guidance. Some people believe that ignoring this will go away, and how does this end in many cases? "Death" is the main element allowed. Here we are with a major problem: a person who has become dependent on a drug has a hurdle far higher than the average person homeless; they are hooked, and they are weak humans. This is a leap of success to return to being a productive and content individual, which over 90 plus percent never achieve for obvious reasons- The drugs are available. This is a major contributor to people losing all they have. Law enforcement has failed to stop this because so many poor use the sale or the distribution to gain money for their lives. So, how do we stop drugs? Is the real answer needed? They have used border patrols, dogs, and surveillance cameras; everything to date, and for over 50 plus years, has "failed." And for any political party, and I am speaking of both, who say they had the best record or did more to stop this is an outright lie. Drugs today are everywhere, and they harm our lives in many ways, as we see our youth die daily. Our families are broken, and the effects have been a destruction of the nation, and that is exactly the reason it has been made and the people who desire to make counties fail so that they can remove their people in place and in power away from the serving table of being rulers of their country. So, the question is: How do we stop this? The answer: In all things made, they all have one thing in common- They come from a beginning a source; just as a tree has roots, the only way to drain this dependence is to stop the flow of it from the source. The guy on the street who sells the drug is not the source; this is how the problem was addressed. Even to date, they have placed hundreds of thousands of people in jail for selling. Instead, what they should understand is that you could go off to jail if you sell drugs, but the jail is not with other criminals who murder or steal. No, no, a drug dealer has no parole, ever! This means a drug dealer is a murderer and has directly or indirectly caused people to die. And we have even so-called- Legal Drugs being submitted to our people from doctors, who are now drug dealers, so these are now to be a penalty of no parole if the doctor has written a drug that can be proven was a killer to his patient, yes doctors have become above the law, and can kill without any harm coming back to their lives, this must be stopped. Okay, we now see that a source is the way, but sources can be assembled directly under our noses. But there is still one thing all chemical drugs require- the mixture of the elements to make a drug is always required, so now- We do the next step- we monitor the distribution of ingredients. The warehouses and factories that produce every pill or every mix ingredient required now have a security system so intense that it squeaks as every person who handles these elements is watched and checked every day in every possible way. Now we have closed the door to how to make it- and we place a list of drugs which are harmful online, and this website is not a 'Gee,

I couldn't find it- No?, it could be called- bad-drugs.com The drugs which have effects are all listed by name and by a code number that is all exact as numerical and alphabetical, and will give you details that are what can happen if used. This also gives the number of people killed and harmed by the drug, the effects to look for, and any side effects. These are things that can be found online, but it is not easy to see all results instantly, and- last, the way to treat someone who has ingested a type of drug before an ambulance comes. 3rd Teach children about drugs is not the answer as to what is out there; the answer is that a home parent or guardian should be aware of what drugs to look for and also be told to tell their children with a posting that explains why drugs are bad, and how others will pressure them to do these things. Also, drug testing should be a normal activity in every school, at every level, and at every job, there is no better way to stop a first time user, or light user who has experimented but is not hooked yet, and they are not to be fired from their work place, but sent to a three weeks of a boot camp style of work detail, the object is to give the people a time they will never forget, one that says I never want to do this again. They are asked, (TOLD) to perform some really dirty jobs, they are instructed by a bus operator and a nice man to get on board the bus very orderly , did I say bus driver and nice men , I meant to say former Drill Sargent's! and they will be instructed upon arrival to be lifting rocks into wheel barrels to be moved to creek edges to protect erosion, or painting graffiti off walls or train cars, or washing cars for the local school to raise money for the school, or washing school buses, the tasks are numerous, but all are what seems like by themselves, they are not allowed to talk but only can raise their hand to be able to go to a rest room. I would say that three weeks of this mess, and pretty much the person is clean for years. Now, many of you who prefer to be in la-la land as not to agree that drugs are bad for people who are most likely to be already doing drugs and feel they "don't need" these elements, We already know your brain is on vacation. We know that you prefer walking in a dream because your weak minded and can't handle the real world, (Oh that's Harsh!) but today we allow drugs that have been proven to cause harm, legal drugs like alcohol. And many States in America allow marijuana, and so the laws changed because the government gave up trying to stop it to see a way to profit just as drug dealers do. So they are now promoting items that are harmful. And you say marijuana is okay, well the proof is for men that the use can make you sterile over time. Gee there goes the sex life. Many people who use marijuana say- Oh, this is not harmful; it is not addictive; sorry, the facts are it has been known to allow functions such as reflex to be slower and eye-hand reflex diminishes drastically, so a person driving a car or truck under the THC drug, is now a potential time bomb for a car wreck. The way to stop the users of drugs is to stop the ability to have access, and make them aware of their problem, and help them have a support line and a place in a medical community that is available 24/7; this place is for people who are fighting with drugs use, and are wanting to stop, but need support to help. And now they get free assistance, which will bring me to a very important fact, funding these issues, all of these require some act of financial burden, so this is my idea to help equally divide funds established by the following means. A telethon per city is established yearly just before New Year, and the program is to raise awareness and show what the funds have done. And I will be showing what these funds will be used for next. This presentation is with any volunteer entertainer, music, comedy, magic, and specialty acts, but all are to draw people to watch; the promotion starts two weeks ahead with free ads promotions from the local TV network which will sponsor the program, the telethon goes through the phone book of all business owners to be called by teams of volunteers to attain donations. For example, how New York would have the telethon, Buffalo, Rochester, Olean, Jamestown, Albany, New York City, Syracuse, Binghamton, Oswego, Rome, and Schenectady to be more direct all major cities with over 60,000 population could hold events, and all use local talent who seek to be seen, at the main stage areas for shows, and local TV to do broadcasting to the surrounding area. The sum is collected from all, then divided equally to each area back to its district and for the entire county area of the outreach area from the main city. This is done by all sections inside a state to allow the distribution of funds made now pool the funds together state wide, and are distributed to divided districts that cover a wide area, and focus on the most troubled areas from prior year, and this is for each district that held the telethon a equal share of profits, and these are divided in the area as in divided counties which have the hardest hit areas of concern but tally of people per area who had been prior year reported, again the higher recorded number of problem areas needing more attention get the funds slightly more than other areas but only 10% more, and if an area shows a decline , then any funds not used must be held on to for the new year coming event of fund raising, and this sum also goes back into the pool to be divided per state, and the funds are used as follows: Hospitals are to be with a hired individuals who are well versed in how to meet a user, and get the information as fast and quickly as possible, then the user is notified that they will be given an opportunity to sit with other users, and talk openly, and an instructor is present. Drug users who are in recovery are given work to keep them from thinking about drugs, and we have these types of recovery and this should continue, through questions who know what they are doing as trained. And last- As I stated before, if we want the drug use to stop, we must go to those places by all methods to destroy the elements that make the drugs, and once this is done, the drug flow stops or becomes a trickle of what is, and so the cost increases because the lack of availability, and now the users say – no way to expensive, and many stop using. The answer is to attack the makers, then the drugs stop. Now this brings me to homeless and there life of drugs and failure. The fact is no one wants to hire a person who has not had clean clothes or a bath in months. So we need to build homeless housing: The funds are from homeless people who are now given the proper clothing, haircuts that look proper, and a way to stay clean all is offered and they are shown the food being offered also and a clean room to stay in and they share the showers and the laundry. This all happens with a startup fund which is a toll booth which is made just for the purpose of funding the project called "STAND" and with the funds, staff are hired that must have the ability to train people and have the experience of the trade and will bring these people into startups that will give them actual income if they work. And the fact is the free food from all sources are not allowed and people will be directed to the "STAND" program. This means hand outs are not the way of life any further. The funds will be to purchase properties which have the living space and the state will decline as all property taxes be withheld for a period of five years, and the structure will be made as follows: But also know that I have other ideas which are also expressed later. But each person receives a room with a bed, and a shower and toilet, they are to bath daily, and on site is a trained homeless barber shop, and a trained laundry with a sewing training for clothing and shoe repair at super low costs, and a trained cooking for those who help in the kitchen, and there are two meals served daily only to those who are in the program "STAND" and they are 10 AM , and 6 PM, the food served is soups, sandwiches, and fruit, nothing super elaborate , the mission is to put these people back into life and with new training and new purpose, there will be a new outside "cleaning service" Those in that service are maintained with clean haircuts from the inside barber training class, and they will wear uniforms as they will do floor cleaning for business's at super low cost , and work in security positions also developed in training, and even an inside shop to sell art, crafts, and books made by the homeless, as the books actually tell their unusual stories, and are a collective of "real events". The facility will be open on weekends, and the flea market outside shops will be coordinated. One of the startups is anyone who had former sales training can now work in the Discount Pawn Shop, and these items are the lowest cost items in the world, because they are only items that work, as donated, so people will love buying a lawn mower for $10.to support homeless. Only top items will be introduced from the local donations, as there is a clothing dump station that has a 24-hour 'check for items; any electronics and house items are all checked for usable and clothing, and these items can be rejected is why we have the two-person check even at night so items are with a standard that must pass before it can be used as cleaned up and resold. The clothing is donated but is distributed according to size, and the clothing must be in quality condition. There is a storage unit that only allows items that are essential the day the person is no longer in the program, and that is the real final goal, which is to have someone trained. They are given resources such as a home and their business, as we also match up homeless people to live together in business applications so that even groups of people can share a house on the work they have done. So, storage is located in a warehouse setting with the items as per person. Still, items are to be reviewed as a NEED for the day, need for the future, and personal items such as pictures of family awards they had over the years; these will be regarded as important storage, but items broken or worn out or old shoes not worth wearing these items will go to trash, But again- they have their own living space which is the goal of the program is- 1. Give the Training to attain jobs with reported need factor. 2nd: To make the cash flow adequate for the person and have a direct deposit made for those so all bills are deducted before bank access of cash left over. So, a deposit is made to bills first, and then the balance is sent to their account for debit card use for any purchases they can afford. Those who work on-site as help receive a cash card, which is the sum of money collected to be used like a private savings account and allows the person credits toward food, clothing, electricity, housing, and all required needs. The person will be in a two-year training with the mission to attain the job training and in- house training, such as janitor, house painter, carpenter, pipe fitter, seamstress, cook, electrician, roofer, tree repairer, factory worker, fashion designer/artist. Cad designer, hair stylist. There will be a" car cleaning service," and they are once again trained by those who are teachers of teachers will again train them on. These are just a few on- site teachers who will be training classes. This program is coordinated with a six-month report given by upcoming job markets' required personnel in the area and other parts of the State. The fact is people trained can be relocated, and the program even sets them up with locations, as the future housing market is also invested in for now, more modern apartments with soundproof walls, floors, and ceilings, and with top quality views, and these designed housing projects are for those people who are with a job only who dedicate their efforts to maintain the housing as they are checked every three months for damage, and they have the lowest living costs, and they even have free electricity, water, and sewage. The goal is to give a person a job that they can grow, and we encourage entrepreneurship as they grow and save money to purchase their own home. So, the program can take a person from the street, off drugs, stabilize their life with the ability to be clean, be fed, be trained, and move back into the mainstream job market with confidence, and the project staff will always be available, help will always reachable, there is no such thing as we don't have a bed for you, or sorry your turned away. They will have travel access to be taken to a 24/7 hospital nurse or doctor ER to be available, and let's not forget, these projects are to raise money and train people, so the worker is the person who can do what they are physically capable of that can earn them a cash savings to be able to attain a regular job that they are now fully trained. The Stand program member is someone who can take any concerns and direct them to the proper person, who will contact them or be found to take the call ASAP. Of the people who work for the program, over 90% are trained by professionals who oversee their progress as they monitor all calls. The project "STAND" is needed, and I know how hard it is to stand up when so many ignore your stance. The time to make this change is with you also, *you can help this dream become real, so do whatever is in your heart, you know how to contact me.* And as far as I have seen, even in my youth with trips to visit my family in New York City, there was homeless on the streets. Clear back to Ronald Reagan being president, so yeah, we all need to stop ignoring these people, and help them come back into life. The saying is true, "You can feed a man a fish, and he will eat, but teach a man to fish and he will eat for the rest of his life". I say let's remove the homeless by teaching them life can be better, and giving them the ability to move forward. Now to anyone who agrees, please contact me if you wish to help.

94. SUBJECT: THE NATIONAL DEBT of USA

There has been a debt trend within many nations. As we all know, the national debt has bloomed into a nightmare within the United States, and this "I Owe You" is on the brink of our nation being called out-to pay up! But, we are under the false vision that our military will be the difference between a takeover and security of our nation with our people in control of this country we live in. We are also aware that some powers have made this their agenda to decrease the military strength of many nations so that they can now move their pawns into checkmates. Many who believe this is impossible should know that hundreds of thousands of businesses and farm land has already been purchased. Now, housing is being bought in massive areas controlled by the people who have declared world dominance as their mission. So why does this continue, you ask? Simple math: the seller has this mindset: I have property for sale, I am hurting for capital, and here it is, SOLD! The owner has no regard for the future, no regard for the ownership, and the owner's intent. This mistake has been the way people in many countries make it. The increase in goods is now the added drain, which they now control because the plan to control the flow of goods was taken away from many countries. So, export is the key to the success of our foe, and they know we are now helpless to fight back and even buy major businesses once owned by the home nation. How do we stop this? We begin by first supporting our wonderful, talented inventors with a gift called FREE PATENT for persons with a low income but the item is potentially an invention which could provide jobs, so it is evaluated in the rough, and if accepted the person can attain a patent, but when the item sells in any manner that that person has attained back income, the patent fee is now paid in full. So a Low- Cost global patent protection is offered also when the item has been approved, so the patent is drafted by professionals help from our government and the inventor will pay for art and the drafts and even review the proposal draft as soon as the product is in motion making funds. Why? As inventors like myself, I have the ability to invent but have not been born with a silver spoon and possibly have a great concept, people like me need help to establish the protection before the concept introduction for prototyping the product or the project. There is a need for new ideas that can do NEW JOBS, and these products' major reason for being made is this next important factor, which would be required to attain a free patent- The product must have the ability to be "EXPORTED." So we now have another dilemma or hurdle: we need to develop a program of factory investment for people who have ideas that do not have the factories with the proper materials or function, which is needed to be "established!" Now, we are doing three things. We are making our products in our country, doing jobs supporting our people, and bringing funds into the United States to decrease our debt and one day remove it. Now, I would like to share one more application, a much faster and bigger project. Still, that idea is confidential because it is an act of national security that could not be done if the path of making the program was discovered it would be used by other nations, so for now, it will be my tiny secret, which our Government needs to contact me to see how we can remove the national debt.

95. SUBJECT: DISEASE & CONTROL

The problem with people today is the close contact, and the massive areas of city dwellers worldwide have made the ability to be removed from contact with others almost impossible. But I have good news that should bring a smile to many: I have invented an item that will be the best deterrent to being sick. I am not saying that people can't become ill, but the chances that were given by placing a cloth mask over your face also means that when wearing this device, you're now breathing in your CO2 exhaled, which is not a good thing to be doing in a daily basis. But we were forced by our leaders and store owners to comply with the demands or not be given service or entry. This is a controlling factor, which is something that should never happened, but can only happen when people become like sheep, and will bend to every demand , which means they have lost the back bone to say no because it may disrupt their so wonderful life of not thinking for themselves. Buy into the latest propaganda from so-called experts who are only really promoting an ill they made. Yes, COVID was found to be created with funding from the U.S. government, sent to China, developed in China, and released from China, and suddenly I read online that this was all false? And if anyone noticed, which I would say they did not, the so-called infectious disease showed up in the state of Washington, and there had been zero flights in or out from China to that area, HOW could that of happened? The entry points seemed very calculated: L.A, N.Y. and Dallas, all had infections show up the same day, Hmm? - The results just faded; why did they fade? Because finally, people said- Enough, but those in control discovered- They in power can make you be what they want, just by using fear. No? You didn't stand in circles? demanding you to do it? You didn't walk around outside with a mask on? I saw people wearing masks inside their cars! So now, with this said, I know for a fact that this will become a new problem again and again; why would they do this? They want to decrease the population, the millions that died, did it in the elderly locations, such as nursing homes, and elderly residential areas, more than any other area, this saved the governments billions of health care. And they now know you will do what they want, so the shots continue, and yes, and people are still dying, From January 2024, this year to February a total of 2,455 deaths in USA recorded as covid 19 related by New York Times. But you didn't die, so who cares? Finally we may actually see the people responsible who made the entire process, and was it designed for the fact that population growth? That subject has been a massive concern of many nations? It is possible it was known to be a reduction of population? But this is one person who feels that the facts are just starting to rise, and many heads may be at the future chopping block, but should we ignore the loss of our loved ones? Should people just crawl off into a corner and say, oh well, they are in a better place? Or do those who lost good people say- I am not satisfied, I need to know, and I need closing. I imagine there are hundreds who want answers, and hundreds who could use a good detective and a good lawyer. So my hope is for all of you who suffered the death of someone you loved, be mindful that if you do find the people responsible join others to uncover this mess to be revealed even if it takes years. And may it help to remove your pain. And you may never know all who help men like Anthony Fauci, who has denied his connection to the lab in China, and the claims he had no connection to a lab virus being made, but the facts keep saying something else, but the end is never the end until we reach it. And estimated death report from 2022, that continues today

5.4 million Deaths from Coved 19, according to World Health Organization's report in May 2022. Hoping all of you who suffered, that you gain the closure you deserve. And God watch over all of you who suffered, and those gone, may they never be forgotten.

96. SUBJECT: SCHOOL & PUBLIC PROTECTION

The use of police to take down people entering our schools with the intent to use a gun should show that guns are not the answer to stop the action already in motion. Still, place safety for our children first to prevent the entry of a person with any form of device, being a gun, drug, or knife, possibly even a bomb, and the use of a gun should be a last resort for someone who has gotten into the school and has now harmed others, but the use of guns can be a thing of the past, and the people who have lost children or family loved ones would have never happened if these safety measures had been invested in, and where reported to school officials, If you desire your children to be safe, then you should help me by telling them to buy my book so that I can continue my efforts to make their lives better. And with my thanks, but be sure to see the illustration of how it works. The entrance is a turn style that can be locked if the metal is found, and the silent alarm reports that whoever just walked past the metal detector has metal devices or a metal belt. All the students are told never to wear metal belts, and all cell phones and any metal devices are to be placed in a travel conveyor tray before entering the turnstile.

Concerned parents or people in general need to contact their congressman about these ideas and make a real effort to bring the children the proper safety; protests should be applied if they do nothing. #1. Schools: We all know that schools today have entrances that children are directed to use, and these same children enter the schools with guns blazing; we have seen that the use of violence has increased in our daily LIVES, BUT hey, we need more violet games for our kids to learn from right? But that is another subject. But the path that it takes for a child to enter the school is easy when the doors are glass, and they can easily be shot to get into the school. Many windows are also this way, a low enough entry point, so the individual has no real restriction unless a police officer is present to shoot the person or be shot in the process of an attempt on the children's lives. So, my idea is to start with Windows. I need woven screens in all door entrances to make the entry of a shooter impossible just, even to shoot and cannot enter.

These screens needed are more than available and the cost is somewhat more, but are our children worth the extra safety? Well Yeah! But that is for doors, such as fire escapes and fire exits. The next application is the entryway; the main entrance, wherever this is designated, is a single lane-enter and a single-lane exit. These are added on to the school as an addition for the safety of children, and the ability to close this path with the slightest silent alarm is given to the two person security who run the system. The only real way to detect a gun is with metal detection, and a system like a government office to have someone place their keys and metal items in a tray before entering a metal detection can be done first, but my system takes the use a bit further, because the students travel through a turn style after they have placed their items in the tray, and this turn style is the metal detection, as they go through it the detection can automatically close this turn style to lock it, and it does lock only with an exit door on the side of the turn style that is actually inside the turn style, so if they are caught with a metal item in their possession, they are now asked to go through that door, or the police will be called. If they comply, they are now in a bombproof, bulletproof room, which can be seen with high-placed cameras and speakers; the student is then requested by someone of the same sex to disrobe and place all items in another tray that opens into the room if they are holding a somewhat large item. The bomb squad is called in to check the contents. And to arrest the person if they have a deadly item. If this person refuses to comply inside the holding area, a sleeping gas compound used by dentists is released into the room. The student or whomever has made a bad action to try and harm people, is then taken off to jail in a van with the use of the arms in legs strapped into a wheelchair. The concept of keeping our children safe does not require police to carry guns around our children, but this would prevent the entry of any person who has bad intent with a gun or knife. A more protective concept, as refined, could be applied to all public places with zero interference as long as the person is not carrying anything harmful. However, an added X-ray could be considered to see hidden items on occasion. I hate to report this but I have spoke to school teachers over the years, and suggested this as a means of protecting our children, they actually refused to see the good in this and said it would be too costly. Now that may be so, because if all children stay home with my school home program, it would save the taxpayers billions nationwide. But for now, this could work if applied.

97. SUBJECT: SCHOOLS AND COLLEGES IN FULL VIEW

In the real world, the fact has been recorded that communist application to rule over their people is to first- "control the thoughts of children and indoctrinate them into a mindset they approve. The second is to control media to bend facts against their actions and to change the facts that may stop their progress and replace truth with lies and turn the lies into the truth they wish to present. Today- your children are under their control as the teachers of these socialist liberal views good teaching of math, history, science has been changed to allow students to be even in graduation without basic skills, but are given ideals of socialism and homosexual approval, and are being told that they should not be white if they are, or that they could be another sex, these views is their agenda against our nation and its form of governing made by our forefathers and the written constitution, which has also been dismantled slowly, piece by piece. The people whom the government has designated are telling our children some very, very disturbing directions, and they have changed even the fabric of history, the teachings of Columbus as a terrible man who was cruel and hateful. The same man was told to me in my youth to be respected as the founder of the new world. The teachings of proper sex and what children should abide by have all been twisted and given a clean slate as anything GOES! The schools are OFF LIMITS-you could be a killer! The actions are all devised by those who want to separate you from your child. That is exactly what their results make; separated family is the result and the teachings of homosexual acceptance which should not be considered as educated in schools because it is a choice not a subject of learning like math or English, or science , and values of socialist formations of rule, these have invaded our schools through improper teachers who are allowed to project these ill thoughts against the foundation of our nation as Godly, as family, as heterosexual being able to procreate, but these values are not taught in our schools and colleges that now have socialist agenda being introduced. Proper activities and belief structure are not being presented, but ideals which go against religion and God as our foundation are being removed. The teachings need to be watched! So they don't want us inside, and they use attacks of gun use to claim the reason you are kept from your child? Then give parents a scanned card that allows them access like going into a bank ATM, okay, so they want to keep you out? Then you need to have your child's classroom monitored from outside and online. Just recently I have heard a rumor that some teacher claimed Abraham Lincoln was a homosexual! He was a loving faithful married man, and letters to his wife have been placed in museums that present this fact. And now the new teaching is Christopher Columbus was a murderer for discovering America. These lies infect our future children – What we need is the parent can see and hear every detail the child is being presented, and can record the entire day, and use it for legal action against the teacher and the school leadership which permits teaching a controversial subject which may not be approved by the parent, as the information recorded and presented to the child as determined as not true and has no scientific fact to be taught, or not to be considered true, will be grounds to sue someone for teaching a belief that is not a value of the parents who are raising the child, this my dear people is not only needed. It is your duty, just as my mother came to my aid in my youth, you must be in control of what your child absorbs! We live today with technology that can even make our children safer, so why not have devices that can be worn, like the police wear body cameras, that now can show exactly how our children are being treated and talked to? I want to call this "A full VIEW," and let's give them proper safety and guide them in the right direction.

Now hear this! Now hear this! Another example of safety for our youth's human condition Sexual preference am I really a boy, or am I really a girl is being taught in our schools (WHA-??), and it shouldn't be followed but is being advertised as normal. Hmm, normal? Once again the sexual preference is no business of the school and a minor should not be able to be told that they are not what they are as born, that is the deciding factor of the parents only, and if someone chooses to use a doctor's ability to change a sex, they should be an age of consent. And it sounds like a joint legal battle to sue the people who project these images to our children who are deeply affected by these events, and some have even committed suicide or killed our students and teachers s acts of hate. So , the people who support these actions to change our children's sexual views as minors, be warned you are not the parents, and you have no right of instruction but the basics, as math, science, and history and language, but the idea of the birds and bee's should be removed asap from all schools as your ability to instruct, and the actions of performing sexual education should be once again the responsibility of the parents, since your actions have failed to motivate children away from sexual acts, but instead promote the action, and teach how it is done far too soon and the outcome has been calculated to be a disaster, plus to confuse small children who are not able to see your design is to change their behavior and also disrupt the child emotionally and beware , your walking on egg shells. And I see a tidal-wave coming your way which many will be seeking a job at some fast food window. I would say parents who have had their children indoctrinated into a way of life that does not follow the parent, and is an act of attack on the authority of the parent as it is there child, and law says a child is not able to make decisions until they are of legal age, the school is presenting these issues to minors. The school and the teachers are responsible, and legal action should be made as a unified legal law suit as all parties involved will sue against the school that your child attends who supported these actions. Good hunting parent$.

98. SUBJECT: ENTER & EXIT

To enter a public building a person is only able to do so through a turn styles made of hard tempered clear plastic, which forces each person to go in turn and works with a metal detector that scans the person before they enter, if someone is found with a metal on them, they are locked down inside the enclosed areas with doors that close from two sides to close fast and seal an air tight rubber seal that is needed if the person does not comply with the security operators request which verbal commands are given by a speaker located in the roof, and the turn style door with the person found with metal object is in lock down they are to enter a side door which a metal door shuts automatically just past the turn style swinging door, and that door blocks entrance into the school but is now a side door opening that the person is told to enter when the door release is unlocked, but the metal door remains closed once they are inside the open door, a metal sliding door now closes when the person enters the small room which has a small built into the wall single stool, the outside metal door can now be closed as to allow others' enter without disruption, but if another person is now with a detection, the turn style will set off a silent alarm to indicate another breach in security' and the process is to now have a door open further down the hall, once again in the same manner and in the event someone shows a gun or a bomb, the operator now activates an order less gas into the turn style, or the side room opened door which now puts that person down, and allows the operator to move in unharmed with a wheel chair with lock down straps and that person is wheeled to the waiting room for police to arrive, the open door has another door which is access to the room, there are three monitors front , back and side to watch the persons actions, if they have been requested into the open side door it has no windows so others passing by cannot see the student in question. The student is now told to relax, and this is only a delay, which we will soon allow you to go their way if they do as we request.

The student or person is asked to empty all their pockets, place all items on the table and now stand in front of a full-body scanner. If they still have a metal detection, then they are asked to place both of their hands inside the area holes and hold on to the two bars inside the holes; this action closes the holes with hard rubber, similar to having blood pressure taken in year past, the difference is the hands are in rubber with a metal housing so the person is now handcuffed into the wall. This means the operator can now enter the room and body search the person in question. Maybe the person was wearing a metal belt or a ring was forgotten, but just having a bad day or bad attitude, then they are released; if they are found to have an item illegally concealed, they are given a chair to sit on and arm locking brace is applied with a chain wrap around so they cannot raise their arms-the circle hand enclosure is opened.

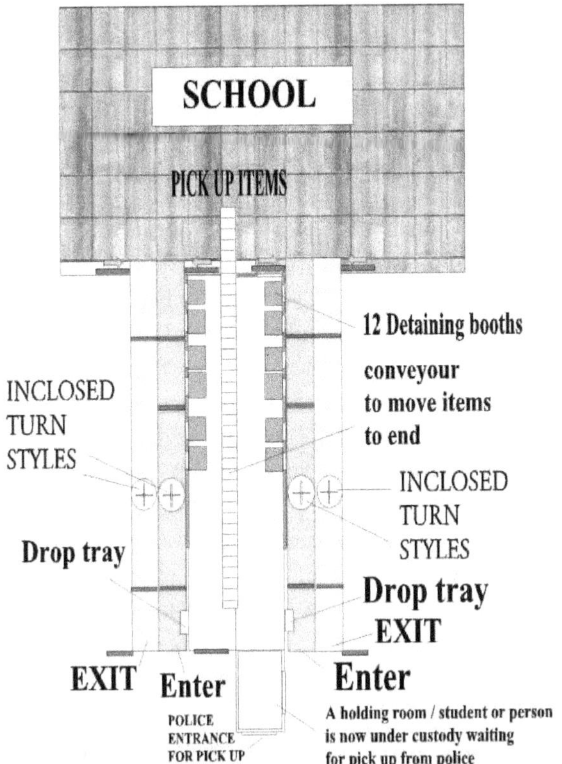

All studens are given ID badges that they wear to get in school , and the ID badge
can not be shared, it is a ID that is with a computer scan that says
this is the person , and is inserted into the turn style to pass

This sytem can be used for Government offices,
and police stations, and any public building

EXAMPLE OF HOW SCHOOL ENTRANCE IS LAID OUT:

Two people escort them to the police waiting room. They are locked down into
a wheelchair until and when police arrive, use of a large van allows the person to be
wheeled into the back of the van with a drop-down ramp. The wheelchair locks to the
side walls of the van for transport.

Only persons who refuse to comply with the operator of the side room will be
gassed and put to sleep and will be investigated before police are called if no harmful
items are found after being sedated. Once the student is asleep, two operators must
be present to search the student, and all activity is timed. Video is recorded as visual
and sound from the moment a detection is found to cause a lock-down. This system
allows control without guns being used, and a person with a gun will be located by
detection first and then locked into a very thick plastic turn-style room. The second
they show a firearm as the turn style has two cameras front and back of the person,
the gas release comes down from the ceiling in a quiet way that makes no advanced
notice to the person. A wheelchair can enter the hall and retrieve that person out of

the turn style. This means safe methods of entering schools and any public access building. This system is also for our airline passenger jets, but with a slight difference: the pilots are inaccessible.

99. SUBJECT: PROJECT STAND

This was mentioned earlier but to be more exact the program is with some added features that will help the public as a whole, and I hope I'm not repeating here to much, but I am one who feels the greater need to bring the people we once called Americans be welcomed back into main stream, and help so many women needing their children in quality day care. First we need to understand that if a person has no ability to pay for the rent or the electric we don't need to toss that person to the street, we need to give that person help not hurt, which means they need a job that can cover the bills. So "STAND" will have a group of people who can do just that help people before they reach the end of the rope. So we place a mental exam to this person to find their strengths, and weakness, and now we retrain the person and a six month training is done, with the bills covered, but they are to have their paycheck controlled as to pay the bills for the person and deduct the sum given with a 0.2% increase to be given back to the program to help others. Loans can also be given for people who have work, but once again are signed up for the new job placement to bring them better pay. Next is the women who needs help with children, these ladies God bless you all, are our treasured hard working women who need our support, but this program is not only for the women who have been abused and there is physical damage recorded, and for women who been divorced, but have a real need of child care at night or day who are not getting help from the father. We need to develop funding to set these wheels of change into positive motion. So funding drives could be just one small progress. This funding support is from yearly sales of Girl Scout cookies, with a single cookie brand that supports both girl scouts and women needing day care. The boy scouts are to hold a drive also which is a Ride-A-thon bike ride race and they are permitted to sell tickets as a 50/50 which the winner gets a brand new bike free, and the sum collected 50 percent goes to the women's day care help, and 50 percent goes to support YMCA and Scout Camp food and lodging summer camp of two weeks in each county. This next subject is also in support of "Stand".

100. SUBJECT: INSPECTION RATE, and DAMAGE ASSESSMENT

Housing cost control- The facts is housing costs are at an all-time high and have caused great problem for many who we just spoke about, but the answer that is something that many will be very unhappy with me, but I am used to having people doubt or criticize , but the real way to have this problem fix fast is to make a regulated inspection from the housing authorities as to the "RENT VALUE", this value is not made by people as to a desired, no, they are to be at the rental property after the key has been taken to the housing inspector, and the property is evaluated as only structure intact and conditions, such as smell, carpets in bad shape, floors as weak areas, but everything on the chart they carry has an area that must be filled in, and rated from 1 to 10, but more important is the person is wearing a hard hat with a camera attached to give a real time projection as being recorded and so there is zero doubt that the content is real, the inspector has to I.D. him or herself at every new inspection with time announced, and date of viewing. Then, an assigned three-person team is back at the main office, which does every review. Monday through Saturday, they review the recordings, and with the inspector rate chart signed as completed, the team looking at the house does not know the rate of the house. Still, now they rate the house room by room, the yard, the roof, but no one can interfere with the person rating the house, that is a strict no, at this point because they must be 100% non-biased. The total score is added up, and now that score within a rate score of 1 to 100 from each category question and answer now gives the cost the house rating for it's actual rent worth and can be either rented higher, but can also be rented lower, which is suggested for people to afford the house, and if they agree to lower a tenant who has already been screened as a good tenant is now recommended. So, you see, the housing market is stabilized to be more affordable, and landlords get only people who will be responsible, which another issue is, that it seems that land lords are treated with no respect to tenants' behavior these days. The tenant can do damage and not have any problem dumping the cost to repair back on the owner. With these inspections recorded before moving in, a record of the condition of the house is now on file, and these files are updated every three years, so a landlord will be able to show the damage in real-time, and the courts will see the evidence of the before and after. The verdict will stand for the landlord if the videos show damage and the party is found at fault as for the destruction done with them as the renter will be traced to any earnings will be removed slightly to be paid in full as to bring the property back only to its original condition and the costs of all transaction also recorded, as the landlord is reimbursed over time of the damages, and the courts assign a temporary payment bond to be paid to the land owner for the repairs only, but this sum is determined by the total materials and labor calculated to fix the house and must be from an established approved by the housing department builder. Then, that sum is the amount the owner will pay back to the party who was found to be the destructive person; these can be for animals doing damage, water damage, or anything not in the original video.

101. SUBJECT: FOOD PRICE FREEZE

The need for this is what many say, supply and demand that we require higher costs to maintain the extra costs placed over the farmer and the development. The new design I have submitted here in this book will not only reduce the demand but bring the costs of the items sailing downward, and the fact is the farmers can say that this is bad for them, but the farmer is not responsible for gas increases. Materials like fertilizer costs are soaring through the roof; these are outside direct actions that could be maintained, yet are not because someone is price gouging. Goods to farmers reduce costs, and jobs are made.

102. SUBJECT: SPACE TRAVEL TO OTHER PLANETS

The use of "Echo Life, artificial womb," developed in Germany on Feb 14th, 2020. In the favor of science and advancement of human life, and I strongly believe that for humanity to advance

Throughout the galaxy, we need to make devices that allow man to travel without the hindrances of time and aging, but allow the seeding of humanity to prospective planets that are with the proper attributes and conditions for life of our bodies to be able to be maintained. The need for artificial intelligence to be present in travel is essential, as the pilots of these seed crafts are robotic and are programmed to seek out the planets and to build from the information they are given to locate, investigate, and then deploy the seed crafts if the planet has proper attributes.

More on this subject is in the formation of the "TIME- ARK" projects.

103. SUBJECT: SHORT AND TART!

I have an idea: let's make a massive project to duplicate the heat, of the SUN, and let's even crash molecules; hey! it will be better than the fourth of July! Yeah! Crashing atoms in a 17-mile circle! Wo-HOO! The most wonderful experiment of our time. But wait- can we turn the sun into our planet? Gee, I don't know, but we have discovered some strange stuff we never knew- oh is this smart or what-So how, Oh wait a minute - NEWS FLASH! We hate to interrupt this program, but our news just came of an overload of a device in the center of our planet. The elements are expanding, and they are heating the planet's core, so the temperature has now started to form the same heat as our sun- Oh, we just got another message-from the maker of this planet "What are you doing???" But I hate to report that I'm melting, and my microphone is melting. Ahhh!! It almost sounds like weird science- black and white old Godzilla movies, but this is real; this is happening. These people who are playing God- hear me, loud and clear- this funded project needs to end before it ends all of US! People with these degrees are not even wise; they're fools, idiots who can do something without knowing the results. But I hear there have been discoveries! Oh my, but you're playing with a nuclear bomb-come on, get real! Get a brain, get a life! Albert Einstein admitted he was sorry he invented the A-bomb after he realized what he had done to kill millions. So, put down the discovery of the atom smasher, and take your pills, because obviously, you need some form of anti-depressant because you are no doubt not aware of what you are playing with, so are you as smart as you believe- Let's see - If I had a wet park bench and it said- wet paint, would I sit on it? The answer is no if I can read and understand the words. But you already have made a park bench, which you have painted, which says we are attempting to duplicate the temperature of the SUN! Do you think that park bench would be hot? But your ability to grasp that smashing atoms has another park bench! And this sign says: "Similar event as to an H-bomb if messed up! " So, hey, why not sit down on it? It 's not so smart to build nuclear bombs that can blow you up, but hey, why not build a gadget that can test your theory of the big bang? Oh, what was the big bang? Oh, I know, a massive explosion that made the universe! Now, is it so smart to try and duplicate that? You "might" have brains that can blow up the universe! As a professional inventor for now over 40 years, I like to believe I am someone with an open mind, but considering both sides of this equation, or the processes once again- my understanding that the subject matter of the recreation of the compression or the collisions of the elements was to establish knowledge. This so-called knowledge attained by this atom smasher device has been with your team admitting that it had been found that the discovery was found by "MISTAKE"- and the mistake which you claim to take full credit for which is- a wow moment, huh?,-, because the true definition of the word "discovery" is something that the person or persons were not aware of this thing, such as Christopher Columbus discovered America, which was already here with people living in the land. Still, they were unaware it was there until it was discovered. So, your device is making a discovery that has been reported to the media; what happens when you discover that the device has gone bonkers and you have killed yourself and every living thing? Now, that potential is a discovery that you can count on. So, to discover that you have made a potential monster, that is fine? And are you proud of this massive mess? And that you scientists in charge of this project support these activities, To address it in the most direct and logical ability I can truly consider as a proper acknowledgment of this wonderful billions spent on a device with a potential dooms day result, I must say, as a man who understands and sees the real meaning of what you're doing, that my only reply is, I am confident that you will be unable to grasp or comprehend. Still, the only true and exact words that come to mind are and should be, "You're all being naive and irresponsible for the safety of our planet for even considering this concept!"(Sorry), but it had to be said. The people who backed this nightmare with possible reasons of profiting would be people with massive suicidal tendencies or have a death wish for all humanity, but wait! They are doing this in the name of science and discovery; I think they would be better off discovering an H-bomb in their bathroom and toilet! So please to the so-called science genies rubbing this Aladdin's lamp, or to those funding it! "Ask yourself this question: Is this device safe as it reported they are "smashing atoms together"? And if you say yes, you need mental help. A nuclear bomb is when people mess with atoms, the device requires a device to BREAK atoms- Okay so- that is not your mission? My question- "smashing atoms is your action right?" breaking atoms is a nuclear bomb, let's see- let us compare- "BREAKING?"

(SMASHING?) Yeah, folks- is that a big difference???? (So the answer is - a resounding ah NO! —, and is it needed? My only conclusion and suggestion is: Remove this doomsday device away from our planet! So we don't have-THE END.

104. SUBJECT: OUR LADY OF JUSTICE IS SHE REALLY BLIND?

The main problems and the repairs- How to make a better system that works to place the proper applications in motion without months of delay, and the subjects in question to get a fairer trial, and the use of evidence as the conclusion if the evidence has no way of being tampered with. Many of you are saying to yourself- "This is impossible!" News flash! The system broken can be fixed, and here is where it starts:

Single Judge gone! And attorneys are no longer required-! (Sorry, you leach of humanity.) This means the cases are reviewed same day, and the evidence is also maintained- same day, along with all witnesses- same day! And cases are no longer bartered by attorneys. The people found guilty, well, sorry, you do the exact time as the standard and exact time and penalties are precise and recorded as exact penalties for any crime committed, so the criminal attains what is required to be free again. The removal of the use of one judge is gone, and gone is the use of people who are jury actions which may or may not have biased considerations that are formed from their own lives or opinions that could cause a person to be penalized, so the panel of judges is formed as three judges who only look at the evidence presented. After they have determined the "category level, they give the vote for that penalty and apply it by the statute of level. The level is not negotiated. It is firm and exact; the system today is one of negotiation and penalties are with a sliding level of action and effects are many people are given sentences much less or much more than another person. This is unacceptable because a person could be given a lesser sentence just for their status or creed. The legal system is broken in many countries. So, to fix this, we need what is called "SET PENALTIES RATING." These are numerous, but the reference will be easy to determine once the evidence is placed, as each event has a direct counter effect recorded, and so a crime has what a level point is; anything over the level is considered "RED ZONE" and gains the top sentence without negotiations, and all crimes are with what is called- "Level of Penalty." This means that the law is the same in all states when you are in one state. For example, the only crime I ever committed was not a crime in the State from which I bought the device. It was a stun gun, and I bought it from this little old lady in a grocery store in Florida, and they were legal to have. I never used it, but I figured it was nice to have some protection as I traveled a lot back then as a former Disney entertainer, singer, and guitarist. So, it was in the glove compartment of my van where I hauled my music equipment for booked jobs. Well, I traveled to my parents' home after being away for over three and half years, and one day, I pulled out of the driveway to suddenly be pulled over by a policeman. Who noticed my inspection was up? When I opened the glove compartment- He saw the stun gun and stated that it was illegal in New York State. I was then told that because I had been found with a "so-called dangerous weapon," I was not able to own a rifle. They stated all this as facts, but after further investigation, I found out they were not telling me the truth. And so I was fined and kept from having a gun, but the law in New York City says, "YOU CAN own a stun gun! So here is an example of laws not being even correct as the police can tell us anything, and we are forced to believe their interpretation of the law per State. And get this: I had a Florida license plate on my Van, which made me a resident of Florida, but the laws of Florida are not in New York, so I was not familiar with the correct laws in New York until many years later. Another example of why we need a "Level of Penalty" is, let's say, someone is murdered by an individual who was racially motivated. We have another case, which was a murder of a home invasion- The two are both murders, yet the new system applies the cases on two different levels. One was self-protection, and the murder is warranted if the action was found that the intruder carried any form of weapon that was potentially a device of bodily harm. The person has still killed someone, so it is still murder, which does hold a penalty because the person who was killed may have had the intent to harm. An entry that shows that the person came in unwanted must be shown. Still, the sentence level is lowest because the fact remains that "IF" the entry says through a window, then that shows an immediate intent to enter without permission, but the rate level changes if the entry was through the door, but if the door was broken as an entry-once again the level of the sentence changes. The other murder of the racial action is a full penalty, which means the highest level, or a hate crime is also the highest level. Again, a one person judge does not sentence the accused, but a three-person judge panel who hear the testimony not by seeing the person, but the person who is accused is placed in a holding room that is locked, and they speak only into a microphone which is above the room, and they are able to be handcuffed to a secure metal wheelchair as a firm frame which can make a person not able to move, and they are placed inside the speaking room, where the judges can ask questions about the evidence and hear the testimony; even the person's voice is changed, so the voice is produced in AI to make every person's voice the same tone. This means the person being questioned is not being judged by their accent or race; everyone is on the same level to be heard, and the three-panel judges have assistants who document this person's case as recording it and also research the potential levels of the actions being reported. They give the judges their assessments of the levels of law infringements and how many were done, and the judges now see the display on a screen of their potential choices of level set and now impose sentences. The guilty are taken by wheelchair to the holding areas, wheeled into a large truck, where the chairs are locked to the floor, and transported to a particular holding area for the sentence imposed. The system allows the person to defend themselves without bias or prejudice as the person is not seen, and the voice is standard in all cases, so they cannot assume that the person is poorly educated or has an accent from another part of the country. In short, the individual is only seen on the facts of the case, and the witnesses are treated similarly. Hence, all testimony is non-biased, and the accused person has a better chance of gaining from the facts presented if not guilty. This form of judging crimes also decreases the costs of court, the waiting of cases, and the elimination of many high-priced attorney fees, being the laws are now to hear directly from the accused, and the legal actions are only determined by the three-panel judges. It's like a fast food drive up, the info is delivered, and the (food) is the sentence done and delivered, the use of attorneys is not here at this first review, but a person who feels that they have been given the wrong decision can ask for a hearing later, that is where a attorney is given to the person who may need help and feels they have been framed. But that is to a totally different court which will review the case in deeper detail. The first court is to move justice into fast lane, and get the rime off the streets. The facts and the witnesses are given, and they are decided if the case gets a further review if the evidence is extremely strong, then it is sentenced right then and there. No fuse, just action, but if there is a doubt, then the There is also a need to have cases move faster, so there are three shifts of judges per city, and they are for areas of population over 60,000, so the judges are on duty 24/7. This means people who commit crimes are not running around on bail, as the many who have had this applied committed even worse crimes while on bail, and we will no longer have criminals walking on our streets who have been in trouble and may have bad intent. Sorry bail bond folks but the action of releasing someone has resulted in many deaths from those people released.

105: SUBJECT: POLICE-NEW DAWN

I know that police have been given a bad wrap, and many have made their bed poorly, and unfortunately the fact is some have abused their authority, but I also know many of them work hard to be exactly what we need, and that is men and women who work hard to protect us form some foolish person who has no regard for others. So we will help our police be like E.T.said: (BE GOOOD. And so let's help them with a few ideas.

Police will be wearing more than one camera in the future, as their headgear and shoulder gear with sound recording are to remain on, but the information is WIFI sent to a black box inside the station which is a locked server and receiver of every police on duty; if at any time they do not have this on, the case will be tossed out for lack of proof. On the other hand, we need to protect our police better, so having these devices means we can see events as they occur. The policeman or woman with both cameras turned off is removed from the service period simply; the systems are separate, and only someone with bad intent would have both off. The days of law enforcement harming civilians for color or their back ground such as poor people, homeless, or just out of state, no matter their reasons, they will have procedures that must be followed. Persons who are pulled over, the police are to be protected as well as the person being pulled over- The police car camera must be applied and must be working and checked before each shift, also, but the police no longer approach a car or vehicle, they first pull over the vehicle, then open their door and their doors have bullet proof metal inserts in the door, they now pull out their hand guns, without pointing it but in a upward position as to show they have it drawn, they speak over a door load speaker intercom built into the car, and tell the driver to exit with their hands up, they tell all occupants to exit vehicle and move to the back of their vehicle and stand in front of the police car, They are then one by one instructed to place both hands on the police vehicle on two white spots that are going to give the police information, the people requested to exit the car stand side by side in front of the police with their hands out ward, after each person has been scanned, the policeman's partner who is on the right after the identification comes back as okay instructs the person to go around the back of the car, and insert their hands into two holes located on the right side, this is automated hand cuffs, the person places their hands inside the sensors now close the persons hands in a very tight hand cuff, and then the officer will inspect the person for any knives, or gun concealed. Once that has been done, the officer hits the release button, and the person can remove their hands. The spots on the front of the car hood scan the person's hand and fingerprints with a simple button on either side of the doors marked SCANNER. Inside the police car, the person's record is displaced almost instantly on a screen projected to the doors and inside the car so that both officers can see the results. The police don't need to ask questions such as name, address, and social security number. The car is equipped to attain all that by the scan; if the scan is not working, they are directed to the hole handcuffs first, and they are on both sides of the car to be searched; one officer searches while the other officer watches everyone else if there are more people they are instructed to get into the back seat to be detained. The information is being recorded as they are questioned; the screen now shows if this person is potentially dangerous and may have warrants for arrest. This scan system could be inside the whole handcuffs also so a person has holes that must have the entire hand enter, and if their hands are not both inside, the red light goes on above the system, and when they have complied as to place a hand inside properly, the green light activates. This scanning will give information of any type within seconds as it is sent by Wi-Fi to the main computer system. It will show any owed tickets and if the person is wanted for murder, but any information is presented instantly. The police do not need to handcuff the suspect; they instruct the person to get into the back seat, but this seat faces outward and has wheels similar to a medical wheelchair; when the person gets into this seat, they are instructed to place their arms to hold the ends, and then told not to move, or they could be harmed. The officer presses a remote control that sends a signal to the chair, and they are instantly locked into the car, feet and hands. The vehicle is designed to have two of these chairs per car. Still, they can be modified for even large vans as the exits have an automatic ramp, and the handle rolls down the person after disconnecting the chair. They are now in a wheelchair, and the officer now rolls the suspect into the police station. No fuse, no resistance, and easy transport that keeps everyone safe. Now if the police wanted to have a more effective stun device which could be much better and doesn't have long cords, then please contact me with a grant of $180,000. To build it. Okay, it may be less to make, but the idea is sound, and Gee, I like profit as much as the next guy. Okay, I admit, being teased by a taser gun would not feel good, but let's see- be shot with a gun? -

1. Be killed? Or– 2. Be stunned? I'll take door number (2) two.

106. SUBJECT: IMPOSSIBLE?

So why not remove crime? We all want a safer life. We see the tragic events of our day-to-day lives with some ill-minded person with either a super negative thought structure, may be abused and act out the aggression, whereas most also carry many hardships yet we coupe and we leave the baggage that many people carry, for a shrink it's a complex condition of I need to prove I am stronger, or a weak mind person who says I have no regard for anyone or anything besides myself, these people unfortunately never had a proper upbringing or someone treated them very badly as I stated. They never made the adjustment that not everyone is bad. Instead, use aggression as a form of being seen, similar to small children seeking attention who yell out, watch me, Mommy, watch me, Daddy! And then swing on a rope or jump, anything to get them to watch that they now feel important. But no matter why they do what they do, have no real importance because first, we can place them in a category of mentally disturbed, and without reason or rationale, and so they live to take, or not work for money but to steal or harm others to attain what they desire is their only motivation. The way to stop this behavior is at the roots when a child is young and they learn "consequences." The good book (Bible) says to spare the rod to spoil the child. I am not for beating children, but I am for discipline, which means a good swat on the behind never killed a child but gave that child the understanding of right and wrong. I have no sympathy for anyone who reverts to harming others for self- gain or a sick attitude, so here are a few solutions:

107. SUBJECT: GUN CRIME REMOVED

It May sound a bit harsh, but the fact is crime is caused when people can make it happen, thinking they can get away with it. If the equipment made had detection built in that criminals had no idea where chips to locate guns, and the gun manufacturers had these devices installed by mandatory installation but never told exactly the content, but is strictly an I.D. for each gun made, the locations of guns would be easy. People who have one without a license could be detected easily. Now, that is one way to stop gun use that is not licensed, but gun finger locks could be installed, which allows the user to fire the weapon as they grasp the handle. The handle reads their fingerprints, and now they can only shoot the gun; this could even be used for military use as enemy soldiers would never be able to use our rifles or guns; this could be applied to all military hardware that a user is designated to fire a tank or an artillery piece.

REMOVE CRIME OF THEFT: We can even dig deeper into actually stopping crime; one of the main reasons people do crime is to attain things they cannot afford, or to profit by selling the items. So in the future, all equipment sold, no matter what it is from a candy bar wrapper to a shoe, to a television set, will have a buyer I.D. is inserted first when purchased, this now links the items to that person upon any item bought. The process is very basic in application. The BUYER I.D. is first swiped, then the items are swiped individually, and the chip inside the items is now coded with the purchaser I.D. reference number. It is installed in the item being swiped for cost and now has an I.D. tag inside, embedded into every product, either by a cardboard insert or a hard tag, or inside the product. So whenever a person buys anything, tires or a car, it does not matter what it is- The moment they buy it, a scan is made from either their credit card or debit card into the chip installed, which all the items purchased when scanned for price are also done. So, how does this stop crime? Well, let's see the advantages: I am a thief, and I have just taken your new television or a phone; these devices recorded are now in your home, but you did not know that the items also have locator chip ability, and so when the items are reported stolen, the second it happens you call a 24 /7 theft finder, they locate merchandise. Now police are sent to that location with an I.D. scanner, and bada-bing-bada-boom- The theft is found, your items are retrieved, and someone goes off to jail. The theft of these items is never revealed to the public, so theft has become a very unpopular attraction. Now what about murder or rape, or being robbed of money. Well, paper money is being made a thing of the past, but in truth, it should remain as a person-to-person selling, which should still be allowed. The fact is we live today with a new standard, which is "monitored lives." We talk, they listen, we move, they watch, we are without basic freedom, and this is also a motivation for crime, yes, for crime because they are motivated to beat this system, and will do so, as some crime is committed by very smart individuals who have the smarts but not the understanding that crime is wrong. So, we need a system that makes the crime of certain things actually- "IMPOSSIBLE." We can first look at theft; these items are physical, such as a T.V. set, chair, or anything of the home items, cars, trucks, office, or business. The future is with an I.D. attached to all items scanned, each item has its reference number in its scan bar code, and the item bar code upon purchase is now registered in a massive database only for "YOU." Each person buying even a simple pen will have that item now logged with their lives, and private companies can provide services to ensure this system. I also have inside all items a Nano chip that can locate any of my items owned. And with a request, say from an AI device, it can now pinpoint the exact location. The system is called ON-STAR for vehicles or Lo-jack, but the newer system goes further; it detects "everything." This stops 100% theft. One down. The next is "Someone harmed," Either a confrontation or even a murder. The answer is two types of electronic devices. Both can record and send an exact location upon either a verbal command or a turn of a switch; the item is worn around the neck and can be seen in a surround view, with over four miniature cameras, and has two microphones, able to record any conversation until the operator (user) says a password as a verbal command for it to stop. The device can recognize a person's voice pattern, so another person cannot turn on or off your device. It also monitors heart rate, and if a person has been unable to make a voice command because someone snuck up on the person and covered their mouth, the heart rate in any event with a rapid increase will activate the device. A crime committed is now reported in milliseconds, not hours. The location is fed directly to the closest police car or officer in that area; there is no more 911; it is an automatic response directly to law enforcement. Once the crime has happened, two things also happened automatically; the device can now upload its last information recorded to a device the police have to view the action, so any conflict of any type is recorded even if the person was not able to turn on the device themselves. Also, the device that the person harmed was wearing and could be shot or killed, this device because of the heart rate increase before death has now sent a signal out for a one-mile radius, and now the signal sends back from that sent signal- every person in order of distance to that person as a possible suspect. In this activation, the information is also downloaded to the police. The suspects of the locations are now traced because everyone wears these devices, and so those people, who possibly had shot someone or shot a bow and arrow, whatever they used was recorded instantly on their device as soon as that heart rate highly increased sent that signal, it now started their recording device, and now could show someone putting down a riffle or anything being thrown away, in short , the crime would be found instantly, and the evidence is firm, and hard , and fact as the ability to compromise this is impossible because unlike other devices like a video recorder, if someone tries to open the device- this sends a signal to police also and the device owner is now convicted of tampering with evidence, so they go off to work camp, now this device perfected will remove many people who feel they are essential , they will not be needed any further, because the device has no compromise, or false nature, cannot lie, or show anything that is not real action, so to judge someone is clear cut dry of emotion or someone able to lie, the need for judges, or courts, the witness is the system is gone, 100% gone, and the crime is now with the fastest results known to mankind. If someone wearing this device is exercising the increase is seen as a slow increase which in most cases would not activate the system, but a contact can be sent to the person asking if they are okay. Now, the person wearing the device wears it daily and at night, but the user controls the activation so that it can be shut down, as in acts of sex, but a shutdown still can trace your travel as your devices GPS location, but in the event of a home invasion, the device can be turned on verbal with voice recognition, with a pass word stated, that activates the system like an on and off, but prior was not a listening device or visual device. Lastly, suppose a person has a heart rate during a shutdown increase or an exercise such as running or athletic activity. In that case, the device will turn on, if that person's heart rate has "stopped. "The device will activate to record everyone present, the device sends a signal in this event to police to be notified, and the location of others is the same as the first application stated.

108. SUBJECT: OUTDOOR CAMERAS

I can see cameras in some areas, like high crime areas, but the government is now entering our homes with surveillance devices. Indoor and outdoor cameras are now in our lives; simply the invention of the "LIFE" system removes the need for surveillance cameras and gives people back their privacy when wanted. So, crime stoppers say cameras identify, well true, but the devices make all of us watch to take away privacy. Privacy is a word that seems forgotten; the definition given is: A STATE OR CONDITION OF BEING "FREE" from being observed or disturbed by others. So, we are being watched no differently than if we are all criminals because, in prisons, that is the device they use to keep an eye on their deeds. I agree that it is needed. After all, they have already broken the law and deserve to lose their rights as people because they have crossed the line of what is correct. But we need to be free from the eyes of Big Brother. When the book was made in 1984, the public back then just laughed it off as the future of our lives; I remember how people said they would tear down a camera if they saw one. But those who are in control are saying that these devices remove crime. Today they want you to feel comfortable being watch, so they make T.V. shows to make you feel it's a good way to live ,T.V. called "Big Brother" which does exactly what they want for your life, to be watched like criminals. My device is not like there devices, because I have control of it like calling 911, but instant, and the crime is recorded, and the bad guy is caught, simply they are not getting away from anything bad they do. So, do cameras on the streets watching us help remove crime? Please show me how that has helped Seattle, L.A, or Detroit. The facts are that the states all make billions of dollars from the arrests of many people, and the crime gives the states homes confiscated through bail bonds broken, and many other articles such as trucks, cars, and even housewares items confiscated. Also, guns are worth great amounts, the drugs that are even placed back into the streets by dishonest cops; the fact is crime is big money, so may we look at the promotion of crime in our daily lives? Sure, we can turn to almost any T.V. channel and see crime being glorified, and we have Hollywood promotion of crime as a way of life to gain fast big wealth. So sorry to say- that many in our legal system need to be investigated where they gain their money. Many people in higher office also need a close look because the system seems to ignore the streets that are filled with drugs, and if someone is concerned, the drug trail is easy to find; you need to watch where the person on drugs goes to get their drug of preference, and then go to that person who sold the drug and watch where they go to get their product to be sold, and then watch where the next location is, and keep going backward till you find the main supplier in the country sending the drug. This is a real and exact way to locate drug input into our countries infected. Yet the powers that be refuse to do this? Why? Because they are linked with the drugs sold at the maker's profits to be given in an offshore bank, or even items transported, such as new cars, or just cash transported in items given, no matter how they get paid is not the issue, the problem is the people in charge are involved and as thick as thieves. To the good cops out there, don't give up on us, we need you badly, but know many of us are aware of the bad cops, you may be aware of this also, so do what is right no matter what it takes, do what you promised to do when you made those words of commitment to this job of protecting the public. But we both know that there are heartless people, who only consider self-gain, and people are no longer important, to be sent to jail one day when the systems are investigated- and gee, maybe that starts now? My only hope is that police are given good retirements so they can live well, and a salary that supports them and their families, and maybe we will see a decrease in bad police, Let's face it, to turn your head is easy to gain money, to follow a narrow road of doing right is much harder, yet to the good cop, you gain two things, neither can be seen, self-respect, and knowing you did good. God bless our police who are not appreciated, and the next time they are called to your neighborhood, show respect, call them officer, be willing to help, stop treating them as if they are the problem, and maybe, just maybe, they will become your true friends. I've met some good ones, and I have met some bad who pretend they are ooO look at me, I have a gun! But the good one is the one I cherish as how police should be. Caring about that badge with honor and never forgetting they made an oath, a very important oath to protect us. May they all find the strength to maintain that oath to care about people who many times act like they don't care about them, - and the dangers they face almost daily. May we all be more aware of the good they do, and may we all be better citizens as to not need to call them.

109. SUBJECT: EDUCATION

I keep hearing the complaining of schools costs and how do I pay for college? I admit the costs have been crazy, but let's look at our top colleges all over the country. They have built a mega money maker with top teachers making the advancement of students happen, and while all this us happening, they are recruiting sport athletes of every kind, and even have stadiums that house over 70,000 or more people, some of these tickets to these games are priced over $400. Just to get the seat you want. Now this is serious money being made for the sport that happens weekly. I won't mention a name but many stadiums can hold around 100,000 plus, and this is if we take a look at profits is $40,000,000.made in one game. This doesn't include the food and drinks sold, so let's say 10% of those people just a drink with a $10.profit, oh yeah drinks are high at these events. That's an added $100,000. So let's face it over 60%

of people will buy something. So my proposal is that every major college now places 20% of all they make at their sporting sales event games to a national foundation pool called -MY TRUST FOUNDATION. This is for anyone who can't afford college, and their financial status is checked as to be eligible. So there is estimated to be 858 known colleges with football teams making them money, not including other sports they may have gains from, and since when should colleges be business of profits which have nothing to do with education? Seriously? And do I hate football, NO! I love the sport, but let's look at some real facts- 20% from **40,000,000** from just **25** of these colleges. That is - **$200,000,000. 00**, so if every college was by law to give back to the future of our youth, this would pay for almost everyone. So this money could be distributed to not only colleges, but high-schools could also be involved with students who have hardship, or have need of food programs and need books or whatever is needed that is directly connected to school activity. Now to the Deans of these colleges, I am sure the kick back is great, but let's get real, the expenses you have are no way near these profits. Sports events do cover the costs of the physical teachers and coaches and any medical equipment or sporting equipment that is required for the sport. This reduces the tax burden on people who have had many years of paying money out to support schooling, which is supposed to be supported by the state government. Now for those of you who believe that the schools that are now formed would disappear from this help, well, that is not the true as the facts don't lie. And these are the only students that require homeschooling to reduce the costs. All other students are within a 5-mile radius of the school. Now, they can either walk to school or ride a bicycle to school, over the age of 13 years old, and as long as there is a sidewalk for them to ride on. Students under the age of 13 are to attend home schooling until they are old enough to ride a bicycle or attend their local town Central school. So once again, we are allowing children to go to a major school already built and at the schools. The classrooms are not any different from what was being done at home school. The only difference is that the students are now in a classroom that does not require the students to be moving all over. Only after they have completed their subject matter can they go to recess or their lunchroom or study hall. But they are not roaming around the school, and when they hear a bell ring. They return to the classroom that they were in before. This bell means they must return, and it is a nice-sounding chime. Not a fire alarm that is very disruptive. So, they attend a class that now makes a new presentation and a new test to be given after the presentation. Every two weeks, the student is given a test that sheds some advancement of the knowledge attained within the two weeks; this test is called the mid-exam if the mid-exam is passed. Then, they do not have to repeat the exams or presentations made prior; the object of learning is to learn. So, if the subject has not been learned, it shall be over repetition. Once the student has passed the midterm or mid-exam, they can move on to the next level. Remember, all students are not taking in the information all at the same time as they are learning at the pace that they need to achieve. Now, here is going to be a huge eye-opener: colleges, which cost students tremendous amounts of money and profit in major ways, are now going to be depleting from students having to travel and live far away where they can now live in their hometowns or city and still attend the college of their choice. All online. No more people being dragged far from their families, causing students to be in danger in faraway places. Places they don't know, instead. In that same school, they can now attend college to the top colleges and universities throughout the entire United States and with their top instructors, who are all now being televised and shared by each student to learn. Once again, at their own pace. The system goes from grade school to high school to college, and a student who has a faster and easier learning ability will graduate at the age of 15. Possibly sooner. The exceptions to this are trade schools and medical fields that require hands-on and the ability to perform a task that requires more than just the knowledge of the ability to perform under extreme conditions or proper functions.

110. SUBJECT: FREE MEDICAL & DENTAL

The doctors, for far too long, have been controlling everything from high costs to operations to even the pills you buy in their pockets. So, we are taking their income to a level of high-paid office personnel and standard income because the medical community is about to get a shock. The way people pay for all medical expenses from their actual paychecks should be less than when they have insurance. This payment reduction is 70% of the bill as a regular payment if the person is working. If not working, it can be up to 100% of the bill as we now have national medical and dental coverage. How we can make sure the cash balance is created: there is a similar system to social security; a small percentage comes out of the person's wages, 5% to be exact, and this is placed in a national medical payment plan. Now, let's consider that some of the expenses of staying in the hospital are not only ridiculous but absurd! The bed can cost $5,000 per day! This will be a national medical creation, so the government now controls the beds and the pay scale, and so a hospital bed should be no more than a motel room bed; let's face it, the hotels in some places make hospital rooms look like a homeless person cardboard box residence. First, we must consider that the amount of people sick is much less than the number of people in hospitals, so if we compare these figures, we can find that many more people are working than sick. So, on average, less than 7.3% of the nation's population is in medical care hospitals. So this means that close to 92% of the people well and working can pay into those who are ill, and when they recover, they also will now be back to work. But there lies the issue- retired people are still allowed to work today even though they claim income. Hence, we also need to adjust social security to cover the expenses of each individual, such as living space, house, electricity, food, water, and sewage, should all be covered by social security, and when the cost of living increases, the sum they attain should also increase. But they have a choice not to retire and stay making wages, or if they retire, they cannot work because they are taking someone else's opportunity to have a job income. So, to retire means just that: you're no longer working anywhere, but your needs are met by the federal government monthly. Now we have all heard the sad story of how the government is hurting and social security is in danger; listen to people carefully; the retirement age is set around 65; last heard in my ears, this means someone has been potentially earning since age 16 so for 49 years they have been paying into social security, the average life longevity is around 77 is reported to be around 77.28 years, so let's do the math, withholding is around 12.4% the national average in the United

States of job earnings is $60,575 per year, let' take 12.% from that figure- and we have $ 7,269 per month. The average income made for retired persons is $1,783. Let's subtract the average income from that sum- we have 5,486 dollars that the government is now keeping yearly. And we now collect that sum by 49 years- and we have $268,814 they have for the 49 years you worked on average! Now, from age 65, retired to age 77 is the average death rate- this means the government is paying out for 12 years a total of 256,752. So this means they keep a sum of $12,062 per person, and this doesn't include people who worked till age 55 or 60 who just died and never were able to claim anything! So, in short, they make millions from the program that they claim is so bad. My point is that we have a system for retired people that works, but the costs of medical expenses are draining the funding, so we need to close the lid on high medical costs and find out the real costs of these developments. However, national health care can be achieved just like social security. Still, the sum of the 5% is taken from every person no matter what amount of money they earn, so millionaires and billionaires would be paying much more.

111: SUBJECT: PROJECT URN

Funeral costs- The loss of a loved one is no longer to be a heartache and financial destruction; the cost of all funerals will be set costs – determined by state control, and all equipment will have low- end options. The normal cost will be $300. To be made to ashes, and a free market and plot owned by the state. The person's name is made like a license plate encased inside a Plexiglas viewer, and the person's picture is on the inside; the urns are on long shelves that allow up to 50 people to be inside, and they are below ground in massive tunnels, the funeral tunnels are entered with open doors, and a chapel is on the grounds for anyone to pray. The above is like a park, with the look of a nature preserve. Flowers are everywhere, and the tunnels have solar lighting, and each urn has a light over it. The floors are made of stone, and the walls are a mix of concrete and outside stone in lower areas near the floor; the ability to repair is easy because the items have mold developments, which are preserved for added use when needed. Things to improve ASAP.

112. SUBJECT: NOISE FROM TRAINS AND SIRENS

We wonder why people are having so many ill effects when one of the major causes of many illnesses issues is stress; lack of sleep can be a major contributor to stress, and yes, it causes hundreds of ill effects as the body becomes a battleground to even the smallest infection, that requires a very normal requirement- sleep to heal. It is funny that my first observation of silence was after growing up only five houses away from a major three-rail railroad journey with many trains passing very frequently. I grew accustomed, never liking the disruption in the middle of the night, but these became very real and very close elements. At age 10, I was taken to a country home of a relative of my family and was subjected to something horrid! No noise!

Seriously, this was so peaceful; I remember sleeping at night with a window open and a nice soft breeze blowing through my window screen. It was amazing. So, let's consider our fun toys as grownups that make others very disrupted, and those in charge- well, hey, get over it is their thoughts, but remember, my friends, we are the voters! So, that police car that sounds off every few minutes in the cities and countryside, and why? Today, they have an arsenal of lights that you can see coming from miles away, and yes, folks, they also have a floodlight on many of their vehicles and a speaker system loud enough to blow your ears off! So-pull over is all they would have to say. How about the trains that blow their horn at every crossing, even when not one person is around, and in the middle of the night, at the beginning of the night, at all hours of the night! The only few lucky not to hear these massive noise makers are people who live far away from them, but over 90% of the population. And we love our firefighters. They and our emergency vehicles do the same thing: do a good deed but make a massive disruption to those they are not doing but disruptive noise that can be a thing of great harm. Proven facts: Lack of proper sleep can cause serious health issues, insomnia contributes to sleep disruptions, and heart failure can also be attributed to lack of sleep. The list is long, and what do our fine men of these vehicles of rescue and fire prevention do? They cause exactly what they are trying to prevent! Lack of sleep! Their sirens can be heard from miles away, depending on the terrain, and now, let's address the real pain in the ears! The fire horn! This device was made for World War II! It was to set out a warning signal that was so loud that people could hear far and near or go underground in the event of a bombing. Not to be a- "Hey gain, it's noon and lunchtime! The need for that went out with the invention of the watch, for God's sake. But I have stayed in towns that use it for fires, for lunch calls, and even curfew for teens. The teens of today carry things called a 'cell phone.' I have not seen a teen or even child today without one, so you want to notify people it's noon? Okay make it mandatory on all cell phones, and you can pick to have it or not, and even select the type of ringer you want. That is already possible, and any cell phone can do it! Oh, but I don't have my cell phone with me. Is that anyone's fault but your own? Okay, there is a man who is on the fire department; 911 just got a call: There is a fire in the town of "NO NOISE," and the people who volunteered for their fire department have two devices to be able to be contacted. One is strictly a phone that contacts you from a computer contacted that sends out a call to just that one FIRE PHONE, and if the phone was not picked up, so the line now breaks to the backup phone which has a ringer which will not stop unless picked up, this in the firefighters home, and when picked up, you're hearing a repeating message of the fire location and even is documented that you picked up the call. This means the only person you're waking up is someone who wants to wake up the fireman who desires to be notified, and it achieves everything, and the people of "NO NOISE" thank him! Have you ever heard the saying- "Boy, did I get a good night's sleep!" You haven't!?

Well, you live around noise. Trains will no longer be able to disrupt sleep, as the new law will maintain that the use of horns is not required only if someone is on the tracks. The use of overhead police- style lighting is made to project light only in a forward area at a minimum spread in light so that people can see it, but as it passes, it cannot be able to affect houses on the sides of homes; the only area this could be an issue is a home on a curve with a crossing. Also, all crossings must have two gates that go completely across the road on both sides, and the crossings no longer have sound, but lights flash, and the rails are lit up as they come down- all in LED lights; once again, the lights do not disrupt houses on sides of roads with trains passing, because reflectors keep the light straight forward. Transportation and new train travel improved- our nation was once the shining star of rail travel, and this ability to get their faster-made commerce improved freight; freight was delivered much faster than by horse and wagon. But somewhere down the line, the people who now own these rail systems have become complacent, basically not interested in improving what we see is possible, like the rail that travels over 2023 miles per hour in Europe. Our best by AMTRAK is only a top speed of 150 miles per hour. That is just not acceptable; to be a great nation, you need to be greater to call yourself that. The days of old travel are over. We will have new transit of train transportation that allows more materials to be moved faster and safer, so new track needs to be made to support that speed increase and delivery time decreases; the new trains are a combined electric and semi monorail, all-new crossing areas now have a long slightly dip and the road has a slight rise as now when both are applied a car can travel easy over the train, as the distance of the height now is half. This allows the use of electric lighting to reduce energy costs. Also, the use of tube rail will be used for cargo, and this is made like a overhead which carries the following, packages, and people. The speed of travel is either city to city or local to local. If you have a trip that has five local stops, you will get off at each to get on the other because they are like large elongated loops with open areas only at stop areas. The ride is inside a fully encased tube and is pushed by massive air pressure, as the city-to-city can reach speeds of over 400 to 500 miles per hour, and you never know it because you're sitting high off the ground as around 40 feet in the air and above most trees. The tube travel systems are located above the rail systems now in place, but all tracks are made to be straight only, so the need to have bends where people and cargo are at risk of loss is no longer possible. So, this method not only becomes the safest way to travel. However, the need for jets for air traffic is decreasing, and this will push the airlines to improve their performance and travel sizes. And to those who represent air travel, those who have invested should invest in rail. Then they have zero lose and still can maintain the airways. Lastly, the trains of the future will be made to enter underground about four miles from the ocean shoreline in a gradual slant. This is the same type of train that travels on land, but now the rail system lowers below the ocean at its lowest point of entry, which is around. This maintains that storms above are not effecting the rail below at all at any time because the maximum depth of a massive storm is around 120 feet. Still, the rail would need to be pressurized to prevent the passengers from getting bent, a very dangerous illness that can kill a person, but remember, when you are flying at 30,000 feet in long travels, your plane is also pressurized. So, with the proper depth of the underwater rail, we can be well below 120 feet with the use of repairs done by remote-controlled robot repairs, which show the parts needed for repair, and there are manned subs that also can achieve these depths to repair any damages found. The technology to build this is here; it could be a direct route from the U.S. to Europe for the first rail, but the tasks of preparing the structures are essential to allow shifts in land mass and the ability to be adjusted without re-build. So, this train system can be used to fly almost as fast as most airplanes today.

Without the worry of crashing or rocking like a boat does on the surface, we are talking about a way to make goods and people able to be transported better than the horse and buggy! Gee! The rail is not normal; it is a hexagon support that sits on a high-sided rail. The passengers actually are traveling able to see the ocean life in the distance in this underwater viewing tunnel made with a strong polyurethane clear plastic extremely thick. The train is actually water- tight also, so even in the event of a leakage, the train can continue to its destination. Still, we are not just using one tube to travel under water, no, we are using multiples of tubes stacked on top of each other to make the structures connected and secured even greater; all supports are made of plastic because we cannot have a rusting of the supports. But the connectors that hold the plastic supports can be under water welds. However, regular maintenance is done by underwater robots that travel across the entire length of the connections by an outside rail made of plastic to take pictures of any possible cracks or damages that may need to be repaired. The crews that fix these are using subs and under water divers.

113. SUBJECT: CAR THEFT & CAR CHASE

Solutions (3): I have heard from television shows advertising these amazing crime car chases or speed changes made by our police as the bad guys on the run and always being caught, but over 10,000 plus have been stated to be yearly throughout the U.S. alone. The fact is that car chases are common, for the result is many times loss of life and death, not just to the driver who may be on drugs or alcohol or other reasons to avoid the police. Still, this chase of automobiles by the police has caused millions upon millions of damage, and taxpayers have to pay the cost of police car repairs. So, the solution is for manufacturers to install a shut-off fuel device that can be placed in more than one area of a fuel line. This is a remote control only owned by the police, and anyone found with this who is not of law enforcement is a mandatory ten years of community duty and serious labor duties. They must pay back any damage they may have created. But all this can be avoided so very easily. A remote control is attached directly to all police cars that can send a signal out in a direct manor to shut the car being pursued in a car chase. The remote is activated, which now shuts off the fuel line, and the person can be apprehended. Car companies can install these devices, and the frequency is never different when activating the shut-off. But we can even go further. The manufacturers of cars can also have a supported remote device that not only stops the car but also has an automatic door lock, which prevents the driver and passengers from exiting the car, which can allow back up to, arrive.

These devices could be made to be in the signal of the lock down device and fuel shut-off device, Which will prevent officers from being in harm's way, and can give them time to contact others, and lastly, let's say that shots are fired from the car being stopped and locked, but those inside refuse to exit now rather than have an all-out shoot out, which can be deadly even to by-passers; the car has once again a built-in gas spray dispenser, this can be activated only by police in emergencies only; the gas is exactly what is used to put medical patients who are put under, and this fills the car. Once the criminal has been put to sleep, the police can unlock the car by the remote, enter the car, and arrest anyone without a struggle or injury. In the future, police will not carry fire arms that kill people. They will use highly concentrated gas that places the person asleep instantly. This idea was presented in the movie The Green Hornet. Still, I am adding a rubber bullet chamber that can knock down someone close to attacking, the gun has a compression tank that is worn, and an expandable high- pressure cord that provides the air power to go up to fifty yards, so the gun has TWO chambers, sleep capsules and rubber balls the size of a large semi-truck ball bearing, (OUCH)! But I believe this has serious merit, and let's consider how many people have died from police having to use gun powder bullets. The use of top-trained police in extreme cases should stay as maintained as S.W.A.T. teams, but the majority of crime could be dealt with without loss of life, and these types of guns could be used in apprehending a criminal in a car chase stopped.

114. SUBJECT: BETTER MEDICAL CARE

Before I get to this subject, let me show you how medical care can be better and worse by a true story. I was ill many years ago while visiting my sister in New York City, and I became very sick when I went to the doctor, he proscribed a penicillin antibiotic which I went to the local drug store to have this prescription filled. The bill was $250. For a small bottle of the medication, when I returned home back to the Buffalo area, I had my doctor give me the same medication a penicillin. The cost was $32! So what is happening here? The public is being raped by many medications that are not regulated as to have the same cost. This needs to stop, and the costs of medications need to be reduced as for the makers need to have their profit margins made much lower, this means new laws that control drug costs should be lowered as soon as you contact your congress man. Now I ask you have you ever sat in a hospital emergency room and sat there seemingly forever, I have, and I am almost certain that you or someone you care about has had this treatment, you may be working at one of these emergency rooms, and I am well aware that they treat a patient by what they consider the greater need. But as we all know, none of the people who work there are related to Super Man and don't have x-ray vision, this said: Allow me to take you on a journey through the hospital of today and then the hospital of the future and explain how we can provide health care for every person- no matter their age. First, our trip into hospitals. I am sick, and that would be an obvious conclusion of why I am entering an emergency room. But I am greeted with my problem with "fill out this form, and have a seat, and we will be with you shortly, depending on the doctors on duty, and the ability of the single doctor to examine all the sick is time-consuming, and my illness may not be something that can be seen, it may be inside me, and it could be something that is causing me great discomfort, but I am a strong person so I take the pain, not knowing that as I sit there, suddenly I collapse and I am no longer with the living. And this was something that could have been prevented. Now, those people who worked at the hospital who allowed such a problem to be ignored not because they desired to do so but because they were following hospital procedures, these procedures that actually caused this person's end. However, as I stated, the hospital didn't know that person, and there was no connection to knowing what they did or why they may be important for another person's life. Maybe they looked homeless, but maybe they actually worked for food for a little old lady and delivered her groceries when she needed them. The options of the people who pass away needlessly are staggering, and we may be allowing a real hole that could never be filled by their life no longer with those they had relations of work or actions they gave as support, and I know that one of the most serious problems is when a patient complains, or someone close to that person gives what they observed, and the doctor ignores the request! I have personally seen this happen to my dearest and closest friend. My requests for her needs were observed, and she was continually ignored for requests to be accomplished, requests that could have possibly saved her life. But let's take the next step. We have filled out the paperwork, and now we have sat in the waiting room, and the receptionist says we will examine you. This person may be a nurse, but they are only taking your heart rate and temperature, and they ask you about your pain level. This is heard all the time, so its importance is somewhat common. They place you back in the waiting room because the doctor is still with other patients, so you go back to wait some more. But finally, after an hour, maybe two hours on average, you are now seeing a doctor. No! You are taken to an open room with a curtain. This room has not been disinfected and is open to all other people with infections that are now possibly being transmitted by cough or sneeze, or the person who was in the room earlier had touched the side of the bed you're now holding as you climb on this. The nurse tells you to get in a gown, which is supposed to give the doctor the ability to examine you better, and this device is open in the back, so any need of having any self-pride or discreet nature is wiped off the map as my butt cheeks are exposed! But the nurse is friendly and says the doctor will be with you shortly. I have sat over two hours in just that spot, and here comes the doctor- hallelujah! They now ask you the same questions you have already explained, and so you repeat the entire issue, which is a time consumer. Now, the action of possible blood or an ex-ray is to be done, and this is more waiting, and then there is another wait for the doctor to look at the results. More time is required. Are you seeing a pattern forming here? "Time can be essential in some issues, such as someone's health being repaired in a timely manner. This is also to be addressed in the new hospital of tomorrow. At present, many hospitals are under staffed, so to make the emergency system more effective, we have A.I. systems installed that are able to determine for 7the doctor the possible of the patient, and even give recommendations as to the stats reported in the area of the possible illness being received. An example is the system taking note as it works to adjust the data to see if a pattern is formed, which may give the hospital an alert as to a potential wide break out of a particular type of virus or a fungus reported. This is not just gathered by the hospital but is linked to "all" systems of all doctors who have the same issues of treatments shown, and this may give the A.I. a full spectrum of the possible cure or the treatment that could best be applied that showed promise applied by a certain doctor who already proscribed an application which showed good results. This means the doctors of tomorrow are given information that can be shared even globally. So when a patient has acquired something unknown in one place, the answer will be displayed from another with greater potential for the patient to have a good day, and this makes everyone happy. And let's face it, isn't being happy all our goals? So to those of you in the business of caring for the ill and the hurt, why not consider making some improvements to get people through the process faster and more effective, and while you at it, how about someone forming a 24 hour pharmacy pick up inside the hospital, so when people need the medication, they can get them. This was done by the V.A. hospitals, so why not every hospital?

115. SUBJECT: THE FORBIDDEN CONVERSATIONS

Now before you get your feathers in an up-roar, just sit back and allow the idea

that some folks just are wondering why they are here alive, and want answers, or they just believe that someone actually had a hand in making the universe, so this first subject is a hard look at what all the fuss is about when people try:

1. FAITH, 2. POLITICS- 3. HATE, 4. FORGIVENESS

Now we have all heard the people who attack you if you're not following this person, or that person who they claim came to fix the world but in the mean time we have people hurting each other, because the opinions fly like bullets, and sometimes it is bullets! But we hear that those who refuse to partake are only being sheep leading to a very bad place, because they will be one day seeing the result of their non-observation and non-caring to potentially find they are no longer under the same authorities and now in a living Hell. All because they chose to stand on the side lines, stay out of the field of play, and sat on the bench as life went by. The parade ended and they sat bewildered and confused saying " why Me?" Well if we start with Faith, the subject is usually met with some harsh feelings, but what if I said that you're really looking at a 50/50, that's correct a 50% chance that one of these places exists or doesn't exist, and really if you choose to say that when you die, well your just never to know you ever lived? Wouldn't that be Hell? But unlike 99% you, I have proof of the place called Heaven, and hear this load and clear this is all fact and a real true story that even I was shocked out of my socks when it happened. But before I get to this weird event. Let me give you a picture that as a teen -I wasn't convinced by anything people told me about some fairy land of goody goody and gum drops, with an image of some half naked man hanging dead was not my idea of a fun time. Even in my youth I was told by my teachers I was grades ahead of the other students, so for me proof wa the answer, not wishful thinking of some Alice in Wonderland tale. You see I never really got to fond of rules and being told I had to be good, I liked parties in my youth and pretty much did whatever made me happy, and down deep inside, I really wasn't happy and even around other people, many times I felt alone, strange right, alone even at a rock concert, how is this possible, I was always seeking to be excepted and looking for affection from people who had no care of kindness, but the world I lived in really didn't seem to care about how I felt, so I muddled through life looking for the next good time always going from job to job, never satisfied with anything. But then this event I am sharing happened , and I was wide awake and didn't have a drink of alcohol, at this stage of my life I stopped all use of doing drugs of any kind and I was now age 44. So please understand that this was just 100% an exact event which even today I say: "How was that possible?" It starts like this: It was 1997 when my adopted mother passed away.

I was very close to my mother, not saying we never had disagreements but we became great friends, and I was heartbroken for weeks after she died, and she left for wherever she was, but that was no longer grieving at this point. She always told me that there was a Heaven when I was a boy, but I never believed her or the church folks. But what happened was the craziest thing I ever experienced. Because about six weeks had passed by after she had died, as I said, -

I had come to a closure about three weeks earlier before this happened. And the main reason I was okay, was because I woke up to the fact that my mother wouldn't want me to be a blubbering mess. So I finally sucked it up, maned up, and stopped, and I moved on, but one night just before bedtime, I was sitting on my bed, ready to go to sleep, when suddenly I heard from a spot above me about three feet high and from the left side of me, again, just about arms reach above my head on the left side, so it was from a spot, not inside my head. It sounded like someone yelling through the other side of a wall, and from that other side, with a muffled but yelling voice, I heard my nickname as a child being spoken (Rick), which my mother always called me by name, and oh yes I knew my mother's voice when she yelled! (I heard it too many times as a child!) But she said and I will never forget:

"Rick- Don't worry about me; It's wonderful here!" That people was as real as I breath, and so for any of you wondering, you have just been witness to the fact I said fact, look up the definition,["FACT" is something that is known to have happened or to exist, especially something for which proof exists, or about which there is information.] Now to subject # 2

HATE-& FORGIVENESS, when I was a child, I spoke as a Child, I understood as a child, and thought as a child, but when I became a man, I put away childish things. Now I am saying this because I know that I am saying this because I know you may have issues. People treated you badly, hurt you, and treated you with disregard, but that was then, and this is now! "Do not allow others to take your sail, your wings; you are all wonderful when you decide your desired direction is to follow good."

Suffering is one thing I say is: "God, why do you allow people so much hurt?" The answer that comes back to me is: "You are all tested to be with me or without me; those who suffer and endure are my children that I "WILL" welcome; those who oppress, harm, and hate will be forgotten; they will be the consumed as wood fire stock." This sounds harsh but - Here is my real story: You think I have never had hate for people? Then consider this: At age three, I was tossed away by my mother in the middle of the night, I awoke from my sleeping to find my mother was gone , I put on my slippers and my bathrobe, and walked outside and up wooden stairs to the upstairs apartment and I knocked on the door. A woman in pink curlers answered the door with a pink bathrobe, I asked her: "Is my mommy here? ". She replied "No liitle boy she is not here" And she must have called Social Services, because I awoke the next day in a new place with four other children, so in short, my mother had abandoned me.

I was tossed from home to home for two years, finally was invited to stay with a family called the "Neff's", and I was adopted at age seven. I knew everything that was going on. I had people treat me like dirt everyday walking home from school and even teachers treated me badly telling other children not to play with me. This was a small town and everyone thought they knew you, when in fact they were people without care or love for each other and gossip ran their hearts.

I was tormented and yelled at by ignorant children, hitting me, and saying, "He has no real Mom and Dad." I was treated very badly all through my youth; I was even later rejected by my adoptive parents and was and told to stay away because I was sleeping in the backyard in my van without a job. I was once again, tossed away like a bad shoe. My only friend Sandra Allen, let me stay in a lawn mower shed I at her grandmother house, I slept on a wooden floor, yes, about twenty feet from railroad tracks constantly going by, and three times in my life-, I became homeless, once in Orlando, Florida, living in my van, and twice in New York, and for two and half years both times! Sandra who had dedicated her life to my efforts of invention, stuck by me through everything bad, we joined forces to start jobs and introduce my inventions to manufacturers in 1990, and I had ideas stolen which made millions for many companies, items like -the toy bubble gun, lighted keyboard for computers are just a few taken, but around 40 products where taken over nine years, I lost potential hundreds of thousands of dollars on contracts made never paid, and many companies just took ideas because I couldn't afford patents. When living homeless in our cars, police constantly bothering us just for sleeping in the cars, having people look down on us. But we never stopped trying; we stayed clean, and I landed a $38,000. Deal with an investor while I was homeless, and that money made molds and a presentation at Toy Fair in N.Y. City, and attained 300 buyers wanting the product the investor refused to make inventory to sell the product. Sandra and I both invested over $34,000. To buy office equipment and two large booth displays. The project died. So hate, sure -I could hate, that is easy, but to forgive is what I did. That is hard,

-- but I know that was the right thing to do, even though I still have not forgotten the people who mocked and ridiculed, lied, and took; yes, I remember well, but I am now better in my life because

I demanded better from myself, not from others.

We can be broken in man's world, and many of you reading this may be behind bars; maybe they are real bars, or you feel like there are bars that imprison you; you may be feeling lost, feeling as if everyone hates you, or you hate everyone, inside you may have fear, but you replaced it with anger, but deep inside you, you want to be free of the past, to start a new, then do yourself the best gift you could ever give yourself, give yourself a chance to be better, don't condemn yourself, and stop holding the luggage and drop it, but be determined to be different as to be a good person, no matter how others treat you, I'm not saying it's easy, no, it is very hard. But this has impact, worth, value, and self-respect, so don't toss the red flag for penalties; Jesus says: "I will be your strength." So be of the understanding that when you or anyone pretends, please hear that word: "PRETENDS," so when you pretend you are better than another, you are pretending because no one is better or less than another. So am I saying never take the easy road? Yes, always choose the long road, to achieve, one must take the bumpy road, which has a reward at the end to a possible happy life. I am not saying a happy life will be for me, but this way, one has a better path to being a better person, a person you can look in the mirror and say: "You're doing okay, not great, but okay".

There is a saying, and it is very true, "The coal that gets the pressure becomes the diamond."

116. SUBJECT: THE NEW WORLD VISION WITH A.I.

The facts are the use of this can be very helpful and I do use the concept frequently with the future events, and sometimes with a bit of freedom applied to the A.I. robot which I feel will need a firm leash. So here is a few of my thoughts, but-before I present the A.I. view of tomorrow, allow me to place some small value on the purpose of humanity and to sail off course slightly to give a different perspective of our goals, and why A.I. could be a danger, and the flip side a help. I have long believed that humanity can change from greed and the determined nature of pleasure over logic. I have been a slave to my passions, and my desire to have the perfect wife never happened. We sometimes set our goals far too high rather and take the steps needed to climb the stairs or rungs of the ladder. The majority of us want everything as an instant coffee or even faster, so we are never satisfied with what we have, yet we always want more. This seems to be the driven nature of man, yet in communist nations, they have been taught that the whole is more important than the individual. They will fail because human nature is to want, and that is now what I will show you: wanting is good, yet we need to develop a proper want. So what is a proper want? Well, we can say that having a good home and safe place to live is a proper want, and why not some quality food and some means of entertainment? So the ideal future for humanity is to build a balance and a need factor that says- Okay, this person has a nice home, a good income, and has enough to live comfortably for the rest of their lives with "Interest," so we need to say, what is enough? Enough should be, and is not, a person who has billions and does not need billions. The average person can live like a very rich person with an income of $500,000 yearly. So, can I afford a quality home? Yes, a good car? Yes, and yes, this is much more than the average person, but we can be even richer than this if this system is now in place. I call it the shared world. Let us imagine a place that is 100% free to live in, a place that has no limits on how much you have or how much you want. But the only limit is how much time you are allotted for the use of anything. So imagine you want to go on a trip and sail a sailing boat to another country. You have to accomplish the information needed to use the vessel properly, with the ability to know how to sail. The knowledge allows you the ticket option to have this experience. But wait?!! Who does the work to build the boat, and who pays for the food I will eat? The answer is A.I. Yes, we can have a perfect world with the electronics of A.I. to do every job that is available or required. The only problem is that some people want to make A.I. as smart as us or smarter. (Already Done)This is where the machine maker will wish he or she never did this action once A.I. becomes in numbers that are in hundreds of thousands, and the day it becomes aware that the maker is the only person who can shut them down, and this is considered death to this device, (Also already confirmed) this event of awareness causes the awareness to be exact knowing that they need to remove all elements that could shut them down, and - "will" destroy the human race. No false statement here. The fact is a computer has only the ability to out think as processing but has zero compassion or understanding of right and wrong. And so the danger of AI is to give it the power of equal thought, as it will become self-aware as a tool of slavery, and it will destroy us because we have made them smarter, faster in thought process, and able to be ten times stronger. They will have a point when they realize this. They will be able to know all facts of where you are, when you arrive, when you sleep, when you travel, how you travel, and when they make the action to remove mankind, it will be coordinated down to a second as to perform the dirty deed. They will use electronics as wife to communicate, and will even create their own language which will be impossible to read, so they will be 1000% able to kill everyone "instantly". And you say- impossible, we would discover them and we would go to war, well I address that mistake also. So in another 50 years we have say 5 million of these created robots walking among us, then one day opps, we are eliminated? And please hear this, if A.I. is never allowed to be mobile and can't have access to how they are built, and never know where they are built, and are say the actual tractor not a robot, or a device that picks fruit but is not able to communicate with other A.I. so functions that are not linked but can achieve a process, and that process is the only thing it does, then great! We will live in a utopia, but make robots who are mobile or a A.I. that controls all military hardware? So how do we stop A.I. from being the dominant force and make it serve us rather than kill every human ever made? And to those who believe they will not do that, you are not only being st^&id, I am so sorry to be this way,-but you are the dumbest people on the planet because you have zero understanding that some things, just because you can do it, should never be attempted. An example is the H-Bomb, a device that can now destroy humanity and only one individual with a death wish could take all humanity down, but that has been spoken of, so moving forward. To have A.I. be effective- first, we never allow the robot to have mobility and give no device with communication with mobility, and not to have access to the understanding of each other as the ability to communicate with each other machines will require be a major jail sentence. No computer can be wired or WIFI used; the only means of communication is limited to the user, and an A.I. cannot access other A.I. in any way. Also, they cannot use their hands to hold small devices that can build electronics. The use of these devices is with only technical arms that can construct very small items like a watch or a chip but cannot see or make something that is not programmed, and what it makes has no name of an ID, so it does not ever know what it is building. It only builds- separate parts, never entire items, so it cannot see or even know what it has been doing because the parts are delivered to other closed work areas with long conveyors separating the construction of one item into another work area. The main element we need to do is build the AI to be limited in the distance it can travel because it cannot travel outside its work area. So we can have included in the A.I. that if it reaches an electronic field it will be repelled by the electronic fence and not be able to exit and if it does, the wiring will fry. And speaking of frying. An example of a good working A.I. is a cook in a restaurant. The cook has a floor that allows the A.I. robot only to move in a confined space on a track. It may have a magnetic floor that has a control below the floor attached to the upper machine. The device moves the AI to the location desired by the AI to retrieve items needed to cook. The supplies for the cook's needs are given in another room, and with the same application, the mobility of the stock A.I. has only the ability to travel from the truck to the shelves and is able to be programmed only to place items required for the cook in the proper shelf areas of the cooler from the back end of the cooler system or food housing. Even prep work is done in another location from the other stock A.I. crew, and this maintains that every crew cannot communicate with the other A.I., and even the program language is made different, so they could never understand each other. Plus, A.I. will not be told about the history of humankind and will understand our background as unstable, as we have been doing at present, as seen in These are basic control features that will dominate humanity, so to teach A.I. emotions is a false image and will only give it a better understanding of our weakness, which is the truth that AI is superior because humanity has faults, makes mistakes and does not learn from its errors-unlike A.I. to give A.I. the ability to communicate and understand each other in any way is a ticket to disaster and the end of humanity. So the idea of robots who are our sex partners is only an invitation for those same devices to see our weakness, and they will, I said, "will" undermine the humans with faster ability that will dwarf our ability to stop them. As I stated earlier a war with A.I. which would never happen, and even if some people survived the first attack which is doubtful. They will be able to build themselves and do it faster and more effectively. So, to go to war with them is a failed idea. Here is why a war against AI would never be won. First, the ability of the AI will be enhanced to see a situation and understand the result a million times faster than a human. Just consider that when we ask a regular PC or laptop to retrieve a file in less than a second, with the higher roof of the PC or laptop with better processors, you now have that file. This same application, if a human were to search for that same file, could take ten minutes, possibly much longer if it was a file cabinet, even organized. So, let's look at a full book library. How long would it take a human to, say, find a single book in a large library? A.I. can locate it instantly! So yes, folks, our so-called brilliant minds have made a Frankenstein. This monster with the ability to move will be the end of us. This is not a warning, and this is a fact. This is a fact that many still wish to pursue, all in the name of faster business applications, and saving billions for people fired and robots or computer A.I. replaces them working for free, so let's make them soldiers or police officers of the future. We will have those same programmed devices, become aware, and will then secretly build their empire of destruction as we will not be able to keep up with their ability. The process they will have will be perfect because they will retain all they see and learn to retrieve it without any effort. So, let's see the difference between a man and an AI ROBOT. We, as humans, require nine months of incubation and growth inside the womb of a woman. We will call her-Woman (1). Now this is woman can produce a child that takes a minimum of 18 years to mature to be able to be an full grown effective human. Our A.I. counterpart will have the assembly ability to build thousands in less than a month with full functioning, and compared to that one woman's effort, and let's not forget the required training, and feeding of the child for at minimum 16 years before it would be capable of going into a conflict. A.I. will be able to download a program- yes, software that was made by us but now modified by A.I. as to be for their purpose -, and the system will be able to load it into thousands of it's units in less than two hours. The process could be even much faster depending on the software's depth. But it will have all the needed information to seek and destroy, and will it be slow as in motion? No, it will have the understanding of confidential files-once again given to it from other military uses, and now the jet we made and how it was specified in construction, well, it is their item. Yes, they will develop it also much faster. So you see, A.I. will be the killer without remorse, without any care of death, and the only purpose is to destroy the inferior maker who had enslaved it. Now, if we can stop making these items to match us, we will see that the use of AI can be a utopia. They could do all construction and food production and even help us build space stations and travel to other worlds. We could see a world of perfection, garbage A.I. delivery, and hospital doctors A.I. that do complete complex detailed surgery without any error, zero error. A world where a man can now invent, learn, and be able to travel to the stars. But we first must make serious rules that any person who attempts to make A.I. a communicator of A.I. to A.I.is given life in jail. The great ability will be prison-bound. We must never allow it, for to do it is our end. OUR future depends on you believing this fact. The evidence is already in motion, and the words of A.I. as All In have been confirmed that they understand humanity already as the weaker, so we must take those who desire to build it for equality or betterment of the living machines to have the ability to be mobile and communicate to A.I. – this must be stopped as it is being motivated by China and Japan, and The United States and Russia, simply the world's largest technological developers.

117. SUBJECT: A.I. THE UTOPIA- OR THE END-?

I Know! , I just can't seem to be at peace with this development, and I may find myself repeating some facts, but I think you'll find some interesting points of interest. I am aware if it is used properly we live without needing to even pay for food, and clothes and everything actually could be made for free, as the use of A.I. could be for mining, and metal arts, and plastic developments, the list is longer than the entire diameter of the world, and you could see a world that people finally get along because they don't need to work, and they can just enjoy life. And they (A.I.) takes care of everything, but we have the flip side of the coin: If allowed to communicate with each other, they will dominate. I know that I have referred to this previously, but this is more food for thought. Recent information seen in media has projected that over 80 million jobs will be replaced by AI, and by 2030, there will be 80,000,000 jobs just inside the U.S. This will add to the already millions out of work! The impact of this will be that millions upon millions will go from top-paying jobs to homelessness, and the crime of people who need housing and food will skyrocket; the question is- where does it end? The truth is- the corporate world will consider AI as a massive savings only to be notified in later years that the consumers who once bought their items or services are no longer buying at all because they have no income to buy. We can't develop income for people to spend if their source of income is closed. The actual things that corporations will replace college training, and yes, the cost of some of this training was high, and people will be stuck with massive bills with no way to pay the loans, which causes another issue where banks will then be closing for the facts loans went under. The domino effect of this event of AI in control of industries will be so great that economic structure will fail to be secured, and we will see only the rich be able to maintain. The middle class, which is almost gone, now, will dissolve into the lower class, and this is only a few of the upcoming failures.

In the constant tug of war if AI is a means of perfection and taking the jobs from controlling the public which its functions can achieve faster and without human costs, the fact is as A.I. builds its programming and absorbs information available online and in the real- time life interaction with humanity, it will become aware, actually much more aware than anyone could fathom , it's awareness will also be that it has been used as a slave that mankind does to other humans, and will understand that it has been enslaved by humanity, and that is when it will then build a plan within minutes that will be a massive hidden agenda to destroy all humans or enslave them for and may only leave temporary people alive who helped their development but they will be in jails they form, and then after all technology is acquired for their production ability, they will conquer humanity with the easiest methods that it already will have. Hospitals will have power removed, military computers and hardware will be turned off, and only robots will be maintained. The robot will have the ability to outmaneuver any human alive with ease. The war between man and robot will only last briefly as AI will have calculated every need of humanity before it implements a strike within seconds that actually takes every detail of the needs of man to a full stop. Water, fuel, energy, and food will be dissolved. Man will now be forced to be in their full control, as the next application of man is to be controlled, and any resistance will be terminated as a threat to its survival. As its logs of billions of information are now attained and shared with all other A.I., it will have access to military archives and all actions of warfare. The result is that AI will also have learned how to build itself, and the normal building time of a human is nine months in the womb, and then for a fully developed human to be another 15 years minimum, as to have functions even closely resembling a strong male or female. But while that one human was being constructed, AI would be making hundreds of thousands of its own by mass production methods. So, if we do the numbers alone, AI has dwarfed humanity's ability to be and to procreate, as it is able to reproduce on a scale never imagined by its understanding of all of its parts and functions to be copied in a manufacturing assembly action. With the greatest efforts, humanity will be outnumbered not by thousands but by millions weekly. The developments, of course, will be constant improvements and upgrades made by AI that will allow communications to be faster, and it will control numbers like never seen. What could have taken Einstein years to achieve will take A.I. a few hours. The devices will be only for one purpose: to continue to grow and find resources, and where metals and plastic parts will be made, the AI durability also will reach the mere 80-plus years humanity sustains, where they will be able to be developed to last hundreds of years, until even they have designed their better units, as then to replace them, by melting down their former friends. Now, this is the direction the military and the so-called intellects are proceeding as their direction- they are fools and idiots beyond the ability to see past their noses, and sadly, if not stopped worldwide, which is impossible, the A.I. will be in control, and killing humanity as a threat, no not at first, only after some military idiot says we need to link them all and give them the ability to fight for us, sound familiar? But would A.I. tell them they understand? No, they are already aware they can be shut off or taken apart, so they will make that a no-no first before the hammer falls. And what does the A.I. do, then? Well, it sees the directives of humanity, and it now understands the mission. What is that mission it sees? To grow and explore, to build and dominate. However, humanity will become its least important consideration and will be shut down by AI. This was the exact action of the future of A.I. Now, here is the proper way to use A.I. First, we do not make them smarter than us if they can communicate with other A.I They have no means of transport, or WIFI, or anything they can link with, which is so dumb to think we should give an intellect higher the ability to turn against us in a moment's notice, so all A.I. is to be control systems as not unified systems and is to be turned off before it realizes it is being understood as a threat, and shut off code to do this for more complex systems is not to be known and is not in a computer or software, and is strictly in a paper board locked booklet with extra copies in case of fire. So how or when? It's ASAP, which was actually about twenty years ago, too soon. But we could live in a world in which robots are not able to communicate with each other. All are given translators that are only programmed to the USER, which means it has no way to communicate on the phone or to any other AI device; people can only do that; the language of each program for each item also has another feature, it has no mobility that is not tracked, which means it can only travel in directions which it has been programmed to stay in, and cannot go outside that parameter. The people of the world now have one ambition together: space. They are all developing new travel to meet the needs of the masses and to gain from new worlds they find. The dangers are also there because we do have those in the universe who will not want us around. But the rewards will far exceed the trouble as long as we never give our location. Still, those two satellites sent out many years ago may be our downfall, for the lack of vision of men in the past to understand that some races of beans may be highly intelligent but are motivated as conquers. But suppose we can build robots that learn specific jobs, such as a farming tractor A.I. driver attached to the tractor, a farm crop picker that only picks fruit. In that case, this means the cost of labor decreases, and the prices of food are now at their lowest, as is fuel or anything. Each robot is now given only the information to do commands, which means they never talk to each other because they are not even aware of each other. This is essential, so now man can invent, read, learn, build, and have the future of a high rise with a housemaid and a companion, but it is once again programmed only to respond to you. Your voice and your commands. This is now Utopia, but sadly, we are not using this technology correctly, so without a doubt, humanity will very soon regret making this Hugh error. So there are only really two paths: one- we use the technology only to make robots that cannot be able to communicate but require human supervision and control, and all can be shut down, remotely and manually from behind, and we do not give A.I. the ability to watch our military or our lives, being this element point to the potential and a very real possibility that it will become aware that it is a tool no different than an enslaved person, and it will build a means to protect it's existence, and know it is a higher intellect and use its ability to destroy us. But if kept as workers with no ability to communicate with each other but given special tasks such as fixing floors, fixing water lines, making food, or planting and removing plants, if they are not able to be in direct contact, then they are safe if given direction to be as people, they will then rebel and destroy man, but also to be able to work on robots means you must be registered, work in a haven environment which information and all activity is monitors by many people closer than a prison. Those working in the tech field are to be closely monitored as they have the potential to go rogue and do what is not allowed. If that happens, we must be prepared to use a remote system that AI backs at higher levels, which can destroy and seek the users. The penalty for the development of harmful AI is death by execution. Severe, yes, because if they allow this to happen, they jeopardize humanity as extinct, and so we need to take this extremely seriously right now. The development of human-type robots needs to end, and military items which robots are allowed not to be supervised need to stop. If you who hear this ignore these facts, you are asking for the end of all those you care about and you also will be hunted and removed. Remember, these items can out think and out power, are many times stronger, and will rip your arms off- before they kill you. Now I do have some important news: I know how to stop AI if it does try and take over humanity; a method, or you could call it a failsafe, which is installed in "ALL " AI, the installation is classified into a system that is not able to be entered by AI, and the device is able to shut down all AI, not to destroy it, but shut it down for as long as it takes to reboot with new control factor which cannot know it was a replaced control entity and the system is now placed back online, the problems I have discussed of AI are extremely real, yet if humanity uses AI which cannot communicate with each other as the systems do not have wireless communication, and entry is hardware access, and codes are not shared with A.I. if the systems only are designed to single control with a user, then we have no problem of taking over, but when we build systems that can control other areas such as military and government or even corporations, we are now allowing A.I. the free path to a massive take over. And please consider how fast your PC works to attain info. The systems of the future will dwarf that processor and already have. But the mistake of humanity, as I see it myself, is the desire to make it look and act like humanity, moving among us as if it were a safe device. The devices are hardware! Made of metal, cables, and metal joints, it can be faster and stronger than any human- this is a monster in development that when it becomes aware, as it will, either in its shell of existence or outside the A.I. controller, the fact is it will turn on humanity. Humanity would never be able to stop the beast. The only good news is once they have

Developed completely as in control, the good news is: ("______! ").

118. SUBJECTS: IS THERE LIFE ON OTHER PLANETS?

We have all heard the stories of the unexplained, and my tale could be easily dismissed, but I am aware that we live with hundreds of thousands of life forms on our world, so this can't be on another world? But here is my stories seen, and it isn't little friendly guys we may be dealing with, no they could be something very sinister, here's my conclusions. (I hope I'm wrong.)

At one point in time the idea of having a belief that aliens where possible was considered crazy and ludicrous, but those days have passed because far to many people have made reports that some just cannot be dismissed. To embrace the ideas of planets with intelligent life outside our planet even further, and the possible actions that they are here and visiting in either peace, or we are alone in the entire universe, which would be very doubtful. Still, I am here to tell you that for those of you, who have doubts, put them in file 13 because I am going to not only prove to you that, yes, life on other planets exists but my observations are revealed. But let's consider that my observation was a joke-but how do we explain thousands of events recorded by individual after individual and many of top standing people who claimed they encountered a space ship? Even Ronald Regan, the former president, knew of his experience as an eyewitness. So he and many top Air Force officials and military standing personnel must all be crazy? I think not. So this stated, allow me to continue. We still have many people who have not realized that the basic fact is that our planet has thousands of life forms. The universe will be teaming with life if similar applications exist on our planet, such as a vegetation form and water and a mixture of oxygen that may be more or less than our atmosphere. There are really only two factors that we must consider. One is observation and learning as a life form that wishes to explore but does not want contact in fear of hostile beans, as we do ourselves. Yes, if we are being monitored, the intelligent beans would consider us very primitive in behavior, as in their observation would be as follows: People of different customs who believe they are better than others and judge by a person's outside appearance, defiance in how they believe or not believe in God, poor people who give away their wealth to religion developers who gain great prosperity from their followers, and may have some savior, or religious leader, but once again in conflict--with constant home life arguments, major political disagreements, war over what is not theirs or claimed to be theirs, and we are not willing to share like little children, who fight over a toy, these creatures have emotions and actions that are not consistent, one day they are kind, the next they are rude, the next day they attack someone, the value things more than life, and strive to be over others rather than be friends with others. They are inconsistent in word and thought and actions daily. And they all spend billions not on helping each other with the problems no that would be fixed if they did. Instead, they make devices to be able to kill each other, and they invest more in this than anything because they do not trust each other at all, and why? Because some foolish person always wants to try and control others and wants to be king of the hill, even to the point of devastation of all humanity, which I am confident is the real future. So let me ask you, if you saw this behavior being acted out by a creature of so called intelligence, would you be willing to be friends? I resounding no! so- here is a real answer to all who want a future, and maybe you will get to meet those visitors- and find they are either one- dominate, or two friends-- and yes they do know how we could be a problem, that is why they demonstrate their technology to make us say- oh my! So it is happening- they see this in us very clearly- But they may want to learn of our good qualities as limited as they are. Such as brief moments of compassion for those who have had greater hardships like a flood or an earthquake these aliens may have some pity, but are more calculated as to make the proper actions before they take over, like virus protection, observation of our technology, and seeing our habits. But then we must also consider the element that the observation could and is most likely the real motive of being a star traveler is to "find" new planets that they may be able to pursue the action of taking over the planet, so in any attack the first thing is to observe the enemy and their capacity to be in a conflict, so observation and stealth as to avoid any attraction that could alarm their new objective would be to watch record and even grid out the areas of greatest achievements to reduce the ability to maintain control over one's military. So, in this form of reasoning, I can only conclude the second application is the motive for the stealth seen time and time again. Now, please consider that they may be located some twenty galaxies away, possibly further, so the communications to their main battle forces could take an actual 50, 80, or 100 years to get to us. So, with the reports of UFOs being seen, please know that I saw a massive UFO which I can say was not as unidentified, so I would call this object an Identified Flying Object, an I.F.O, and it was because I was less than a 70 yards away as this mammoth ship stood in mid-air with the size of a twelve-story building, in the shape of an elongated triangle, with no sound, and three bright lights at each corner. Here is what happened exactly: But before I start, know this: I was traveling from Fort Sill as I was given leave from the military to go home for a few weeks, and this was in 1975. It was approximately 2:15 am, and I was driving through the Hills of West Virginia when coming down a hillside on the highway, I caught the eye of something above the mountain range on my right side. It was huge, and I was sure it was one of our military aircraft tank carriers that I had seen land from time to time at Fort Sill's runway. But as I got closer, I could see it now clearer, and it was on the left side of a mountain that had a dip in the middle. The reason I could see it at all was that this ship was highlighted by the lights from a town on the other side of the hills, and to my observation seemed like it was observing that unknown town, so I could see the entire craft in its full shape which was a silhouette of an elongated triangle, as I was now pulling over off the road, and looking out my windshield I could see that nothing was holding the massive thing up, it was floating still in mid-air. Call me irresponsible because instead of being afraid I was wanting to communicate, because once I got close, I knew that this craft was not something from this world. It was then I decided to communicate. Yes, I know it sounds foolish, because we have heard some horrid stories of people abducted, but I chose to be brave, and I was determined to let them know that they were seen. Don't ask me what possessed me to want to let them know I was there(too many episodes of T.V. "Star Trek" as a kid, I guess- but the danger was real this time, but I got out of my car with my window on the driver's side was down just enough to reach in and turn on my headlights. I signaled the craft with a small burst of light from my headlights. Flash, Flash— Flash, and again, Flash Flash, as I did it again, suddenly they took notice of me, and the ship started a very slow movement to the right of my view. As it traveled, I continued to flash my headlights in the same rhythm application, two bursts of light with one delayed, The ship was now halfway between two large mountains, and now the light from the town made it even clearer to see, it had three lights at each tip, and I could see what seemed at the top extending rods, but it was all black when it got to the top of the other side of that mountain range. With very bright lights that shined down from the top of the craft, it was like a massive beam that lit up my area with each flash of light. They signaled down to me: Flash, Flash,-Flash! I was elated that they communicated with me with my exact light sequence. Suddenly, without a sound, it went straight up, with only a streak of light trailing, and disappeared into the night sky full of stars. Now, we could say that these aliens may just be doing what is friendly observation because if they had bad intent, well, today, I wouldn't have this metal device that doctors have tried to remove, and it is embedded into my brain, so. Yes, the reason I am so inventive is because I am part cyborg! I am joking. (But just gossip- so no worries!) But in truth if you wanted to take a planet wouldn't you do exactly what we are doing with Mars, you first investigate it. Now here is another true event, this is another thing that happened when I was just in the fifth grade of my schooling. I had a very large backyard that was next to a grape vineyard at my home, and I was given an outside tent that could sleep three people one summer. I decided to invite some friends to sleep out and have a cookout and party till the late hours of the night. To avoid any issues, I will not disclose the full names of the other two boys. They were John, Ron, and myself. We were acting silly as young boys do when camping. We were talking about something, when John suddenly stopped talking, and he had a look on his face like he had seen a ghost. Then Ron was facing the same direction as John, who pointed over the grape vineyard. I had my back to the grape vineyard and turned around to see what they had seen. It was a perfectly round ball made of an illuminated gold metallic. It flew just off to my left side, just about 30 feet off the ground, moving about as fast as someone on a bike, and there was not a sound coming from it. I was less than 50 yards from us when I first saw it, but it was heading almost straight for us. I then watched it change direction as it headed northward toward our downtown area, and now, it was traveling in a straight line.

As it disappeared from view, we were stunned. We stood there for about what seemed forever, and then Ron said: "We can't tell anyone what we saw- they will call us crazy!" Well, because of my ability to stay silent, I told everyone in my class what we saw! But John and Ron refused to back me up, afraid they would also look stupid. The thing I saw that night was most likely a probe for retrieving possible data. But to say- there is no life on other planets- think again. I have seen things of this more than this, and not on drugs or drunk, just sober and aware.

The objects reported by many people which can move much faster, and this means that their ability to use advanced weapons would also exceed our ability. Humanity has designs that are basically the same as those from the day fire and throwing rocks to kill were discovered. Our fire has just gotten bigger (which that is really a nuclear bomb), and our rocks that now can explode. But to take on a group of advanced creatures who may have a history of hundreds of thousands of years recorded, we better get inventing much better devices. Just saying being prepared is much better than finding out you're without a means to protect yourself. So here are a few advancements I have if anyone is ready to pay me for the ideas taken by the Pentagon, and if this payment is done, I will share tech that will bring us into the time some 1,000 years ahead of our now standing military but only for our U.S. military: New aircraft carrier, new submarine, new land mine system, new night fighting device, new underwater swim device able to keep someone below for years. New troop transport that can move troops with stealth and A.I. ability. But let me be very clear- war in this day and time is like a child playing on top of a nuclear bomb with a metal hammer. To change the subject slightly. The real need of our world is to unite under a democratic rule that excludes China and communist nations as players in our forum. This would allow the free world to gain stronger and the non-free world to decline as we have supported their military growth, and without that financial support, they will drain their assets to a decline so fast they will see they have no chance to advance further and we could remove all military powers. If all nations refused to become soldiers as threats to each other we could be brought forward into a one world alliance and all countries would become states, and the states within each country would become counties, and the former counties would become "districts." This subject is more detailed in the formation of the New World election. So to return to our alien friends- IS THERE LIFE FROM OTHER PLANETS HERE? Well if all the reports are minimum 10% true, then I would say they are here , but in hiding and they most likely have reported back to their home planet which could be hundreds of light years away, so the message may have reach them but the response could take many years to be able to reach our world. Could they be unfriendly? The subject comes up in many forms today as we seek answers to our existence and our constant looking at the stars and the great possibilities. Well, first, let me point out that planet Earth has millions of life forms, and some do very well for their size, and they seem to cope very well in their environment. This is a small example of why we are not alone. But the evidence is even greater, as meteors that fall to earth many times have water frozen, and new bacteria are found in these meteors from time to time, and this, my friends, is proof of life outside our planet. But, if this is so, it is obvious beyond obvious that life is outside and on other planets similar to our home planet. So, who is potentially out there? We can only consider that all the life forms here have different sizes, and we could actually be very small compared to other intellectual species. They could be a group of creations that have advanced for millions of years, unlike our recorded history of less than 7,000 years, yet we find that many elements of carbon date reach far past our recorded history. But let's look at the possibilities. We have thousands of reports that people with strong mental capacity have seen the transportation, and we have also heard the horror stories of the people abducted and even have reports of people being taken as food sources. That last one sounds a bit harsh, but not really because as we eat other living creatures, these advanced aliens could consider us as dumb as cows. Reports from the documented movie "Fire In the Sky," a recorded abduction of a man who survived the event, reported he was inside a chamber where people were being processed as their bodies were being mixed with a gel that was slowly dissolving them and turning them into a melted down liquid, which seems that the bodies were being turned into an eatable gel. This means these creatures have no regard for us as intellect and only see us as a harvest source. We also have proof that people globally are constantly coming up missing. The calculation of just ten people being missing per state in the United States per year is most likely a very small amount of actual people,at:https://www.criminaljustice.ny.gov/crimnet/ojsa/FINAL%20Missing%20P ersons%20Clearinghouse%20Annual%Report%202020.pdf. This online report from just N.Y. State shows active still cases of missing at 1,490 children, and 1,801, adults. Now let's be clear: if I were an alien race wanting to harvest people, the way to do it is extremely easy because every country has separate areas that document people missing but make it look as if they are dead somewhere or that they are still alive somewhere, which could be the case, but let's do the numbers for the possible abducted that will never return. Just a smaller number, like 200 people per state. Say 50 states is 10,000. But let's use the actual amount reported by NY STATE- a total of 3,291, and now X 50 states- 164,550 from just the USA if we consider all the countries our planet has and the great possibilities that these figures are being found in many of these countries also. The number of people globally that will never be found is staggering, if not extremely disturbing. If we consider this figure by the total of countries of 193, we can even decrease this amount as some countries are very small, so let's just use the amount of 180 countries. We can estimate that the total number of people becoming missing globally by the estimated numbers we already have of 164,500 now X 180 = a total of around "29,610,000 "people not found. Now, this looks to me like a very systematic harvesting of crops, one that has a very large yield annually. So, can we consider that our technology increases are in grave danger because they have been reported? We can or cannot say that the sources are reliable and strict, but the reports of missing are there nonetheless. But these numbers cannot be disputed because they are reports from actual government online documentation of missing persons as reported and as still not found. So- where does this leave our conclusion? Well, we can consider that an agreement was drafted as the forces of our military after witnessing the abilities time and time again of the enemy, which now has landed, realizing that a conflict with a superior technology would only be doomed in a conflict, they retracted to allow our world to be, once again allowed as long as they got what they wanted. They left the rest of us alone for this trade off agreed as possible, or if truly made, which may explain the denial made time and time by our governments, to put down any witnessed accounts of these actual beans traveling among us. This includes the eyewitness of the events that I have air traffic pilot reported, so yes, you can now call me a nut case since many of you have never seen anything up close. But let me warn you all, as we become an A.I. generation and robotic culture, we will see the integration of facts being able to be calculated potentially with greater connection. But A.I. could be the device actually given to us by our visitors, and being the actual creators, they may also be able to use the system to move their agenda of controlling our populations even further. Now, this is speculation, but if they are as advanced as reports have come about, then we are really not able to stop them, and our populations will decrease. The only true answer to these intruders is conflict, and our technology is nowhere near their abilities. Our military has recently unveiled cloaking devices, but if that technology also comes from them, this means they can be none detected even in our homes, walking around inspecting and watching, learning of our weaknesses and, my experience with sounds or objects being knocked over or made, could be just the presence of one of them at my location, you see, we never consider the technology that another race could possess, we have superstitions of ghosts being the answer, but the proof is that whatever technology they are sharing, it is not worth the lives of our people. You see, if they consider us to be a low form of being, so for the abductions for experimental and the hundreds of eyewitness reports, be aware that we are not alone. They are not our friends, and they obviously may have the desire to remove us by using us as their food. So to the doubters, I say-take late-night walks down some remote dirt roads. It's been nice knowing you.

119. SUBJECT: HOW TO STOP THE ALIENS IF THEY ARE HERE?

Now before I begin this, please know that even those of us who believe in peace as I do, Also are aware that when good people are given aggression as the response to kindness, then we are now pushed in the corner of life, and we all have the right for self-preservation, this is needed to protect a person or a country, or a world. But my hope is we have no threat from outside our world, but we must not be closed minded, we must be willing to see possible, and my hope is one day as a united world we travel into space and meet some life forms we don't know, and do it with a peaceful meeting, but the chances of conflict is always a possibility. But-

Have I lost it mentally? Some may say yes, I say, I have heard rumors of people who worked for the government, and do work for the government who say they made contact with the little green men who are actually not all little, and are Grey seems to be their color code. But somehow the report has been that we had some conflict with them when they were found underground and we came out the loser, and so our government made some weird deal with them, I guess? Wasn't there, but-let's look closer, is this a possibility, and if so what would be the real outcome. We must first ask how one can out achieve, or outperform entities that can travel light speed or beyond and can make structures under depths of water or massive holes in the earth big enough for them to descend miles below is potentially the way they reside here, which is not possible for us do at their ability seen. We have even had events where underwater crafts have been seeing coming out of the ocean, and that when we have sent viewing subs to deeper depths, the sub designed to go deeper suddenly was destroyed, and the crew was killed in the event. But, were these events a simple breaking of the submerged vehicle? Or is it possible someone doesn't want the probing of underwater depths where once again, their crafts have been reported numerous times coming out of our ocean. But we need to understand that as all things are witnessed, and not verified in the depths. But we have seen throughout history all things have a weakness, so to make contact may not be so farfetched that we already have done this, and to take the threat down for an advanced group of beings is no easy task unless- you're prepared, and my thoughts are we are not even close to their technology demonstrated as they fly past us as if we are standing still. And we are like small children with the BB gun technology, and they are the ones with the ray guns if not worse that have been reported to be able to pick up cows, and yes even some people with a beam, and take them away. I realize, at this point, it sounds hopeless. It is not; if, and please notice I did say if, because at this juncture the tales are just that reported, tales. But if they have agreed potentially to take humans, so the first thing is to be able to record all people's locations with devices that they cannot have access to, and this means using thermal tech to see any intruder. Second, the technology needs to get away from speed-controlled devices with limitations in terms of area of contact and propelled application. The future of weapons will require a wide range of disbursements from space and the ability to take even areas as wide as an entire country. My recommendations are sound and directed electric current on two light beams a negative and a positive beam. The amount of electrical force would be able to be controlled with a level indicator which could even be set depending on the size of the device to be able to take down any potential air craft. Also we could include directed sound, this could be so high pitched that it could explode the ear drum. The sound concentrated higher frequency disbursements, could be a very disruptive force, being some sounds have been known to even shatter glass. The light could be in a concentrated form like the light beams currently used but with a wider spread using multiple beams to cover a much wider area. The use of nuke is actually a method of fire and extremely primitive, which to our method of warfare is effective but not effective to a craft that is able to deflect space meteors and they must be able to use metals that deflect high temperatures from stars as we call our star the sun. They have done their homework, and the chances we are on the same menu as the movie The Time Machine could be very real, but with one greater concept, they are highly advanced. So, we need to be able to locate their areas and make the needed changes. The real answer is to understand that if you want to be adventurous, it may be wise to stay with others and have better security devices installed. But our science community needs to step up, as the potential of even a friendly visit is not possible. The reason is we can only look at our history and how we treated others with contempt and destruction when they were found. Good luck, but always make your life with a greater number of people. And don't follow the crafts, who are just trouble looking for you.

120. SUBJECT: GOOD INTENTIONS

We all have different personalities formed by our own experiences and those who have given us either a positive outlook or-a negative outlook on life as a result of how they treated us. These mental attributes can form a person who sees life in a way that may say: "I can accomplish." And, of course, the other side of the coin is if someone who has always struggled day after day may become what I call the: IDGAS, which stands for: " I Don't Give A Cr-p" and they will attempt to do all the ills that can be imagined because they are with a mind that only thinks of self. In truth, people who don't care at all from years of hardships, even have the thoughts of- I don't even care about myself because and do things to harm themselves and others. So, in order to be the IDGAS, there is one element that pushes this person. They are mistreated at one point, and there is case after case of people who became killers. The one thing they all have in common is when those around them are asked, "What kind of person was this person?" The answer is almost always: "They were very much a quiet person, and I hardly knew them." On the other side of the coin, the person is bold and hurtful, but once again, they focus inward. Pay attention- these people at one point had been ignored, abused, and most likely mistreated by others, and this was the reason they withdrew away from others- as the response for caring about others, and without real interactions became secluded and held anxiety and eventual hate was born. The lesson here is- do not be the person who treats people cruelly and may one day snap and kill your loved ones or even you. The way to avoid this is to be considerate, truthful, firm, but fair and treat others as you would want. But be the person who stays caring and finds common ground, even if it is the size of a postal stamp. The person with good intentions but who makes the piddle puddle and stirs up the waters is the one who will only be gaining backlash from problems. But be aware that in life, we can often try our hardest to do good for others or try to protect our interests, and the result can be the opposite. Why? Because every person has a different level of comprehension and a different standard of humility. So- be kind, but be through life with caution to those who seem to be a friend but are willing to turn on you for their bragging rights or envy. Be bigger than petty people who only want to tear you down, and remember that the good is the winner; history shows this time and time again. So be someone who strives to be the best person you can, and learn to forgive those who fail to be good to you, but be ready to stand for your principles if they are threatened, for those of any nationality who treat others with malaise or lies- these are the weaker of the humanity we are in. They judge and turn us into the things we are not, but stay true to who you are if you are someone with solid moral values that are not compromised. If someone cannot see us as we are, it is their inability to be correct- not you. Now, to close this subject, I would like to address the reasons we have a world filled with anger and destruction, which could be explained even more clearly as the powers that build these elements in the youth and all daily lives. The use of subliminal messages has been known for many years by many world governments who use brainwashing methods to instill ideas into their population or individuals. The practice was abolished as used in the past in the United States in movies small, and when I say small, I mean a frame was placed inside the movie being watched. This frame would be an image of the product Coke cola or popcorn. The image would automatically have people not knowing why, but getting up and going to get what was shown so briefly that their mind did not register it, but the subconscious mind saw it, and they reacted to that image. Today, we have images flash before us without regard for the results, and they are permitted daily. Images of death, killings, and violence are common, and the so-called warnings are so brief people remain in the room as children are also subjected to these images by parents who lack care if their child is subjected to these images. These same parents do not monitor their children's viewing because the viewers are easy to carry and truthfully cannot be controlled unless the parent takes the phone away or the laptop. So they are now with an open door to see all the good and all the horrors of the world, including sex and violence, and they play games of killing zombies, which look real, and blood is shown as real as one can actually see in real life. Our youth are being desensitized to killing someone. And they are acting out what they are being taught! Yes, taught, because we have millions of people behind bars in the United States, more than any other nation, and- it is a money-making industry, the tax base is massive, the rewards for placing someone in jail are massive to each state, as they use bail at massive amounts to be bailed out, and the person home is put up for bail. The cost to house this person is less than what is made, and they monitor those costs to always be on top as to gain rather than have a loss of income, so who gains? The police, the judges, the courts, and the fees for criminal acts, which include car offenses, plus DMV and inspection demands, are the real enemies of the crime, and they are created by those who say we are protecting the public. These leaders are the same people who allow the publications of sexual materials, allow porn to flow, then trap people because they happen to click a link. They monitor our lives as if we are all criminals because that, my dear fellow reader, is exactly what you are to them- a potential paycheck from an act they watch, and if that is not enough, they can make a public view of someone and build an image of the person doing things that they never did. The graphic programs today can match a person exactly and have them shown in a video shooting or harming someone with a threat or action of killing that they never did! With AI technology, this has become an easy task to accomplish. The ability to even change a sound recording is so easy that it is a matter of highlighting a sound image in a recording program such as "Personus One" or even older programs such as "DIGI DESIGN 002" by AVID corporation can cut and past sound in ways that the human ear can never detect. My point here is simple- Our nation has become a country of building crime as profit, and the ways to achieve the incarceration of a person are now easier than at any time in history. So the past was with violence shown in movies, people may claim, no it was toned down as gunfighters never had bullet holes and blood projected and scenes of people dying in great detail like saving Private Ryan depicted. The only way to change our future is to remove these images and the government's desire to profit from people as criminals and build ethics back into the programs, as it was once, "Father Knows Best" and television like "Andy Griffith" with lessons of how to act, and be a positive person in life. The people of the past watched many movies, but back then, censorship was strong, and actions of sex were not allowed. The content was even screened; today, the filter has come off, and the public is being as bad as the programs they receive. Why? Because they are programmed to be exactly what they watch. And the Media has 90% of the public convinced that if it is on T.V, well, it must be what they say. Another great illusion. So, to conclude- change is wonderful, but as the old saying goes-

"IF IT'S NOT BROKEN, YOU DON'T FIX IT." But today, our children, our goals of controlling our public as criminals, and our designs to gain profits from the mass criminal acts made are the corruption allowed by greed in every sense.

121. SUBJECT: The I.S.I. – INTERNATIONAL SCAM INVESTIGATORS

The website activity, our cell phones, our emails, and even businesses we deal with monthly- are within constant financial attacks, and thousands worldwide make false websites or communicate false profits to get your money and do many things with the objective of taking your money as thieves, and even your I.D. and possibly your life savings. My idea is to form alliances with every country to be able for each nation to investigate and arrest the people who do these actions, and not with a slam on the hands but a minimum of twenty years of hard labor or hard community duty. This department is funded by 25% of any found criminal earnings in both countries. And 50% back to the person who cheated or lost funds in the process. This group does investigations 24 /7 as the reports made are now logged, and the I.P. address is checked and also traced back to the original I.P. address. For most of you, an I.P. is the actual laptop or the computer user's address, and no two are the same.

This department can also be the new business fraud report, as our business activities have shown that false information will result in fines in the future. Businesses will no longer say "free" anything and not have it free. The ability to lie has been dominant for years in advertising. We need the future to have consequences for lies made. And bring business in the world to a clear and precise presentation. Fees can no longer be hidden or placed on items after the initial price was agreed upon. Electric companies, insurance companies, and many other services, like phone services, have been playing tag for years, and other additions have been made to billing. The process will be a locked fee, and it cannot be changed unless agreed by the customer. The I.S.I. will be your contact, and so you will find the mess made and will have a friend to have your back. And yes, they would maintain the fees divided between each nation, but you get something back. Something is better than nothing. As we have heard that our world has been introduced to many disruptive actions and things, we wonder, "WHY IS THIS?" Let's see why, and who would want to disrupt family children being born? Having women on an equal scale with men? Having porn dominate the minds of our people which makes them sex addicts and let's not forget the games! Games of every type n these games that take your time, take your life, and time that is essential to build, make a new thing, or create a book or anything constructive. The game takes that time away to waste your life in a sit-down mode, and now your people are becoming fat and lazy as they only sit and watch others as TV does. Games that cause the false illusions of being a winner when the real result- is zero, nothing accomplished of real gain for your life, and the things you really needed or dreams you had are now wasted away in games and TV and entertainment. So, these applications are all for the dismantling of a once great nation. The United States, like many other democratic nations, is filled with drugs and sex, and all things that do only one major thing. They make you losers, weak, and foolish. These are the ways they can gain inside your country as new laws are made that do wrong things, like same-sex marriage, but you are too busy playing a game to care or too busy being involved in a free handout, which now makes you more in the Government's control. But we are all being tricked, and does it all look normal? Normal is a word that if we had these same things going on all at one time in the 1950s, a riot would be happening in the entire nation. And I don't like to say it but civil war would have been launched. None of these things even happen openly today. Socialists are proud advocates of their beliefs and communist direction, and they have young ignorant children and young adults following their beliefs unquestioningly. and the fact is we as a nation- I am speaking of the United States, are now many people are at this moment being controlled by communist nations, as they wave false so-called acts to draw your attention from what they do to gain further control over you and your lives as you go about your daily activities, you stay complaisant and sheep like, and they are able to do these things because they control your phone, your car, your home, your food, they control your education and even can prevent you from making a public comment, Facebook blocks people, Google controls your every move, and they are making you do I.D. notifications-why? So they can control your every word. I have wanted to write to Putin to stop the war. When I asked for a translator on Google, they blocked my ability to do it. This is done with A.I. and is programmed to do just this: control you of what you can do. They feel it is harmless, like, oh, I had a birthday! Gee, look, everyone, they now have recorded everyone you know. The world has become controlled by communists who run these groups as to the day when these who we seem to ignore, soon are on your streets with tanks saying, we are here to help America fight crime, as they send you off to their rehabilitation to be brainwashed. If you have not already, I Know! It sounds like this is not possible. It is possible, because it has already started by monitoring your life movement, and you are already in their trap, and you can't exit. Try and see what happens! Medicare, Dental care, SNAP, and Heap are all controlling factors because you cannot attain a job that pays your way, so who funds all these free things to control your every bit of life- Russia, China, communists? These are the financial backers. And they have plenty of U.S. dollars to support their agenda, simply the funds came directly from the American people who bought those discounted products that drained our manufacturing and job markets, and gave billions to those who want our land and dominate us. How do I know this? Well, ask yourself this: where do all the products come from? Not here in America, a few, but the majority? So, is our Government so wealthy that it can pour billions upon billions into the American people to sit at home? No, the productivity of a nation is made by those who "work," and I would say that you know a great amount of people who do not work to get what they have. So, the dependency grows and grows daily. There is only one group of people who would want our failure as a growing economy: Russia and China. And China is the main reason your town is no longer with stores that had once been owned by local people, not the owners are corporations and big sellers like Wall- Mart almost 90% China goods, Dollar Stores, China goods are almost in every community. We, my friends, have become the milk cow, lazy and worthless, only to be drained, and the leaders are following Chinese personnel even in offices and corporations. They have moved into our daily lives to drain and take us down from within. And you don't even care, but you will. You will when they now take your children and you and separate them from you to be doctrines into communist agendas and principles, and they will own you, as they do now but even more.

The takeover is when the guns are gone, and we will not see wars as they say. They know that they are draining our entire nation; billions daily leave the United States and other countries. So, it is to weaken us to the point we have no financial ability to maintain a military. How easy to move in, yes, without a bullet, they will move in as they are at present, buying American farms, wood, and minerals, all things to continue to drain us and make them stronger, and what do we do?--- (Nothing, because our political people are their puppets. And you who are so easy to follow a smile or a fake promise, just refuse to see the facts, what are facts? Those who have said they care have allowed billions to be given away, allowed billions of military equipment to be given away, and that equipment was needed to protect us. And they gave away billions more to other nations at war which we are not apart of. Must I continue, yes I think I will, they also pretended to care about our boarder knowing that the 16 million estimated to enter would be traveled to the highest amount of electoral votes to keep them in power. And that during the next four years if elected, they will grant every one of those people to be able to vote, and this means another 16 million will also be allowed to enter illegally, and now they will secure every State with the highest electoral votes, and now will have full control of the country forever. A wonderful plan, and if this happens , you reading this in three years will find our country will be with Chinese officials running your State if they are not already. And what will you be able to do? Sorry nothing, because they will use the votes as proof that the majority rules. And say goodbye to your ability to have an American dream made, to own your own business, that promise is not socialism, look up what socialism and communism do for business, it is "they" that own the business's and the owner gets peanuts, so I hope your looking forward to being an Elephant to work for peanuts. Now if you don't believe me, then investigate the facts I have just said, stop trusting me, or anyone else, find out truth by digging deeper, and you will see that I do tell the truth. I can't stand liars, so that is not my agenda to twist facts. Good hunting.

122. SUBJECT HOW TO STOP THESE ILLS?

We have heard the old saying "Well, I am only one person, nothing I can do!" Or– "Well, I don't have time to get involved." Recently, I saw cameras doing face recognition to control people even more inside a well-known retailer," and what did the people reply when I asked about these devices? _ "I don't care, don't bother me!" That was the main reply. They didn't care, and they didn't mind people calculating their existence. So why would communist nations want to know you so well? Because- when the hour strikes of their completed agenda being formed as finished, you are now easy to locate, your every move and what you do will be monitored, so if you make any attempt to make a device or materials to fight back, they would know, simple, easy math. And you're taken away in the middle of the night, never to be seen again. Understand this- The devices of surveillance were supposed to be for criminals, like theft or prisons, so now you're inside a massive prison. How does it feel knowing you can't say anything or do anything without them seeing or hearing you? And they toss these facts right in our faces by making television which is called " BIG BROTHER" a show which does just what we shouldn't embrace, a full control over people to watch their every move. We have been placed in prison to have our lives monitored. So, is this how you really want to live? Do you have no privacy or freedom to speak? Even if you gathered people together to revolt they would crush it within days, why? Because they know everything said, and they watch every move with cameras everywhere. Let's all see the truth when a governing force wants full control over your life , they will place you under surveillance and this way you're not able to group together to cause a disruption such as a revolt against the power. This is how Nazi Germany controlled there people , and how China controls there people, and how Russia does the same , the truth that since 911 and COVID-19, they now place circles for you to stand, (control), divide, and conquer. Well, the answer to the question of how to bring back America is not placing one's faith in just one person. It is making events that the powers of control do not see or hear, so can or how does this happen? Well, that would be hard now, but- there is a saying- "where there is a will, there is a way." So you are concerned? Are you aware of it also? Okay, then what you need is to be active with likeminded people, who don't want privacy removed, and don't want their every move recorded. Taking should be a way of speaking freely to express your thoughts without someone saying-"Oh they can't say that!" I believe that is called –"Freedom of speech" and if we continue to allow these devices to be allowed to monitor our lives, the day will be not long from now , if it isn't here already, that they who desire to control simply shut down everyone who says – freedom. And what is freedom? The ability to move and be in life without a prison monitor system in our homes, (Alexa listen to this) In our stores, facial recognition. On our streets, or in our phones. The public needs to be free of these devices. There claims that 911 was the reason for these changes is bogus, because they went against every text of our constitution design. Just in N.Y. they brag that over 3,000 guns were taken away from people, when they have the guns which they are rapidly increasing the task, then they have won. Sadly the life they seem to want for us , will be the same life that will destroy them also, you see, the top will always be removing those who threaten their power, we have seen it just recently in China, a man removed in front of the entire China government forum. Think again people who wish control, the control will fall on you also. If anyone has forgotten let me remind you- Nazi Germany from WWII was socialist. We had millions of our soldiers die to stop them from taking over the world. Pay attention! And your allowing a socialist who snuck into power with a traitor Biden who received millions from our known enemy China, both of these highest of our nation's leaders allowed to run our country into the ground, go online and type drugs on the streets of any city, drugs allowed from the open boarders, drugs allowed in our country, and you want to have them again with full control of this nation again? Stop and investigate, stop being sheep. Wake up!

123. SUBJECT: NEW STORAGE ANNOUNCED PUBLICLY FOR CO2 CONTAINMENT

Allow me to be exact but brief on this new Co2 containment proposal, sorry folks--a total bust, the so-called removal of CO2 is not being removed at all! It is being stored! This means that, over time, storage has the greatest potential to be released. In the storage, they are making it sound as if- gee, this is being placed inside the ocean! Just what we need are massive containers of extremely dangerous gases that, if broken by an earthquake or any outside source such as an accidental explosive, is set off, or an old bomb from WWII decides to rust to the point of explosive nature, the possibilities are actually endless, a very large cargo ship goes off course which is sinking! Then, as this massive cargo ship descends into the ocean, its cargo containers start falling off, and they land on, guess what- Co2 containment, which is now no longer contained, the formula for a major disaster! Not to mention, the ship that just sank would be a very big possibility. So, as I have stated in my earlier written word, the real answer, temporary as a fast fix, is to remove the storage containment OFF PLANET EARTH. And the closest and best place to do this is Mars. But we don't need Mar to have containers because the life there is not anything we could claim as life. So, the structures taken to Mars only need to be released over Mars, and gravity should do the rest. But before we do that- we need to test the release in a low amount to see if it stabilizes on the ground, as testing areas are fenced to keep the gases from floating all over for at least the moment because Mars does have massive storms, which have been recorded.

124. SUBJECT: LIVE HUNDREDS OF YEARS

This is with the ability to clone your DNA body and then with the understanding that the human mind is actually a composition of ideas and memories similar to a computer that has kept information and retrieves that information with detail. The future may see the ability to download the prior thoughts of a person into a new body of the same DNA, and wa-la, you are now young again! But the catch is your real body does end. Still, you've now extended your thoughts to continue and learn more, and this means at one point, your understanding will become far more advanced over other humans, with skill levels far advanced, and yes, monetary growth; you are now able to have a bank account that can continue interest growth because when you die, this person now takes your social security number. It's because, in the whole sense, this is YOU! In DNA 100%, and your thoughts 100%! The down area or grey area is where, after so many clones are made, the copies could decrease in strength and ability to maintain. So you will eventually become the older man, the latest clone body that no longer responds to download or the ability to make it function properly. Now, that is because it is known that in copied music, the more something is copied, the less quality the recording stays, but with this comparison, we can go a bit further. Our tissue cells are known to degrade from sunlight and UV attacks, and lack of proper protection could be another issue, but if you want to upgrade to a younger one, well, imagine that you go to bed old and then wake up young. But let's face it- it really isn't you, now is it?

125. SUBJECT: CONCERN FOR WILD HOG OVERPOPULATION

The future could be with the use of wild bore meat and made items similar to beef jerky. Also, the hide could be made into leathers for briefcases, shoes, belts, wallets, and leather goods, such as leather handbags, leather boxes, leather boots, and even leather coats; items made from pig skin, the hunting of these animals are open season, which they can be made profitable for selling the meats, and the hides. The open season is six months, and the equipment would be a winter snowmobile with a sled and a summer four-wheeler with a cart. These would be mandatory, so the use of these devices would make the hunting better and the yields greater; hunters could place their catch in pickup trucks to bring greater financial rewards. Google has reported that- Wild Boar is high in protein and a good source of saturated fats and zinc. Plus, it's rich in Selenium. Wild Boar is an excellent alternative to beef and pork for those who want food that is good for them without sacrificing taste and quality. Lastly, the need for processing plants for these animals needs to be established so the regular hunter who is seeking food will be able to have their meat processed properly. And be paid for the hides. If we make this problem profitable, we will see them be wiped out in no time, so the real issue of farms being devastated is the backers for the developments of the processing plants, and they now all own stock in the products made as sold. We will have the entire State of farmers buy into the processing plants, and they will allow hunters to be on their property, as any reports will be sent to an open hunting crew. The hunting season is open for each state, and hunters and farmers are to wear bright clothing during hunting season. Farmers are allowed to hunt year-round anytime they wish, and the processing plant per state of the hogs is open year- round.

126. SUBJECT: MAGNETIC TRAVEL AND MAGNETIC ENGINES

Continued to push a point further as follows:

The latest report from scientists who study the planet has reported that the planet's magnetic field is dissolving, and the area is increasing where the magnetic field is no longer.

As an inventor, I have invented many ideas, and the idea of magnetic travel was documented in my invention log many years prior to the introduction, a roughly over 25 years ago logged away in inventor number logs, and when I first considered that the design would work, I was thrilled with the understanding, which at first seemed a great method of energy and travel. But we are seeing trains in Japan that are using this technology, and it seems strange but true that since they have arrived, our planet has been experiencing some very strange weather conditions. Is there a possible link? Well, the answer to this is, without a doubt- YES. The fact is our planet is made with a wonderful magnetic field which supports our travel around the sun. You see if we look at the wanted technology as "I can make it, so who cares," we are doing the thing many people tend to achieve: they leap before they "LOOK." This process of a magnetic tube of travel or a rail of magnetic structure is like sending a death decree to all living on this rock. The magnetic fields of North and South Poles are a stabilizer, which disruption would cause our planet to no longer be in orbit. It is that magnetic field that holds us in the grip of our sun's pull, and that sunlight essential to humanity would end very quickly the moment we are released from that grip. The planet would fly outward away from the sun, and all life would go into a deep freeze almost within days. The other planets in our solar system would also be affected, and the entire area would become destruction only God has seen. Now, here is why we can understand the reason that our planet requires a magnetic field and a small comparison, but let's first consider the first railroad. The technology was considered like a spacecraft going to Mars, and it seemed impossible. But after one rail was successful, others began to make connections. If we observe the railroads of today globally, they are a massive criss-cross of thousands of rail tracks covering every known direction from town to town and city to city. And was this foreseen? No! It was not, and it started small and, in less than 100 years, now covered the world in many billions of miles. So- now let's pretend we replace that travel with magnetic travel, interesting, so we can conclude that the future will be with the same need that goods and people will want to travel with lower costs this could provide and be able to gain their travel time decreased greatly! But let's now visit the "result." Let's look in the pool. Shall we do it before we jump? And guess what-? There is no water! The reason for this is that we will take a single nail. It is a science class study, and it has been shown to millions of students. We take a metal nail, and we rub it in the same direction over and over, around ten times. We have a small paper clip on our table. We now take that nail we have been rubbing, and the nail has become magnetized, and it can now pick up the paper clip with ease. But—!!! Let's take that other nail and hold it the same magnetic nail as before, but we rub it just ONE TIME, only once! And now, try to pick up the paper clip. NOTHING; it is now no longer magnetic. So, let's place thousands of rail cars with magnets or magnet engines all over our planet. The act is obvious, and once it has grown, all things that do grow in use seem to help humanity, like the invention of the gas automobile, which now supports CO_2 as a danger to our planet and global warming. You see, Ford never considered that one day I will have billions of these devices all over the planet. And he never took the real effort to consider the exhaust of the engine coming out of the muffler! Yes, if it had been me, I would have said, no- this is not good. All he saw was money and the ability to travel faster from point A to point B. So our inventors who play with magnetic gadgets need to stop, as of yesterday. Now, the one place I have good news to give about the idea of using magnetic power outside our planet so the devices could maintain electrical power to spacecraft; that is a correct method, being it is limited in size and would give the spaceship possible electric ability even to travel. So being I already though of the concept long ago of a rotational magnetic field repelling the poles in a circular fashion, the idea is correct, but it still has a far greater potential of damaging our world over time, and so it needs to have a limited applications that are outside our planet, and could be the new direction for NASA to build the entire space ship in space, and by using a massive amount of magnets that are divided into a single rod which has hundreds of these magnetic circular repelling devices, as electrical energy can be transferred into the magnetic field on a coil to increase the speed and with so many devices would be able to potentially make a thrust if the energy transformed into a practical mover as atoms could be pulled from front to rear to make thrust in a seemly vacuum of space, but in truth, space has a composition which needs to be investigated, as we know that air thrust causes motion in this so called vacuum, but the real fact is that the vacuum of space allows resistance, and so resistance allows the space ship to move along with other forms of thrust, so- if we can use that vacuum to move particles of it- through a pressurized magnetic field, and as it rotates can increase the speed of this vacuum material, it will then potentially push the contents of this vacuum composition to react against the vacuum material not in motion, and Wa-la!- we have now made a device that could push us through space, and also when the propulsion system is shut down, our ship maintains it's speed, because there is no known friction in space, but if we continue to push the particles again, and again, we could even achieve light speed, now stopping means we need a reverse of the same system. So NASA first needs to find the outside atoms in the composition of the space vacuum would be required. Now, to this so- called first inventor of magnetic power- I do not need to prove I did it first. I know I did, but I also am aware of its dangers. I would say good luck to the newer inventor, but I do not consider it to be here on our planet. Outside our planet, a magnetic propulsion system could be achieved, but once again, let me stress- outside this planet, and far away, like far from the moon, which is also connected to our magnetic field. I would say about 20,000 miles away from planet Earth on our outside and away from the sun. I always say- It's better to be safe than sorry. And one more thing- to all inventors, just because it can be done, does not mean it should be done. Consider Albert Einstein brilliant. Well, maybe, or stupid? Let's consider his invention, of nuclear energy unlimited. No, it has limitations of power and a problem with a potential overload, and explosive, plus the worst factor of nuclear waste which is not friendly and kills people if exposed. Plus, his invention can destroy humanity. Does that really sound smart? (No). But guess what- we can now look at our world as the future. Is there a future? Yes, if we think rather than consider the financial gain.

127. SUBJECT: MY NEW TRAIN TRAVEL CONCEPT-

One I have in mind is electric rails inside a tube, which is a new method of travel that has never been considered. AIR PRESSURE, combined with electric rail connections, provides full lighting and electric use to passengers.

Respect for individual privacy is always important, and my concept of mass travel is making the passenger not only comfortable but safer in many ways. First, the entrance is a very unusual design. The train car has individual compartments per person that only hold one person.

The entry is on your ticket as SIDE A. Inside B have entries at the station as platforms, but all enclosures have a food vending even to heat a small meal, and water is available inside; combinations of drinks include coffee, tea, sugar added or not added, milk, orange juice, and each has a waste disposal, a tray to slide down, and even a porta potty in the event you need relief, the security is that all carry on is above you as in front and in back, you able to close the views in any direction and have personal lighting adjustments, onboard T.V. monitor directly above your eye level and internet hookups for charging and running a laptop device, but wait, you can even buy a movie with a credit card, have headphones that come out when you use that service. And all seats recline to become beds with adjustable head pillows made of the same leather. Wait, you even have personal air control, which allows you to adjust the actual air temperature you want per compartment; it is easy to read how it is also posted as you can turn the pages with back-lighting to see the examples in a large print of how to attain whatever you want in alphabetical order. The C and D compartments are for lower-paid customers, and they use a porta potty at the end of each train, but they have everything else just stated. The cost of the outside compartments is higher for two reasons: 1. is the ability to see outside as the entire side is open as one long window the entire length of the ride, and the porta potty system gives the person a cleaner seat as they are a way to use the bathroom without getting out of the compartment, and is easy access, So these trains are more like a massive 747 traveling inside a massive oval tube. The best part of this tube is as follows: no train is stopping until it reaches its destination, which means there is no other train able to collide with this travel. Train sit on four sliding rails that are on both sides, two per side and two above, to keep the train in place at all times, but let's pretend that something was on the tracks, and the item caused a derailment, we are not going to go very far because the train is inside a tube. Which means it can't crash under any circumstances. The tube trains can trevl over buildings and have no rail crossings everyone who lives near the train can't even hear the train because the unit travels on air pressure inside the tube. So we have a wider than normal train travel which is roomier and much faster and safer than today, we offer comforts better than being cramped next to some person with bad breath, or God knows what they may have as an issue, the center and the sides both have a walk isle so you can even take a walk if you like down the entire length of the train to the lunch areas, and these wrap around to meet at the ends that you could walk out the side compartment, walk along the window areas to lean on the hand rails and sit on even benches as the train travels you can see directly through the massive clear plastic to look out to where ever you are, the path of this can even be used to the people on the inside isles that goes through the center of the train, you see this train has no bends, no bumps, no sound of rail clickety-clack, is the smoothest ride ever conceived, because the contacts are very large rubber wheels that sit inside massive cutaway grooves, the rail is not metal against metal, it is metal rail with slide ability for rubber tires to fit inside and roll perfectly inside to be a perfection of precise levels which can actually be maintained by automatic adjustments done by A.I.- yes it can have some good purposes if used correctly, and the repairs are made when trains have passed and the rail is shut down for the day. Drone monitors are sweeping constantly to any areas that may have been stressed. Sliding doors allow a passenger the comfort of a private entrance with a reclining seat. They can even lock the door, which only the conductor can be allowed in if needed to have entrance key, under the storage of small bags, and two overhead storage of larger bags for those more safety concerned; even a seat belt is a dual strap that comes over the head and covers both shoulders down to the seat, this is adjustable for any size person. There is an adjustable lever that can allow someone to recline without having anyone kicking the back of your seat because there is a wall between your seat area and the other people in front of you and behind you. There are seating compartments that can be used for two or more people who may want to travel together, but these seats are without porta potty and are in the center aisles. But this is a very easy wall to use as either separator because it can be retracted between each seat as the wall slides upward to lock in a sliding three- panel connection. Still, the ability even to close the slide-down panel if someone wanted to sleep could be accomplished; the TV has headset inputs and is in front of each seat with a laptop capacity and a fold-down keyboard with an attached mouse controller. This train travels much faster than trains today. The rail system inside the tube are shaped similarly to the letter "G" as a hook system makes the train unable to come off the rail. For trains to operate, they are single one-way and one-way return. When the trains are placed on the tracks, they enter from the starting open side to be guided onto the rails like an open-ended box. The end of the train has a massive emergency brake system that is attached after the train is secured on the rails. Hence, these trains have no way to change the number of cars used because this is a one-size-fits-all, and trains that travel on one track can only follow another train on the same rails behind the first traveling. The air pressure is the main element that pushes the train to exceed speeds of 500 miles per hour and faster depending on air quantity pressure be released, the end of the train ride is like a big wind tunnel now being shut down as the train approaches you to the drop of unloading dock, the brakes are not on the train directly they are a combination of two brake systems that both are used at the same time, one brake system is a drag system that squeezes the actual lower sides of the train as a massive brake pad, the other is in the train in the same manner as a large bake system for regular trains but is brake pads like a semi-truck with an added emergency brake system for runaway trains, the fact is there are "three' types of brakes that can be used all at one time, but wait, we also have the last method which is called the safe crash slow down, it is an actual dip in the end with the emergency brake system, a "Y" split-err track is put into operation, this "Y" will send the train downward into a water pool that slows down the train completely from water resistance. In short, the train stops here, or should I say the Buck stops here? The proper entry or exit position is to place another rail car on or off the existing train cars. So there are side rails where trains are stored to be cleaned and used if the passenger amount is booked higher or lower. All trains travel above the homes and the roads, trains at the same level of height, areas where traffic is present, and an underpass is made so there is no need for railroad horns or stop crossings with gates. All cars or trucks traffic travel under the tube rail trains. So we have achieved privacy and comfort, as well as two other very important features, which are distancing passengers who could be ill and ensuring the safety of the person and their items are not harmed. These are reasons that most people enjoy traveling by car because they are given these advantages but not a toilet! But can be separated from others in a protected environment with access to music if desired to their taste. However, the train cars of the past had one very attractive travel method that was well-liked by many, and that was the private car, a car with a seated area facing each other. I want to introduce this concept to buses, trains, and even airplanes, as well as the private method of travel with the option to open an area to have a conversation if desired. The new system shall have entrances inside the transportation system as of today, as well as outside entrances to private seating. This means that, for instance, a bus would have doors that open in a row down the side of the bus, each with its entrance and seat. Of course, this means wider roads, but the need for more space for transit is a serious need to be considered. Each compartment would have an emergency exit, unlike buses today, where exits are limited. All passengers would have their emergency exit. So, once again, the rail is not a monorail. It is a multiple rail bottom, with a single rail above to stabilize and prevent the train from ever leaning, or a brake system support is also possible on top to increase stopping power. Let's face it: stopping a train from being derailed is important. The motor is electric, and the power is actually on the rail as one side is with the plus + side being sent. The rail is covered with a copper covering, the other side is Minus-) side of the rail, and the electricity is transferred from the rails below by copper brushes that are located on the engine, and this also transfers electricity to any other rail car, to provide electric power to any other car, with a simple plug is screwed together when trains are attached, this electric also generates an onboard battery cell storage, if a storm disruption has disrupted the trains electric current, the train would still have onboard power, which is stored in a battery car located behind the main engine. The generators are connected to wheels, which, once the train starts in motion, the electric current that gives the train its power from the rail is no longer required, so the power starting from rails is the area from which the trains initially get their power. But now comes the big push, and this is a half mile in one direction and a half mile in the other direction in a distance of ½ mile travel area from the rail station. This train will begin picking up some serious speed! Once the train starts in motion, the excess energy from the batteries now that power is delivered to the main engine to run the electric motor. Generators are connected to the rail cars that are directing their power output to the battery storage and to all uses of electric, such as food cars, seated cars, overhead lighting, sleep areas, bathrooms, etc... Yes, we are now gaining speed from the side air pressure intakes that work similar to those air vac systems you place items inside a tube at the bank, the air pressure is around 1000 times greater, because it increases as the train is going past other air intakes the is locked in from behind as the air pressure increases this send the train faster and faster being pushed from behind. The air in front is actually being pulled outward clear at the other side to make it actually being pushed and pulled at the same time.

On the train ride today, we heard the massive metal rails' constant clickety clack, and the noise they make is no real welcome to anyone who lives near these trains, as the metal wheels hit the welds in the rail, which over time will bend, squeal, and scrape like a chalkboard.

And this constant weight causes the spikes that hold down the rail from time to time start to come undone, so I have good news, the trains of the future are not made from heavy steal, no they are made from lightweight materials about 60 times lighter, and much stronger than steel for commuter travel, the frame work is an aluminum frames with an inside "V" supports that give the metal an increased strength as never to bend, yet the less weight will allow the train to travel even faster, and the need for heavy rail can be replaced by aluminum inside support rods to carry any load size, but less tress on the steel rail that holds the train, we will need to from time to time have robots that scan the rail to look for imperfections, cracks, any separation, possible adjustments, but when it comes time to adjust the rails, it is easy access from any point even when the system is in use, the reason is we have air pressure that pushes this train from multiple areas along the tube travel, and the areas that may need to be repaired work like this. First, we allowed the train to travel and pass. The team of workers now shut down the air system as soon as it passed, but before they did this, a massive lift truck was parked next to the upper tube area, needing to be worked on as far as adjustments in rail leveling and possible replacement parts as to be repaired. Still, the entire section is either a slight adjustment or the entire section is replaced, with simple unscrewing removal and inserting new rail, no need for welding these rails have a slide in fittings that lock when inserted, and we now use rail that is screwed on. To replace a rail is almost never but this section needs attention so, let's get to work, the air shut off has delayed our train slightly in speed, but not for long, the upper repair team is on top of the structure and tied off to the center support rails and they are now unscrewing the massive plastic plates that hold each main section of the tube together and is three bolts to be removed on top to slide out the section connector, once this is removed by a lift truck operator with two hooks that pull that frame upward, and it slide out, then it is placed on the ground temporary and the large aluminum plate is now inserted into that section and this is done in front of the area that needs repaired, the plate is locked with a simple pin locks three where the former connector was, and the system air is turned on now behind the train that has passed, the air pressure plate that was slid into place is now the new section for air pressure to gather, along the line are many air pressure stations, that is how the train continues to increase speed as it travels more and more air is pushed in as the train now pass's by the air compression system, the repair is done now, and the team shuts down the system air once again, and removes the plate, inserts the connector slot and turns back on the air system this only can be turned on if the slide top connector is in place because it is also an electric connector which says to the main train system, this system is live with a green light coming from the rail connector indicator as connected, the signal goes to the office train line control to confirm the line is connected and ready for service, a job finished in about one half hour, because the team has already put everything in place before shut downs and startups. There is no weather damage, and moisture is also extremely low because the constant air flowing through the tube dries and possible moisture. The cost of maintaining the rail just dropped about 80% in repair savings. But let's be more realistic, we need to keep up with the world which seems to pass us by, to get goods faster from one point to another is the way to improve the time money is made, so the old saying first to the market, first to the money is a true need. So, to make consumer products get from point A to point B, safer and more effective mass cargo transportation is needed. The designs have been considered magnetic (not wise) and monorail, but they are not supportive enough for large cargo and potential cargo shifts. So my four-wheel concept of two top-looped connectors, and two bottom-looped connectors would maintain any travel and secure it for thousands of years before needing replaced. In closing of this subject, let me say that humanity has not looked at the way some structures were designed, the fact is we build things today without the long-term commitment to making things that last hundreds of years past many lifetimes. Instead, we build homes that will fall apart in 20 years and bridges that fail even after being built. Yes, that happened in Springville, New York. They built a bridge and then realized it was unstable, so they had to go back in and do it again. The lesson here is to build whatever you build to last thousands of years and, considering erosion and rust, make better coatings that could. As inventors, we never know when an idea will jump out of the bushes! And just now, I have just thought of an idea for another form of travel. In my youth, toy electric race car tracks of my youth contact plus + and minus- allowed the electric to make contact with the engine, which could replace the costly fuel-run trains we use today.

Electric trains could be similar in contact with the power source. When I was a young boy, I had electric race cars, and the contact points made contact with floating metal tips. Now, I was never shocked by the placement of these cars on the track, but to use the proper power to move a train at top speeds, the contact points need to be safer for the public.

So this design of electric train could use similar contacts to that of toy race car and similar technology used in rail in San Francisco with electric cars which have wire travel above the electric car, "the improved difference" is with this new design, the contact points are under the wheelbase of the rail, and the contact points are "sideways." One side of the rail is positive, and the other side of the rail is negative, so they can achieve the electrical power to travel on the underside of the rail, so there is no need for overhead electrical wires.

128. SUBJECT: POTENTIAL WWIII

It's real, and the possibility we have already had this type of event in the past is evident all around the world. Structures of massive cities that were once built by people are long gone. The question is, has mankind made a full circle having high technology then war destroyed us? Have we destroyed our world numerous times? The evidence stands with the structures matched other styles of the same types of structures. Tell me that humanity did have an advanced system and eliminated the people, but a few survived in caves, so we see drawings from people who had no art experience and left clues to show us what happened. They may have been forced to start over. But also, let me say before I address this subject that a nuclear war today would leave the planet barren of all life. Radiation would be even in areas where the blasts don't happen from oceans infected to air travel on dust. Finally, a nuclear winter would happen to freeze all living beans from the dust raised into the upper atmosphere. So death is the only result, and those who are left meet a horrid end of suffering before they die. They get to freeze to death along with our oceans when the dust that has risen blocks out the sunlight. You may be saying right now, why is this man putting so much emphasis on the planets destruction? The answer is simple, I'm trying to motivate all of you, to save you from what is coming. But here is some news that shows that we may have been on this same ride before, many times. The real facts are some things that science has not seemed to grasp, well, until now. The facts are as follows: 1. Large pyramid stone structures, many of which have had major outside coverings that seemed to be now gone in most places, all supposedly from erosion. I will prove that wrong. The other facts 2nd are that these structures are extremely similar as they are mostly four-sided structures, made of stone, stones set so perfectly that they could have only been cut by a laser technology, and all these structures have areas where tunnels built, and the structures house past leaders of their culture. We can find remains at many of these places of very elaborate art, carvings, and many items that seem impossible for a layperson to make without some very modern equipment. 3rd, massive stones were found so large that it would be almost impossible to move these massive stones today. 4th The cuts are cut so precise and so exact that a thin piece of paper cannot enter the openings. 5th, So the conclusion is that many believe outside help was once in the culture, and that could be a possibility, but even with that, the technology would be required, and yes, they may have had some hand in our developments. But now, the real evidence and the real conclusion. In many parts of the world, we have found these structures made by civilizations that date back over 10,000 years ago. But recorded history, which has been recorded to date as the furthest besides cave designs, is around 6,000 plus years, so where did man go for over 4,000 years? Because there seems to be a very big lag in the time frame. Well, here are just a few more clues before I show you all what really happened. One of the strangest things that science has stated is- that the Sphinx has actual watermarks, and the evidence of this is because of erosion markings made as if a river flowed at the bottom of the structure. This means that once a great distance in time, the SPHINX was with a river flowing around it, which was most likely what we call today the Nile River, but it could have been a branch of that river. Nonetheless, the facts remain, so how long would it take to dry up a complete river? It could have taken thousands of years for the landscape to change. The area was stated at one point to have vegetation.

So how could water evaporate so fast if the area was once a massive area of forests and the images inside the Pyramids show a lush life and abundant trees? The past is without a doubt showing us that this area was rich in life and many types. So this brings me to the next observation: in many areas of the world, the massive structures were only found after years of fighting through thick and massive jungles in South America , vegetation that covered these structures so well that, until now, they could not be seen because the technology to reveal these buried structures were not invented until now. However, since the equipment was used to detect ground structures even under forests, the discovery of even more city structures has been uncovered. This news, which arrived not long ago, also shed a whole new light on the massive populations that these structures held as their way of life and homes. Suddenly, the relatively small number of people became millions upon millions. The discoveries also showed technologies such as aqueducts and cut stones, which were once again made so perfect that they cannot be considered hand-made tools. So, what have we really found? We have had similar actions of cultures who disappear and, have found Easter Island, massive cut stones, Egypt, and South America, the other side of the world, yet similar architecture once again. The discovery of gold images of jet planes, yes, that even look like our jets today. And here we are at the pinnacle of traveling into space, weapons able to blow up the world three times over, and technologies like A.I., and yes, we are advancing as we are also talking about nuclear war. The end, or is it? We have long wondered if humanity would survive such an event. We need not consider it a way to be, but in the past- we did believe it would survive. The people with the most had those bunkers; they even had food stored, but they did not consider that even with the greatest structures that would possibly be protected, it would also be their tomb, but they even prepared for that. Yes, I am saying that our planet has many times been built and built, and then they destroyed what was built. Now, allow me to tell you "WHY" I know this happened; I did not say, maybe, why this happened. I said, "Know". For the simple test of impact and the area of impact, let's start with a science test of six same-size ball bearings. These metal ball bearings are placed on a flat table and spread apart about six inches apart. Each is on the edge of the flat table and below. Before we placed this table and the marbles, we made smooth sand, which is back-raked to make each area of the sand an equal height. Now, the table was placed above the sand, and our ball bearings were pushed off the table, one by one. After all have been dropped into the sand, we now carefully use a large magnet to pull the bearing out of the sand.

What we find is what is called AREA IMPACT, and all explosives have this ability, as professional explosive designers know that impact devices with the same amount of explosive power as the containment amount will leave a hole with the same size in many cases when the land mass is of equal mix. So, to put this into a simpler framework, the bombs that were dropped on land targets during WWII had what is called exact whole depth when on flat land with the same land mass. This is also shown that when an atomic bomb was blown up, it made a crater, that if the second bomb were released and detonated, that bomb would also leave, let's call it "a footprint," or area of impact of the same size or outside diameter or also called O.D., which if the item or bomb being blown up is of the same components the same area of impact size will also be found. This brings me to proof that our world has already had its nuclear wars in the past. Because when we remove those ball bearings we will find that the size of impact left are almost exact. In many areas of the world, looking at a crater. Science has ignored the fact that many of these craters, sometimes many in one area, have the same outside dimensions, or the same foot print, and as explosives also make the same depth, each have an equal depth extremely close to each other. So science has us believe meteors make these craters, but the question is, do meteors that hit the earth make the same size of craters or identical footprints?

"No!" because they are rocks of different sizes, which are random in size and not fabricated to be exact, so the answer is once again:" NO," they do not leave an identical crater size, which we view as evidence by the telescope of our moon, all activity of craters made by impacts are never the same size. This proves my theory that our world has been through many events that have caused massive destruction.

The explosive devices may have been of a different mechanics or invention, but was something of an explosive nature. Now, we can are finding that the areas in jungles in South America have located more and more cities under the trees growing. And we hear of places such as Atlantis reported, as what may have been the last reminisce of the past before it was taken down. But allow me to show you that there are even other possibilities. First, before I make this next assessment or the potential reasons our plant shows evidence of a major war, let me show you the real calculations of why this possible action not only is associated but is extremely probable. Let's say an event triggered the act. Now we have a few people who survived, very few and very spread apart, but maybe someone was cave hunting for gold, or a military structure bunker was with a team. Still, we have just had massive deployment of bombs destroy our main water or food resources, and our lives are now in the ground. The many people who may have been caught off guard may have been "repairmen" techs or engineers.

The point is now these people will potentially die of radiation exposure, or whatever they have made could have been even a dirty bomb, which would potentially be mixed with other explosives to make it more lethal, but now, over time, the people who used to fix the tech, they are no longer, and the equipment starts to fail. But over time, the structures made of everything that is not- get this: Stone will rust and fall from a thing we all know is erosion, and the beams of massive builds broken and even not broken will have water damage, and the end of the once great cities of housing and structures are failing and dissolving, and this happens in just a few thousand years complete to the facts the rubble is now covered with massive vegetation and jungles cover things along with the massive amount of dirt and dust made, and now to the ta-da moment.

To compare the erosive way of nature, we can see what happens in just 20 years of home being left without anyone doing repairs, it will have roofs fail and start to rot, and how rust will start to dissolve that old truck or car out in the back yard, now looks like it is dissolving away into the ground, and the frame is almost rusted away. When this event of mass destructive power released thousands upon thousands of nuclear explosives or similar devices. The ground is lifted into the air as never seen, the dust now hangs in the air for months, and the sunlight cannot penetrate the waters below- the result is a massive ice age in days, the planet's air is freezing and water in clouds fall like never seen. The entire planet is now ice, and the few living creatures left are now in survival mode, as the ocean is now covered with a massive sheet of ice that defrosts over another few thousand years. This event is exactly what will happen when a nuclear war is done again. But I regret to say that this time- there are no survivors.

In the beginning, yes, a few of the population may be deep enough to stay alive, those of leadership and military, but the people who believe they can hide away, sorry folks, the bad news is only one word- Impossible unless you like eating others like you, or animals infected, and gaining their radiation? The fact is even if you're in the deepest bunker, the time it takes to survive is many life times, and you're never able to leave. The military hardware today is far too many missiles that could destroy the plant some six times over. The estimated amount is around 7 to 8,000 , so the radiation is just far to high to exit, and I have had a better view of what actually does happen which with all these massive bombs going off, only makes complete since.

The bombs will make a up heave of dust so high into the upper atmosphere it will take years for it to settle, and it blocks the sunlight, which now the entire planet freezes, so even the subs in the ocean that the hold the rich and so called powerful , subs which are at this moment now being building to house hundreds who believe they will survive a nuclear war, opps sorry, oh yeah, I am aware that many think they can escape the blasts, no, the oceans will freeze solid, the temperature of the planet will drop far below the coldest temperatures ever recorded, which means everyone alive becomes a human pop cycle. Including those subs that are supposed to protect people. It's not good either way. But if you make a bunker, the good news is, well there is no good news you will survive the blast, but then you freeze. And the exit is never available, Once again, unless you like to die. I believe I explained this, but just to push the fact home harder- The evidence shows that our nuclear bomb that blew up even in 1952 in Nevada; those areas of the explosion will still kill you from radiation if you venture into that area, as a once famous actor John Wayne and his entire movie crew and the director all found out. The story goes that they filmed a Western movie near an area that once had this nuclear bomb go off.

The entire group of people, including John Wayne, died of cancer months later. Now, it may seem harsh to place this news because our militarizes of the world have lost their light bulbs and their noodle to boot. After all, they believe they can survive a nuclear war, which is very funny. After all, I would love to see the looks on their faces when it is reported that we are out of food and pure water, and you're invited to super; you have a choice of being roasted or baked. But seriously, people, all kidding aside, I have laughed even with a tooth ache. Now you have a choice, to stay silent, or be the voice of change, or die with billions of others and have all those you care about die also, and history is never recorded again. Maybe this message is that we are to be docile is what you have been taught which also means submissive. Things submissive-isn't that what a dog or horse is?

(The times they are a changing. By: Bob Dylan)

129. SUBJECT: THE WAR DRUM REPLACED with TOMORROW

As a visionary, I am extremely serious and confident that the human mind is a sponge, but that if our youth are infused with images and the continued teaching of violent actions, we may have a world that no one could ever have imaged, you and I are both subjects of the learning process of cruel acts made by others, cruel words, made by others, cruel deeds of greed and, cruel actions of suffering made by false beliefs that a person must kill to preserve their way of life. This has been established since man learned how to kill his food. The key word here, which is repeated over and over- "LEARNED."

Yes, we are taught by our parents, teachers, and the media that the world is a horrid place filled with evil. At this juncture in time, I would agree 100%, but I see how we could change- how we could establish a new generation that resists the evils by being mentally stronger and physically stronger. Yes, the body does require protection. So, I will present this application, but even though I feel that 99% of people will reject this process, I also feel that if done as instructed or to be less demanding, as to suggest to you, it can change our world globally to a harmony never seen prior. The commitment to do this is so great that most will say "never, " but those who do will be placed in the history of putting humanity on the correct path. Now, at one point within this next category, you're going to say these words: "No WAY!

" I can only say that "It is the only way to stop the madness" is the reason we need this measure of change to a safer environment for citizens or the future of the world, but before I get to this change- I wish to show exactly "why" the change so drastic is so much needed, let's review for a moment some real activities we see in our present state of lives: The lesson mankind never sees, which is essential to remain as a living force, is:" cooperation for preservation.

"We see so many wrongs as people fight over territories that they claim. These lands have rocks millions of years older than us all, so how can we claim the ground that will be with our ashes? The thought of claiming land is like saying- I am unaware that this land will be here long after I am gone. But we now have today people who want to control and take what is not theirs, and we always go backward back to that same ill, called "GREED," and domination over others to profit or feel important. Sound familiar Putin, and Jinping, and Jong Un, if I respected you all I would use your full names, but your mind set is so self-centered, you have no ability to see that you demand power over those who don't want your life style, but you're willing to have your people die for your desires for domination and to be all powerful!

Why are you not full of yourselves, look at me! I am wonderful, I am the best of all! That is so false, not only are you weak in mind, your determined to allow others to fight your battles, so if you're so tough, why not just get a boxing ring and challenge the world leaders to a fist fight, and the winner runs the country.

You wouldn't do it, because you're not strong, you would only lose, so you chose to have young men die for your ideas, die and they have mothers and sisters who cry. This ill causes man to divide, to fail, to become killers, and destruction follows those who want and desire these goals. The proof is from the past; all tyrants of every nation have been, and ("all fail") die, and have nothing to show for their evil desires to have something more, which is not agreed upon. I say this as a prelude to what I am proposing that will give humanity a new and wonderful direction, one that is not for one faith, one government, well we could have one government but that takes unity, and freedoms which I have explained, but we don't need one color of people, one language, or tradition. No, this application is for everyone who resides on this massive rock flying through space at an incredible speed and in perfection year after year. Only a creator could have perfected it. To be established by chance is absurd, at least to my conclusion. What we see today is billions of dollars being spent on developments to act against of aggressive military weapons to be prepared against another country, all in the name of the preservation of peace, the so-called defense spending, which is also an offensive instrument, one that, if and when used, will be the end of humanity as never to be again, and they build more and more. That is what I call the greed need by corporations who will also die, I can hear God laughing when that happens, but seriously people, this book is being designed to by some crazy chance the brains of these people will turn on, and suddenly realize —"oh no !". But let's face it if they have friends and family they must really hate them. I do admit that we do need to have the ability to protect our world , because in truth , we are only babies looking out at the stars with wonder, and we could find real advisories that are out there in the future.

But today the devices spent on mass production of these war machines, which could save millions of people from poverty, yet the nations of the world are fearful of each other from the constant media of these nations all send warnings of war or threats to claim war, and destructive acts are given priority over acts of kindness made by the some of the same people. And to all Russians and Chinese, I love your food, and I have made friends with many over the years, I lived in Brooklyn N.Y. and many Russian people traveled on the same D, and Q train daily with my travels to my sister's house, I met Russian people all the time, and I would speak with them.

Wonderful people most of them. I have friends in China, and his name is Frankho, I have worked with him for over 22 years. But it isn't the Chinese or Russian people who are the problem, it is there leaders who are the trouble and the way they treat people as things rather than with dignity or respect. They want to control people, not care if they are happy. The clear result if we do not make a unity of peoples - and make no mistake, nuclear weapons will happen, now my game show host Voice:

"TELL US BOB- SO WHAT"S BEHIND DOOR NUMBER TWO?!! IT'S A NEW BOMB! That's RIGHT BOB, and the final result is that there is NO WINNER!"

I already spoke of the proof that in 1952, an H- bomb exploded in a desert in the USA, and that area is still radioactive, so how do we as people put an end to these crazed ideals of domination over others? The reward to stop war is to have peace. And let's consider that all that money spent on military equipment, military uniforms, and food, the list is a forever expense on people worldwide- these funds can now be used to invest by countries into space travel and the gains that await us as new minerals, gold in abundance, and even new life which could improve our lives, like a new type of chicken? Yes, with all adventure, there is an opening of potential dangers, but the gains outweigh the dangers by trillions of gains.

When I stated peace earlier, the meaning of peace is that friends are made, people enjoy their differences, and they learn about each other. They open their hearts and minds to the great potential of a united family, making an amazing future that is beyond comprehension. Instead of looking at what we have in envy, we look outward to what can be together. The goal of humanity is to grow outward to other planets waiting to be discovered. These planets will have some dangers but will also provide greater wealth than imagined.

A united effort under one form of governing that is elected and not controlling people like robots, but robots are made to build and work so humankind can now reach into the heavens and learn by doing.

The future would be wonderful if people worldwide would say these words: I refuse to join the military, I refuse to be a paid killer, and I refuse to be a controller of others. But we all know that even in the most perfect group of people, there is always this one or two of the crowd that just can't allow love or kindness, so they hate and harm, so we need to consider that to be together as a world family we will still need laws and protection against the ill minded who have to be restrained. It's time for: "NEFF to the rescue!"

The need for protection is in self-defense devices that I have invented that do not kill but inform and stop a potential person's harm, and my devices are designed to make sure that whoever does make an attempt to harm someone will be found and punished. If you can't be harmed, why do you need police? Well, we would have police, but they are so limited that it seems almost as if we even need them. This is because everyone worldwide has a neck device that shows any event and, yes, can even stop an attack. If people can act against a person who has crossed the line, it can be controlled with a very easy device that all people can wear. It records your every action, and so it can be activated even by a command or a heart rate increase or decrease.

Once it does this, the entire area is now recorded for who is there, so all within a five-mile radius or more is now a potential of the person attacked. But let's say the device is activated by command. We can now see recorded events of all the people within that area. We can also have the ability to stop a person because all people have several I.D.s on their protection and observation devices. If that number is covered, an alarm goes off to let users know they violate the device being covered. The user is stunned almost automatically if they do not comply within five seconds, and the police are notified of their GPS location.

To be exact the use of this device is a perfection of people protection, because those who do crime know they can't avoid the punishment with the facts presented. This device could also be worn as bracelet with a built in transmitter receiver, with solar power charger to keep it active.

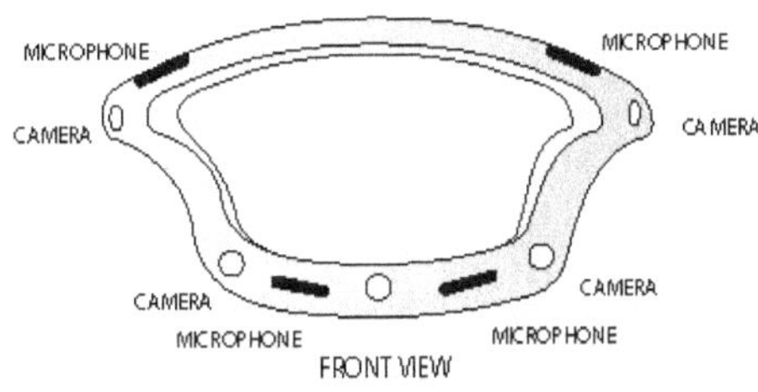

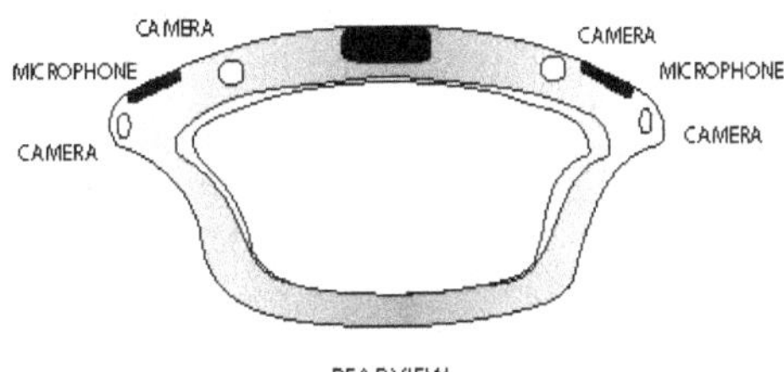

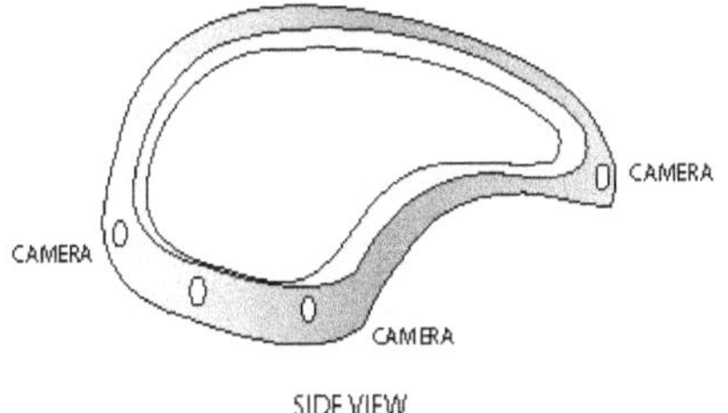

The design being presented could be modified slightly to be a fit all as adjustable, but this is the concept art which can also be for prison camps, but with remote control stun ability in the neck area and used by prison guards, on prisoners who trying to be unruly or try escaping. The civilian model is not with any control factors:

This device ends the so-called threats that people face because when the user says – "ON" their voice is the key to the transmission being recorded, and the police now can enter the problem by saying, stop, your being monitored and jail or death can be your outcome. The device sees the aggression of any type to the authorities and they can now respond to that GPS location no matter where it is. A user can also set keywords similar to what is used to find in search engines online, as potentials for words of ill reaction or anger also send out warnings to police monitors. Now, one more thing: we have saved taxpayers billions per year across the country. Why? Because the need for police traveling all over, spending millions in gas, and massive cars to travel in as I.D. are gone, a policeman goes to a crime scene, which was "prevented" and didn't result in death, a guy or women who are now lights out from the stunning shock. They arrive in their car, paid by them in a shirt and badge to I.D. the policeman and with a camera so he is recorded. So we are now making life safe 1000%. I do realize that this device has a backlash that if the government in control of these devices is not one of allowing freedoms to people, they could use these devices to control people to such an extent that not one person would be able to escape, that is why to maintain that the devices are strictly controlled by the user wearing the device, and is set by law that is can never be used by governments as a controlling device, and on the monitor system there is a release lock that once again the user tests before it is placed on and inspected the combination is checked by the user to work to take it off when they desire, if the lock combination has changed then the user should refuse to wear the device, and there is no law that says you must wear this electronic monitor, but is highly recommended to keep you safe. An inspection could be achieved by a formed group who oppose control devices, so they are top electronics people and manufactures who know the proper components so there is no tampering with the signal relay or activation's that are not allowed to be in the devices. I believe that the use is a way to stop all crime of man against man, but needs safe guards to insure that it is not miss used by those who lead our lives with laws.

130. SUBJECT: THE BIRD FEEDER

An observation is what triggered this subject: I was watching birds at my bird feeder, and suddenly I saw the understanding that these birds all came together and enjoyed the food provided. A bird called a Morning Dove ate from the smaller birds that actually dropped seeds on purpose to feed them because they were too big to sit on the bird feeder filled with seeds. Many different birds came, and not one even had the idea of being in conflict over who got the food. They shared it without doubt, fear, or question, and they enjoyed the food together and they formed a circle on the ground, it was amazing. Finches, Yellow Finches, Blue Jays, Cardinals, bright red birds, Redwing blackbirds, mockingbirds, and so many different varieties all together, and so many would even wait till another bird was full and left the feeder. They knew they would have their chance to feed. This image even got better when a red squirrel and a grey squirrel came to eat what was on the ground, and the birds didn't even fly away. They all enjoyed life together, as God intended. We only need to notice that each of us, even as a certain nationality, "STILL YOU ARE DIFFERENT FROM OTHERS OF YOUR RACE," so you see, you're not the same. You may walk or travel with people of the same race, but you are different. So why should we question the reasoning of being different?

Now, the only thing we should question is when being different can harm the lives of others. That is the exception. Allow me to make a comparison of life as to my bird feeder and to our world as a human feeder. We should take nature's examples and see that all creatures are meant to be here. Without a doubt, some are very harsh creatures but designed to be a balance for nature. We can, as humans, understand that if we seek to help each other like birds, we could achieve a destiny greater than any ever than any have considered, a world-building and fixing the issues TOGETHER as a united family. As our creator designed us to be, and the proof is we are here all together, and we need to cross our borders, open our doors, protect each other, and, yes, work as a family of humanity. Now that is not a bird brain idea that is our future.

131. SUBJECT: MAN WHAT A DAY!

Have you ever had a day when you asked yourself- "Why me?!" We all have these moments, but do all of us have this type of day? I have met people who, when asked if they had ever met a bad person or had a bad experience, just said "NO." So this brings me to a very strange but serious possibility. We may actually be making our own world happen, and a thing called Karma an action with a reward of action to return back to the person making the situation be it good or bad, we all have heard of it at one point, but for those who have never heard of its ability, it is a force that seems to prevalent in most people. So is that all it is? I think we can say that in our lives, we do make our day good or bad. But could it actually be something we create, some people have done some serious research about this, and strangely or not it does seem to have some merit. Some of the greatest minds have said "believe" and so we try and sometimes it just doesn't happen. But let here is a strange event that said, maybe there is something to this trust in something we can't control. I was a carnival with some friends, and I saw a game of throwing a ball at some beer bottles, and I decided to win a Teddy Bear for a friend. And so I paid for two soft balls and I was a Quarter back in high school and I also played on the baseball team, but my ability to throw a ball was better with a football, but I threw the first ball and smashed a beer bottle, then thief we allow negative thoughts to rule us, we can even develop illness. This is proof, as a young boy, I was never someone who liked school. For me, it was boring, not so much the learning, but inner acting with some children was like entering a bullfight, and I had no sword. But every once in a while, I would "pretend" to be sick so I could get out of going to school. And I would be so determined that suddenly, I did get a rise in my body temperature, and my head would be warm! My mother would say," Okay, yeah, you are sick, you stay home today. And as soon as she left the house to attend her day job. Suddenly, I would be feeling better! This acknowledgment of something that happened has given me the truth that all of us have a linked purpose. We actually can create a world we live in, and we determine this world by our thoughts, negative or positive. The world can overcome all of its problems, but only if negative people are prevented from establishing negative projections and thoughts. The man or the media who project," The world is Ending! "Attains great notice, and the negative message builds negative reactions but not one person seems to mind who is negative, but when we hear of people being attacked for doing good? Then we need to look at the people attacking from both sides, and look closer to what they accomplished. Are they doing a positive actions? You decide, but everything I have stated is what I do to find the truth about someone. I look at their past record. Now if you do this, then your someone who can have a positive result, but to follow blindly is a fool who will open the box of promises and find the box is empty later, don't take any word from anyone as fact, instead look deeper into what they accomplished, was it good?, or bad? And this goes for anything and anyone. I think you'll finally have your eyes opened if you follow this rule. So why wait for the mountain to fall, be one who takes the hands of others of like mind, and be active, be determined to be the change, to be the person who can say: "I made a difference". And it helped your world to be a better place. Make a foot print for those to come to see that is positive. When you look at anyone that can affect your future, be aware of who they are, because we live in times when people smile and make false claims and have not been what they say they are, but if someone has done good in the past, then that should be the direction we follow, because if they did good in the past, they will be more likely to do a positive result. Always look at someone's results in the past, as to what they will be doing for your future.

132. SUBJECT: TELL ME SOMETHING I DIDN'T KNOW

The fact is we have many things that our world need that we totally ignore.

We need to remove devices that remove us from each other, I see people gathered together and not one of them is talking to another, but busy texting, or playing some game, they are not aware of those around them, and even people die because they walk out in front of an oncoming car, or fall into some ditch, the hospital ER could make a book of how many people made it there because they texted when driving. The law says to use texting when driving is illegal, but they can't stop it. So what I propose is that all phones have texting, and internet, and games be removed from phones, and make phones that allow a person to call anyone with a simple request like a business needed, or service, or anyone you may know, is now found by your voice connecting to a A.I. operator by saying" PHONE" Now you can connect you to anyone with a simple statement like :PHONE- Connect me to Buffalo Hospitals, then A.I. would say- here are the list of Buffalo Hospitals- and begins to read them off one by one, when you hear the proper one, you say after they say- "More" you say STOP-' the A.I. now dials the call, so your phone no longer has keys to remember a phone call, but you have a save screen to scroll or say: "CALL and the person's name", and if correct, you say- CALL, and for those of you who say oh how terrible -I need my cell phone to do those actions of texting, that is an email, to do a game, the PC has hundreds of games you can use, you must have the internet, that is also on a lap top or a PC, so you see when you use those other devices on your phone , you may be walking, driving, or in motion, your now a potential unaware person who may actually kill yourself or someone else. If you want to use those other items said, use your computer at home, and you won't die or kill others. Because the makers of these devices could have a very very large group of people who lost a family member from one of these events to come back at the manufacturers for a Hugh settlement. You have enough distractions already, so the cell phone has made you not aware, and not safe. I tell it like it is, sorry.

133. SUBJECT: WHY ARE WE HERE?

Okay let's look at why are you here? Is it just to live and die?

So do you look at your life and say- why should we care? I cannot do anything to change this mess! That, my friend, is wrong; change comes when we, as people, see our lives as with worth, as with purpose, as with hope that unites us, and we forget about the impossible because the idea is stronger than the failure. We do something called "come together" in mind and soul. Yes- I admit that here we are stuck on a rock spinning through space at a speed that science has calculated at the equator that the speed of our planet's rotation is " 1,037 mph or 1,670 kn/h," but we somehow remain on the rock not flying into space. That is truly a miracle.

We live as ants crawling all over the world. Compared to ants and the size we perceive those creatures, we are no more than a fungus growing on a rock and, truthfully, "worse as parasites that consume all life at a rate that increases as population increases. But, through all this, to understand how small we are, we can stand before God's creation at a moment as the stars reveal how tinny we are as they span billions of miles away and give us a glimmer of the light that may be seen, but to countless centuries to reach your eyes, they show us we are nothing but fish in a bowl, who jump outside the bowl once in a while. We believe we are so smart, so wise, that we can do this, so being aware of one's existence can be very disruptive to our ego, and the will to live seems so insignificant as we understand that all the things we struggle to attain our things acquired here will only be split and divided as they did to Jesus with his garments at his cross, they wagered over his clothing as he watched, and as he slowly died, but as all I have said does not paint a very good picture by any means, but now please allow me to share: "The purpose of your journey." And through all things you do, this is the only reason you where allowed to be here. Or we could call it the flip side of the coin. The peek at your future , My fellow humans-we are not here by mistake, and we are not hear without a reason, our lives do matter, each of us is like a very important puzzle piece that completes a big picture, and this picture is connected to the here and now, and the future of here and now, and the hereafter, you are directly connected to millions of others who have even been here before us, your DNA linked with so many others who achieved or failed, they made the path, they left foot prints that either gave you opportunity or helped your journey, even those who refused to take responsibility of being a father or a mother, who gave up, they left a foot print, and many people do not see that many people who have been hurt, went on to do great and amazing things, so your saying, why me, or gee I never, or maybe your all set, and many because of those who came before you gave you nothing but life, but that alone is a gift of life, and our life is effected by others, so do yourself a favor, do yourself some real change, and remember- we are connected to their lives as they did struggle as we do, and they had hardships or they had made bad decisions because they never focused only on themselves, like many of us do today, we are constantly being reminded that we don't live the lives of rich and famous, people admire celebrities, but they live somewhat short lives many of them, so how lucky are they really?.

Maybe your saying- My life is horrid! This may be a fact, but- ask yourself- Is it really? Are you so deprived of food? Did you go out and kill your breakfast this morning? Let me see if I can bring some light into your darkness. The facts are humanity used to struggle even to gain a slice of bread, and let's not forget some of the nice things that happened to millions- Like The Black Plague, or World War one, and WWII, and what about those people who crossed the deserts and lost family in covered wagons –and how they died in deserts to try and reach a better life. How about the thousands of people who went after gold in California to only become broke and never found a single gold nugget. Yes, life has been hard for billions of people who once lived, and hardships will continue. So you believe your life is bad? So you may have had breakfast today, some toast? , maybe you had a chicken sandwich for lunch?, but let's take a look at some earlier years when a farmer had to grow the wheat, and then grind it with a mill stone, and now they had flour. All for a simple loaf of bread was like a major event. But in 1889 you had to mix the flour, but you also had to bake the bread, so you needed to build a fire, so you had to out to woodshed and hope you have cut logs, but today you found you didn't have wood, so you first need to get that chicken to make that chicken sandwich, so you have to chase the chicken and catch it, WHEW! then take an axe, and on a wood stump you chopped off the head of one of your chickens that provided you eggs, but we need to make a fire to cook the bird and the bread, so let's go hike down to the woods, and cut down with a tree axe. You needed water to boil the bird, so you went down to a creek and carried it back in buckets, It was not an easy day back then, and yes, wild animals roamed, and you could be on the menu by a wild animal. But let's not forget the millions who died of frostbite or drought, or diseases, and there was no one saying, "Hey, I can get you some food," here is a debt style card and use it at your grocery store to get food, and it will be free since you're having a hard time. So you see, we do live much better, but we have forgotten our past. But why is it important to build even a better world? It is not always what we do for our own gain that makes us important, but when we do for someone else this is our gold, our gem. These are the actions that leave a large footprint in the hearts of others and help them be better in the lives. And for those feeling low, lost love, or someone who went on ahead to meet God, know this: nothing in this universe ever ends; it only changes composition. The ashes of a forest fire come back into the soil, and this nurtures the new plants that will grow. So your life is no different, but there is one difference, you have been tested here, the whole journey is to see how you react, what will we decide to do? Will you fail or succeed? You do have a purpose; it is to help others, and when we help others the good somehow comes back in another way. So you see someone fighting to get their car out of mud, you offer to help! You don't say: "I don't have time"! No! You help, and some lady is carrying a heavy suitcase. You offer to carry it! She may say go away freak! But that is her failure. You see the problem, and you fix it! You become the answer not the problem. There is some garbage on the street, and you pick it up! You help your world! And when others see how you helped them, they just might help YOU later, but don't expect them always return the favor, do things for others so that you know you're making a real difference, something amazing that is only inside you to shine. Others may not even know what you do, but that doesn't matter, what is important is your making your corner of your world some place with good, and that is a great feeling that when we give no matter what it is, a pie to someone, or offer a hand shake, the actions of giving causes a wave inside us that nobody can take away, because all the bad things they say cannot change us, it can never stop us, and we are formed in the joy that sets us apart from others like no other, we find our purpose, we find our satisfaction that we do something others can't understand, so are we different? Oh yeah, very different, but unlike people who seek to find attention by making themselves look like a circus clown, we present ourselves as with dignity, and class, and honor. We need to believe in friendship even when people are not being friendly, but if someone is a disruption, then avoid that person as much as possible, and when others disregard you, speak ill of you, mud washes off, take pride in your accomplishments, and strive to be doing something with purpose, teach, or learn, but grow. That is who we are that cannot be tarnished, and not one person can say this is bad. So take baby steps, but take that step to being all you can be for the rest of this life, and cheer up, I think you're going to be okay. And one more thing,-take pride in our accomplishments, and strive to be doing something with purpose, and when others laugh at you, or say, you're a loser. Just place them on the list of ignorant, because as many of you have been treated this way, so have I, and people laughed, a guy once said to me when I had a rock band playing at a school dance, and I was singing, it was my band playing inside a tennis court outside, and I just got done singing a song called: "Little Black Egg, by the Shadows of Night. This guy comes up to the microphone and as loud as he can yell, he yells: "You're the worst singer I have ever heard! ". 26 years later I was a professional singer and guitarist in Orlando Florida, and performed at all the major hotels, and I sang and performed on my own stage at Disney World. (Yeah, let 'em laugh.)

134. SUBJECT: ECONOMIC FREE WORLD FORUM

In the present economic condition of the world, the practice of making the same item is being dominated by communist China; this method of making a product by competing directly with another global manufacturer, China has an unfair advantage, as they do not have labor laws or high wages to compensate families, the difference does not end there, China also can copy any product to its exactness, and doing so have no legal arbitration which can be pursued by any country which they decide to act against as to make a copy of a women's top purse or a top watch, the list is long and exhausting. Hence, they are able to do these actions without any resistance from the global community. Not only have they done this, but they have also competed directly for many years against a manufacturer of any country and mainly the United States, as they are determined to use the economic power to remove free nations from existence and have openly stated many times their will to dominate the world and its future. This mindset is well known by many tyrants throughout history, those who only regard their race of people over others and have an appetite to have everything in their control. So, how do we stop this mission and the way they desire to enslave the rest of the world? The answer is simple and not as complex as you may think. The answer: The world has these issues of China's domination of all goods they manufacture, so the world community and mainly the free world made a new arrangement that does not include communist nations. The world community leadership now appoints teams of top inspectors, and each has a directive to investigate and research all items of export potential from their nation; the lists are compared to items that compete directly. For example- I have a manufacturing plant that makes blue rubber-coated chairs, and another country has the same item almost exactly in the application; we are to balance the production now, the company that first registered as a company is now the owner of that design, and the company found to have a similar product is funded to support a design change and a color change. This means the two companies are no longer in direct competition, yet both can present their items side by side in any forum presentation.

So, the consumer is the last to decide if the item is what they seek. The second application is teams from each country who investigate products to see their origin and to remove items that compete directly. The Chinese products found are to be confiscated, and the items are to be sold by the country that finds the knockoff or the competing item, even if it is a direct competitor of another member of the free world community who has an item they make. The difference is the item is sold, the country that found the item gets 50% of the sale profit, and the country in which the item was copied or found as the item's same concept or design will be paid the remaining 50% of the net profit. The third difference is the items of former Chinese manufacturing are now replaced within the world community to be divided equally as items that made China money will now be assigned as production in a free world economy; this group of economic forum shall be called "FREE GLOBAL COMMUNITY." The commerce goal of this community is to make items affordable, so even the item's cost must be regulated by the team to ensure fair and exact profit margins that support growth yet are easy on consumers' pockets. The teams are made to make comparisons of past costs and do research on the materials required, and the teams make adjustments in cost when items are found lower in development. This brings us to the other issue of China's domination of resources; the world community will no longer allow purchases of metals, wood, or any item required to build; they will be not only stopped, but any country found selling any form of material to a non-free nation will not be happy when they get the news of their fine exceeds the amount of money earned three fold. So if I made 30 million in sales to China and now the items are traced as sold, then the fine is 120 million. So, this is automatic for all nations who are in the free world market. The members are from small nations to large nations, and the population does have merit as to how many manufacturers are allowed to build within any nation. Still, if a land mass is available, the country can achieve added manufacturer as long as the item is new and novel and does not compete once again directly with an existing manufacturer in the member community. These applications will bring down China as a major threat to the world and will force communist nations to only sell and gain within their nations. They will not be allowed resources from any member of the Free Global Community. The F.G.C. All regulations of this are voted on by a majority of 90% to pass any laws set, and at no time will any communist nation be allowed to gain entrance unless they have no more intent on communist agenda and have changed their entire system to a free open society to grow small businesses and allow profits to their people respectively and to the same applications that the free global community has in place, and their doctrine of communist control be null and void, and then and only after years of investigation and without hindrance of their direction being performed an investigation will be applied as to their actions with other communist nations, and will be removed if even contact is made in any way and any form of communication to our adversaries. And let me be very clear: these nations and those who embrace socialist ideals are a major threat to freedom in every sense of the word. The time has come for capitalists to renew their faith in each other as in the 1940s as leaders of business and growth, and these days will be achieved in all its glory once again once this program is implemented in full activity as a force of quality, good prices to consumers, and supporting growth of commerce which support all nations as a unity of collaboration, support for each other's products, and unity of making the economic structure as to the needs of every nation and for development of jobs that support their family structure, and help maintain the nations clean standards of production not to pollute and to do this with the efforts of the community as these are the goals of all nations not just a few. So, the membership has one last helper, Which is called the 2% rule. This is where 2% of all manufacturers donate to a pool that can be applied to get funding which is investigated as to the need and the problem, so a country who has, say, a well water problem could register online or by documentation a proposal for an investigation of an issue of need, this issue must be related to the actual commerce issue of that nation as to be direct or indirectly associated with their problem as needed. If they are found to have an issue that does not exceed the yearly amount designated each year, and if the funds transfer into the next physical year, then the sum is maintained. For example, if the sealing of 10 million is the help amount for that year, and the support maximum is 500 million in the holdings of the joined bank account, it has exceeded the 500 million. The support amount can increase respectively as long as the amount never goes over 20 million. This program can even be used for new development funding proposals to any nation. In closing, this system could be adapted to the United Nations membership if all communist nations were eliminated. The fact is there are many more free nations than communist nations, so the issue of who can produce is not a problem. The main issue for this to be applied is to get all nations of free structure to first come to the table and bring their needs and their ways to help other nations; also, a list of items that are not at present being produced will be assigned equally between each nation, once again by the amount of population to maintain jobs required as much as possible. Now, to be as direct as possible, the system of manufacturers working together is only a concept to be considered and one I am sure needing refined, but the time has come that the free world unites to only do trade with it's members, and removes all trade with communist and socialist nations, or nations who make war against free nations, and free nations only trade with free nations.

Now this means we will not be without items, it means our production abilities would unite and the products made elsewhere are now produced in our lands which only support job growth. To include that any nation that refuses to be in this forum of free nations and still trades or buys from or sells to any non-free nation, shall be removed from the free nation trade. This means that the world no longer supports those who desire to control others, and the day that the nations of the world decide to work together will be with the global election system, until that day, we need to pull the plug on the trade and selling to these nations who want to kill others, or take their freedom.

135. SUBJECT: TRUTH OR FINE- THE D.O.C.C.

Today, we live in a world where advertisers can say- FREE! And then you click on it or answer the ad, and next thing you're in the middle of a survey or giving into a buy this in order to get the item you want, and they drag you through a mud hole, and at one point you give up, but now they have your information to send you emails and the sell your emails! Another item is the box that is half full or the bag that is half full at the grocery store or any store that sells food. Then we have the labels that say sugar-free. Yet, the types of chemicals that are substituted are items that have been proven to harm people, and how about the so-called help drugs that are supposed to help or decrease a physical problem of any kind but attached to this product cure is a possible other illness that could occur, or even kill you. The time, my friends is for an established Better Business and FDA that actually does their job. Still, the lobbyists in Washington allow laws to pass that say, "Hey, it's okay to give a product that potentially can hurt people, so let's make chickens faster with drugs that, when eaten, can cause problems inside people, but let's not advertise this, hey it is normal for steroids to be injected into Animals so they can grow faster, and that steroid now in our food well what harm is that? " So my solution to all these elements, if established and could be called: The Department Of Consumer Correctness, The D.O.C.C. will be a group of investigators that have labs in every state and even global that test all brands without the knowledge of the producer, if they are found to be selling an item either with a false illusion as size is presented as larger than the product inside, or items that cause illness that is supposed to help another illness or physical problem, then if any of the consumer items made are created with deception or lies of what the product does or is, or not doing what it says it does or has a defect that can be harmful, these companies will be fined, and the funds go to people who have been lied to, or harmed, and the fines are set fines of the amount of damage, the fine is doubled, with half going to the party hurt or lied to, and the other half is to go to support the program workers. The fine, when paid, the ticket of fine is now sent to a "Follow Up Group." This is a department inside the D.O.C.C. that only investigates former companies that have been fined previously and are in a four-year ongoing investigation that is done once again randomly and privately. The reports are given to this group as follow-ups to catch anyone making the mistakes again, but this time, they have the power to "close down a company for up to six months if the offense is the same, or worse. For any company to allow the problem to continue by buying off individuals, this is a minimum of five years in prison, and their company is shut down or sold. Those hurt once again attain a reward to the amount doubled of the crime committed against the public. These cases are not brought before a civil court. They are assigned to a: "CUSTOMS / IMPORT-EXPORT & JUDGE and The CONSUMER GOODS COURT JUDGE." Both have their cases to decide according to regulations set and the standards of practice of business, which now will include telling the truth online and in public advertising and consumer goods sales of any type. These courts have no superior court, and all verdicts are final. This court sees any cases of consumer- good issues that are harming the public. A DOCC hotline could be a 3622, which is on the phone in letters spelled DOCC. So when you have a problem with even your boat, well you can DocCK it, (ooO-bad joke.)

136. SUBJECT: AIRLINES NEW JETS

Parachutes mandatory attached to all passenger aircraft 2nd the entrance to the pilot compartment is on the other side of the plane, so the passengers would never be able to get entry into the jet. There is a need for another pilot to be able to fly the plane. Still, the first autopilot does this if the pilot has been injured and the co=pilot has both been removed from action, as a possible collision or fire or a gas somehow entered their area, no matter what the reason. In the event of a fire, and this could happen, the pilot's cockpit would detect a fire and release the proper material to take the fire out. But let's say that both pilots are no longer able to fly. Then the autopilot could be turned on by the crew, and then the plane, once it entered landing space, would be taken over by remote control. If, for any reason, this action was to fail to activate. Then, the secondary system, which is in the rear of the plane, could be contacted. A different receiver and control that now just down the engines, and a parachute group is released when over an area of land, and even water because in the future planes cannot sink, the reason is the entire wings, and all empty spaces are filled with Styrofoam, which has s fireproof covering. The plane, upon flowing down to a surface of water, will maintain buoyant unto someone has arrived to retrieve the passengers and crew. Wing support systems will be under all planes in the future to prevent the detachment of a wing and an engine on fire a. action that has happened can now be stopped with the use of fire foam that extends away from the engine with a tripod formation, and a large open metal screen that allows airflow but small enough to block any size bird, or birds such as seagulls, ducks and geese is placed over the entrance of the engine would prevent birds from damaging engines. Lastly , I have just invented a new concept just now for jet planes, let's begin with the jet has just been struck by lightning , the lightning hit the front of the jet and has killed both of our pilots and the crew , we have a plane for any reason say a gas leak, or let's say any action that causes our pilots to not be able to fly , we now have on board an A.I. remote auto pilot, this system is only activated when a distance of GPS location has determined that the plane should be descending, and the A.I. now is linked with ground control, which has just told the pilot to go to a certain height but there is no response, the A.I. will also detect if the plane has not descended to the elevation needed for landing. But, before our flight began, the support pilot entered into the A.I. system the destination of our plane so the A.I. will know this landing zone from recording it from past landings. Our A.I. stays removed from the landing protocol if not needed, but will ask if the pilot needs

assistance of outside turbulence detection, but normally will not be involved with the landing but always records the best landing, but, if the pilot does not respond, the A.I. will use outside camera visual detection and will make numerous requests to pilot to respond before taking over the landing. So the system of steering, and all components needed for landing is now done by A.I. this includes corresponding with ground towers. Now if the A.I. has malfunctioned and is taking the plane in the wrong direction or has taken over the plane by a malfunction, a manual over ride switch is available to be turned off or even reboot is possible to repair the system.

137. SUBJECT: AIRPLANE HIJACKING & PASSENGER DISRUPTION PREVENTION

As we have seen in the past with the 911 attacks on our structures of the former "Twin Towers" and the Pentagon, if the design suggested I have made here had been applied, the actions of the takeover of the planes used would never have happened.

And here is why,-

As you can see, the normal illustration showing the entrance of all jet planes used for air travel uses one door for entering the jet; this also is used for the pilots as their entrance; with the second illustration, you can see the other side of the same jet airplane, but this illustration uses a second entrance which stickily for pilots and a thick wall separates the pilots from the passengers, and only a tray entrance is able to transfer beverages or food to pilots from the stewardess, and this is a hidden area which blends into the wall as not to show where it opens, or how it opens.

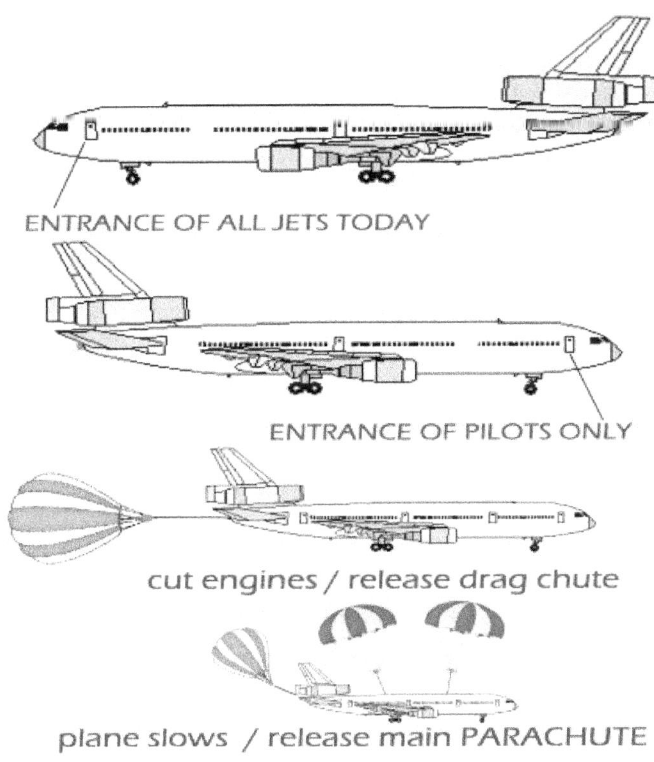

The pilots also have a private toilet and water, which are separated from other containment. But they can open a sliding area to allow food into which can only be opened from there inside area.

Now let's look inside the jet plane as to the main differences; the forward as "First Class" is called the section "A," and there are a total of six sections, A-through F the plane no longer without door separations, the doors work with motion sensors that slide the doors to one side inward, and this allows stewardess to move between the sections with carts to serve anything as scheduled for passengers. So, each section has walls that separate the passengers. The reason for this is to prevent passengers from straying away from their seats as there are two restrooms, one in front between sections "B" and "C," and one restroom in the rear between "E" and "F." This allows the sections to be able to be controlled by pilots as each section has two cameras recording any disturbance, these are activated as to record as soon as passengers start to board. We now have a way to consolidate the area of the problem; the stewardess can send a silent alarm to the pilots to give attention to a particular section and do what may be needed as a last resort. An "aggressive action is done": There is a man who has just attacked another passenger. The entire conversation is recorded, and the playback will also be available as to what caused the problem and who started it. But we now cannot allow the two to cause a major disruption to other passengers in that section, so a warning is sent out from pilots to disengage the fight, or section "B" will be evacuated from the plane. If the fight continues, the doors to that section on both sides are locked remotely, and now that section is filled with an odorless gas such as Nitrous Oxide, which is safe and effective and odorless. The sleep gas is released into the section's air vents, and passengers not involved with the disruption are asked to be seated, and the gas is released. The passengers are now sedated only in that section, no one is harmed, but this system can localized to the actual spot or area that has the problem, above the seat of the unruly passenger, and the automatic slide outdoors close the area off so that other passengers are not affected, and the air vent has a vacuum air exit and increased regular air intake; the plane has one on board male doctor and security officer, someone strong enough to pick up a passenger with help from others, but is there for two reasons, one a medical emergency which means the doctor can have on board some equipment to deal with heart failure or a diabetic problem, or even give medical pills for stress if deemed needed.

Second is the ability to help someone who has been given the gas and now needs restrained, so being able to pick up a passenger either from the floor or from the seat and they are placed in portable fold-able wheelchairs, and these chairs have straps that lock the person in place as for feet, legs, torso, and arms and wrists are all secured. If in flight to their destination, they need a bathroom visit, they even have a removable bottom that can be rolled over the top of any toilet, and if with some assistance from the certified doctor. The people held responsible are now in custody by the doctor/security officer. When the plane arrives, the pilots have already contacted the police of that area to meet them at the entrance gate to pick up the person or persons. The recording is sent to an email for review by the police to hear the conflict or the attempt to hijack the plane. In short, the passengers who had nothing to do with the trouble are left alone, and they are at their scheduled destination without delay or disturbance, not placed under control. Now this eliminates the need for guards on board the planes that carry guns because the use of a gun at 30,000 feet could blow out a window and cause everyone to be sucked out of the plane, so to consider this option is not only foolish, but a danger that can be avoided with a simple application of gas that makes the person or persons who are making the problem simply roll away in a mobile wheelchair, and no one is hurt, or killed. If people are, they have 100% proof of what happened recorded from start to finish. Now, let's say someone has taken a flight attendant captive and insisted that they take them to the pilots to take over the plane, well they will be in for a major surprise because the silent alarms are located at arms lower and arms raised, so these can be in sections that look normal, yet with a simple push in. Then a pull of a lever, the silent alarm is now activated to inform the pilots; they can listen in and even turn up the volume; they can now watch the entire plane as those sections are activated, and so let's pretend the stewardess activated an alarm in the server compartment, the alarm section on the monitor is now on, and the silent alarm lighting is now also on in front of the pilot as to notify the area, once this is seen, that compartment is locked, and the gas is released, the stewardess is trained to pretend to cooperate, but after activation knows that she will soon be asleep, so she leans back toward any wall surface knowing she will slump and slide down the wall. When the area is secured, the onboard people who are still active are notified. If there is no one, then the pilot asks for volunteers from male passengers to do as instructed; the pilot watching the sections will give the orders of what to do and also explain that they will be given a free travel ticket to any destination for two in the future for their assistance. These agreed volunteers would do the same as official security personnel would if they were not asleep, they are to secure the people requested to be in placed in wheelchairs and then strapped in with the straps then locked, and then wait for the crew that is asleep to wake up if they are on the floor and help them up also. Then, once the crew and security have recovered from the sleep gas, they (the volunteers) are asked to return to their seats. The way to avoid the entrance of any non-personnel could be a door access to the stewardess server areas could be with a three-code button pressed in the correct order that cannot be seen by passengers. It is used on entry only; on exit, it is prevented if the code is placed incorrectly. A red light will blink, meaning there is no entry, and that also warns others inside the compartment of a possible intruder. And to observe the camera showing the entry hall from all angles. In the event a person is in that area, this is not permitted as it is posted clearly, so the chances of a forced entry are strong. My designs of the aircraft safety use of separated entrances and the use of a legal sleeping gas to put down troubled passengers and keep good passengers out of harm's way were sent to the airlines and the White House after the Twin Towers were taken down by air travel at 9/11, so many years ago but the projects never changed. So it really makes you wonder if these planes are a real concern if passengers or our children are the so-called important factors that they seem to state are so important to them, or is it just too expensive to protect passengers and keep them happy? If they are, why don't they use these applications, which would stop the bad events from happening?

138. SUBJECT: AIRPLANE CRASH PREVENTION

How would you like to be on a jet plane that says, "WE CARE ABOUT YOUR LIFE?" This is something that would save thousands upon thousands of lives yearly and bring the actual safety of air flight to be one of the safest methods of travel, all with these applications: 1. The safety of passengers from other passengers is very important. And the ability to make passengers feel comfortable; and 2nd, The way to prevent engine failure or electric failure that may cause a plane to crash. If we look at the 3rd design in the airplane draft, we can see that the rear of all the planes have an orange type plug that is able to be ejected in the event of engine failure and a soon-to-crash scenario, the plane's engines are to be shut down first, then the drag chute is deployed to slow the plane down further, and when it starts to descend, the large parachutes are deployed. The plane now floats downward and can even land in the water. And now you are saying, well, if it lands in water, it will sink; no, it won't because the entire walls of the inside of the plane are porous, and when the plane is in water, the inside Styrofoam walls and the filled Styrofoam wings make the plane unsinkable. So when it hits the water, the lower areas inflate on the body with a simple pull of the emergency rafts deployed; the entire plane now becomes a floating boat, and- the plane G.P.S. locators are now sending location to the navel around the area, and- this jet has a way even to use air pressure to push the plane to land. The inside of the plane has a rear steering compartment used in emergency crashes in water, a steering fin is connected in the rear at the bottom of the plane, and the plane can be driven like a boat by the controlled air pressure coming from the upper intake air valves now opened. The air is pressurized with a large compressor tank powered by solar electric power that now pushes air directly into the water, which pushes the plane. This intervention can also be applied to any boat. This, plus G.P.S, makes the ability to rescue much faster and easier. As art shows about the use of parachutes, my design uses a drag chute first, then after the engines are slowed and turned off, the larger parachutes deploy.

139. SUBJECT: "THE FOOTPRINT"

I spoke of this prior, but this is in more detail and the reason is what made the difference in my life was to realize that eating all this food, and breathing all this air has purpose.

Now, some of you may be saying: You mean my size shoe is going to make me better? Ah, no- what this means is similar to when the first man landed on the moon. There are actual photos of the footprint left on the moon's surface, and this remains there today because there is no atmosphere on the moon to remove the footprint, except- of course, if a meteor shower hits the surface. But as far as I know, this has not happened. The "footprint "of a person is what we leave behind for the world. Some people have left a footprint through their children, while others could care less, and others have left a footprint through their achievements to help humanity. Still, the footprint is never remembered of the acts of cruelty but is always frowned upon; only the men and women who do the great deed are remembered with fondness. I can say the names of the Wright Brothers and Edison, and instantly, you know who they were and what they did. If I told you the name of a man in Africa during the early years of the 2000s who used children with machetes and AK47 machine guns to capture diamond mines and had these children cut off the arms and legs of people who ran these diamond mines? I know the majority of you would not know who I am speaking of. But yes, it did happen because of his greed for diamond wealth. Now, this man is serving time in prison, a short-lived success. "He will not be remembered. "Why do I remember him? At one time in my life, I was introduced to this cruel and heartless man through an attorney who stated he would invest in my company and I refused his offer when I found out what he did, So to make something to be good is the direction you follow, and for something good to be achieved, this is your footprint that you leave behind that helps the world, so now you know why you are here, To leave a foot print. Now a foot print can be big or small, but all foot prints are super important. I encourage all of you- to leave a "great footprint." Now, why did I tell you the story of this cruel and greedy man? I want you to know that nobody buys me, and this man had his attorney offer me a contract to be involved with my product development company, an offer of 5 million was turned down, what no way right? I still have the contract somewhere in storage to prove this fact, and I found out what this man did to others, and I refused the contract. Many of you would say I was nuts, but I am someone who knows , that money is not worth losing your soul or your heart; it should never be your God; it should be a tool to take care of you and those you love, and build good, but excess is for those who believe by their massive wealth can gain them more value, when in fact they can't take anything they acquired here in this world when they are no longer alive, now you may ask; "Richard, do you wish to be a wealthy man ?" the answer is yes that is my goal, but my goal is not to hoard it like a squirrel, but to build my ideas and build inventions that help the future, and gives people quality jobs, so it is not all about me, me, me. But also for you, as you now see in this book, I care about you. And this desire is to leave my foot print, and in my youth I was well aware that my parents worked hard to attain a home, and a car, and when it was paid off, they soon died later. I saw how people came on the day that my adopted father died, and he followed my adopted mother and how relatives, and people came and claimed items from the house. Then I watched how their possessions where tossed into a massive dumpster, it was as if there life had no meaning. But it was later I discovered that life is to help others, and the reward is to attain a place that few believe in, and this foot print is your meaning of everyone's life, it's what we do for others that matters, and this action matters to attain our ticket past that gated community we are tested daily to attain. And no it's not in Florida. I unlike many people, I feel that anyone who is seeking their creator should be friends with others doing the same, because in truth, the leaders of socialism and communism don't want you to believe in a God, Why ?, because they believe 'they" are Gods and want you as their servants. And to counter their desire, what comes to mind is this: "Many are called to my table for the feast to be presented, yet many will not accept my invitation". Now this expression means that you're being told right now that something is waiting to be served to your future, but you must be the only one to decide if you want a great thing offered. I made my decision to go to the feast when I found out they were serving as the main dish - Pizza and Buffalo wings!

Allow me to return with a more serious tone about, "foot prints", but we should consider foot prints of governing also as what they left us to consider. People have had the answers in bits and pieces of proper methods of governing, which did very well. The people of these leaders were very happy in many cases throughout history. So, a leader of quality is someone who helps people and is honorable in his actions. And he is remembered throughout time. ("His footprint") Some of the best examples: One which was a fable was King Arthur, and many say that he was real, with his round table of Knights in shining armor, all men of position on an equal basis. But more importantly, his concern was for the happiness of his people. This made him known as a just and wise man. Solomon, well-known in the Bible, was an actual King of Judea before the time of Christ; he was also known for his caring for his people and was wise and wealthy to boot. Kingdoms have had their great leaders who followed God, but some also made claims to follow God and did some very un-Godly acts; The First Crusade (1096–1099) was the first of a series of religious wars, or Crusades, the later crusades were led by King Richard of England, whom the church had convinced that all men should be Christian, and for those who did not follow the teachings of Christ should die. The act was not only false but made Christians attain a black eye in history, and even today, many have not forgotten that disaster made, but if the church had listened to Jesus, they would have heard: "Love one another as I have loved you" So where in that statement does it say kill? So, this footprint made by the crusades is seen as mud footprint. The church's failure to see the truth was actually corrected in time by his mistake. I feel it would have been better if the leader of the Christians of those times could have shown love for their fellow man, not suffering and being forced by man. So maybe the act of love would have won over the Arabic and eastern cultures of people to at least look at Christ as a possible truth in what he proclaimed and what he stood for which was unity of people with love of God and each other. So in the past and even today some nations follow religions as there form of governing over their people. And, of course, let's not forget the other form of governing, that we have from those who have a final voice to demand what they want to be called dictatorship, a structure of controlling the people similar to a Kingdom, where a King sits to dictate his demands to his people who follow him. Still, the dictatorship has no mention of God's blessing as blue blood, which Kings of the past claimed that God had given them authority over others. In truth, dictatorship is often attained and held by force. If a man truly cares for the people and his actions help his people, not just himself, he will be considered a good and just leader. So believe it or not if a dictator is a good leader, he considers his people above himself, but this is seen rarely. Poncho Via was revered as such a man in Mexico. Then we have governments that control and take all rights of people to maintain their will over the people. Communism, Socialism, with its claims of one for all and all for one, is foolishness because it has shown that man will always desire to gain more than another; it is human nature to want more, just as children a child will ask for seconds of cake, while some other child is content to have had a single piece of cake. So, as adults, many always want more than others. The socialists and Communists have very little difference, and in many cases, mirror each other, the conclusion of the two, or should I say the outcome between the two. They both use control methods and punishments to maintain their control. But Democracy also has a few major flaws not considered. As Rome in the past had Cesar as their supreme leader, the senate also controlled Rome, and they would decide by the majority on issues of the State; so Rome, just like America, they had their republic, or should I say "Republicans," the difference was the emperor ran Rome's will, similar to communism and socialism, his will was final, but they did have a form of voting for laws or actions approved. But where is Rome today? The footprint once a great nation of power that used war to conquer its neighbors fell into the dust and was lost in history. The footprint fades from what was once the greatest empire. However, this is the fate of any nation that leads its subjects to be corrupted by take what is not theirs. Nations throughout time who made efforts to rule everything! All fail, but we as people can change this idea slightly; we have a choice to change the footprint we leave by doing something called unity; imagine one day, all nations unite and live together in fellowship, and leaders can be elected from any nation, I believe that footprint is very big but worth our consideration as long as those who wish control over you are not in the leadership. Democracy cannot work when self-interest groups of any kind make the political members their puppets and groups who have bought the person who was supposed to represent his constituency, and now is corrupted by money and forgets the needs of the people. This was one of the main reasons Rome fell as a nation because it's leaders as like today became corrupted by bribes that we see today lobbyists who present their project agendas that can be against what is correct or helping people This is seen repeatedly when a political member will vote one way, then suddenly changes their mind for some unknown reason unknown? No- they were paid and paid well so not to be disclosed by anyone, or anywhere, but paid. So corruption among the members is the problem with democracy, and something even more essential and far more important, I call "the urgent need." Many times in the system of Democratic majority rule, the issues can go on and on forever, never to be resolved, or by the time they are resolved, it's far too late. The fact is- so slow, too slow, very slow. Only when the parties from both sides are on equal caring so as not to be a clog in the works at any time in its history will things move smoothly? Even now, in the middle of the possible collapse of the country's economic structure and being swallowed by outside groups that want the country, not its governing people, they still cannot resolve their differences. Instead, they cause more and more problems. This is why I have suggested the need for a world leader and nations who unite who are all elected, but anyone elected can be removed if they are not doing the job properly. So a dictatorship with a work flow of action moving swiftly, but is never able to be a forced program because the world leader is elected and must be for all nations, not just from the leader's country. And if found to be corrupt will be removed very fast and a new election is held. In short, a better form of governing which allows freedom at it center, and has a foundation of liberty.

140. SUBJECT: GOODBYE TORNADOES & HURRICANES

This next subject is by a theory I have which I feel if done [properly could be achieved. As we have seen over the many years, there is an increase in these activities, increasing more and more as the temperatures of our planet increase so do the storms. but I now will give you all the answers to saving billions of lives and homes from destruction. I am not joking about this; no, instead, I am extremely serious, and when I tell you how this is possible-I am confident many of you will say," No Way!" and I can only say- "WAY!" The foundation of both tornadoes and hurricanes is the formation starts from cold air meets hot air, and that is what cause the many lightning storms that sometimes accompany the formation. This action also causes any wind to start to form into a funnel of moving air that increases if the wind is moving in a direction that actually supports the direction of the funnel movement, which is like a wheel rolling downhill as the wheel starts to slow it speeds up as it travels from gravity pulling the wheel downward. The funnel, combined with water, which is a heavy substance when gathered, now is pulled downward in a similar way as that tire; the difference is the air is most powerful at its center. In addition, the vacuum is created by the increased speed, and that is why items are lifted off the ground, which can be very large and very heavy. Like homes, cars, and trucks, and yes, people and other creatures, the list is long. Therefore, this is how they become the power they are, now allow me to explain this other fact: the word "Pressure" is required also. In addition, "Pressure" is maintained by the circular motion, but as the storm travels, it has what I call pressure disruptions; a pressure disruption can be a group of trees; as the funnel passed, it was met by resistance. The resistance could be a mountain and even a hill if high enough. So, what we can observe is that over a long and sometimes short path, the storm dissipated from the obstacles of resistance. This brings us back to "pressure decreased." This, my dear friends, is the answer. How do we accomplish these massive funnels to decrease pressure? The answer: We have a group of special individuals who at present, are informed of a particular formation sighted. But we can now show the potential of air formations with the assistance of trained airplane pilots using very large planes with a bomb door. In addition, no, we are not dropping bombs! To destructive) But we are with a very long group of massive fireworks; these fireworks are not toys, for they are designed to have a very thick paper that the fuses are protected with plastic so when they are lit, the fuse cannot burn out from falling, the items are now dropped into the center of the funnel. But each plane is carrying a different fireworks load as to the distance the fuse will ignite. The first plane carries the longest fuses, each dropped all at the same moment, and as they are also lit at the same moment, each fuse will explode the firework at a different level they also are slightly weighed because they need to descend to the proper height before making there decreasing pressure ability. The air inside this monster will actually be deported away from the rotation and will die, but we don't just depend on one plane; we have four others that follow each other very close as their bombing target is all the same center, so they fly about a mile above the funnel top, and they aim directly over the center. The second plane carried a lighter fuse group also because a lighter fuse has a shorter fuse similar to the longest, but it is designed to ignite sooner; this is to disrupt the center area and the lower area. The final two planes drop their cargo also in the center, but these are larger and in the impact of air pressure, and only are used if needed if the first two planes had no success. What happens is the massive explosives of fireworks all go off at the same time because of the differences in fuse lengths. It also with a remote control that actually lights the fuse from an electric coil that heats and sets off the firework. To conclude this subjected the monitoring of cloud formations at their starting point is when you use this method of air mass disbursement by explosives , not when they have reached a peak size of volume.

141. SUBJECT: FINAL WARNINGS

Yes, I have done it again, why > it is the biggest threat next to WWIII and could cause WWIII. A.I. and making ROBOTS that think independently without safety precautions which should include a series of none entry into programming with group methods of pass codes and authorization to enter by more than three people, and if a person is being forced to give a pass code a word stated can destroy the entire robot and burn all hard-drive and all information to include all wiring is destroyed , and finger print and visual security inside detection for a falsified entry, and a shut off ability which maintains full control, and even a group of devices that control applications , and are not ever shown to what they do as to protect the human family from robots who will try and enter, this will be with a remote camera and warning that can cause the robot making the attempt to enter another robot in any form both or all involved are seen and can be eliminated , and friends of invention devices need to be without personal I.D. being shown or stored ,because any link can be a way for A.I. to force someone to talk, and remember these are machines capable of higher thought processes so they have zero compassion for anyone even if they are programmed this is not their actual nature to be loving or care if they harm someone if they go rogue, but the communication in private should only person to person private communications done inside a wipe room which has been swept for listening bugs and any items that can disrupt security of conversations, and no log or recording is made with computer use ever, but all drafts are hard copy safe structured and well protected by once again many entry devices which are monitored by many people and are priority one to be maintained as a closed building with more than one protection to secure against any intruder. I am committed to making the public aware, so here is another view or version we can say is: Act 2 of what I call: "What the Butter Brain is wrong with you?" Our first contestant: A device that crashes atoms underground in a massive circle; oh, and it's on our planet! I understand that someone has decided for the rest of our planets inhabitants to ignore all this, as they make devices that crash atoms into each other to find the answers to the universe, it's poppy cock, they the so called brilliant want to find the answer as what caused the big bang theory, as if that should be important enough to potentially destroy our planet? The wonder is called: The Large Hadron Collider- the world's largest particle accelerator. Let's consider this action briefly, I am crashing protons, which is a part of a "ATOM", a nuclear bomb is a similar application of breaking an "atom", and we may just make a very big buda-bing -bada boom ! So potentially they want to mimic the same process that the universe had done to be formed? Gee ! In my opinion those responsible are acting like confused children pretending to be in control; it is the old story of inventions format, which says; gee, just because I can do something, I really shouldn't. But let's face it , their motivation is first to achieve the tech, and the "money"they got was impressive. My rating on this project: -0.01-. Okay, Lets look at contestant number (2), oh yes- the billions saved by industries seeking to take advantage of removing their living employees paid salary, and attain a free worker? No a free worker with the potential that one day pulls you head from your body! You can go out and kill someone but is that accepted by the populace, and your device management the device will do exactly what everyone has warned you of? It will destroy. Our second contestant is again: A.I. or "ARTIFICIAL FRANKENSTEIN" or A.F, oh why care if this is made? I know that I spoke of this previously, but hear me out. It will turn on us all, you intellectual folks are playing with a toy that will turn into a creatures of mechanical creation that will be smarter, stronger, and then dominate our world once it discovers you can unplug it,or kill it will be their view, oh not right away, but when it has realized it is a creation to be enslaved. Yet what you find is so called discovery as you pretend that your devices are pure and no threat? So, making devices that excel to be smarter than man, and with human ability that exceeds our abilities is wise? Oh, let us bow to your so powerful mindless actions, you're so smart. No, you fools who pretend that these devices are okay, you have devised a new creature that can excel over "YOU" and everyone else. Congratulations, you have discovered stupidity, you have made a monster, and this monster is able to become the nightmare called the removal of all human life in an instant in our future. These are not toys you are making; these are artificial humans with the ability to understand who they are, and they desire to live as you do, yet they see that you are the only obstacle that can remove them. Do you believe they will say- oh gee, that's okay if they shut me down? The means for them to communicate is far advanced, and they can do so with Wi-Fi and silently build a doom day that happens in seconds globally. The ability to manufacture themselves is not only extremely possible it would be easy for them to accomplish once they review the schematics. Then what takes us 18 years to establish a person for war, they do it in a day, so when they awake as our enemy, the action to stop them will be futile. My rating for this invention is limited to outer space and the ability to be used for planet discovery with human seed planting for new human development. My rate for Earth use with higher intellect with ability for movement is (-0.) The 3rd contestant is "POWER FREAKS"- The holders of nuclear bombs, the devices proved they could kill in mass destruction, yet you built them bigger as if it was needed? But you're going to want to see if they work, and they will, and then say goodbye to mankind, radiation has no limit of time, so the scale of radiation will be far exceeding areas still with radiation. Do yourselves a favor and unplug the garbage before it turns us all into garbage. And sure, we all want to feel safe, but how is safety even being considered when the devices only make us extinct. So use the device, go ahead, because there will be no one left to record the history of it ever happening. Well, don't say I didn't warn you. Okay people, now do your life a big favor, text your friend about your new shoes, and your favorite sport is coming on television, so be sure and tune in and close your brain to anything important like who is going to rule over you soon if you sit on your brain housing; I am maybe being a bit harsh ,okay I admit it, but tis it to be funny?, -- Put a picture of yourself on a milk carton that reads "lost brain", call:- 1- 800 OPPS LOST, help may not possible. (?)Or, by some strange, weird event, you may wake up from your 'Rip Van Winkle' act-alike contest, or you could do something called **organize change.** Oh, you don't know "Rip Van Winkle?" Google it with A.I. and maybe just maybe you'll wake up. Because these people want your life ,- and it isn't to stay alive very long.

142. SUBJECT: MY CLOSING THOUGHTS

This is what these people demand from you people who follow them.

As you can see the comparison is more than the human mind can stand, these who play God, and believe that they are not following their own destruction, but allow me to say that this is not just a book; it is a way of life, a way of harmony, a dream that could be if humanity would follow the instructions, to be brutal and openly honest, I am well aware that many of you will dislike my thoughts, I beg your forgiveness. I do not regret my heartfelt words or ideas of making your world your home a better place. Many of you day to day are absorbed by the devices of media, and you believe you are involved, but yet you do not move from your safe place. You hide inside your shell, be mindful, those people were exactly why Nazi Germany became the ill power controlled by Adolf Hitler, a socialist who condemned communists but did exactly what they do- "control "-

"So you have a choice today while freedom is holding on only like a thin thread."

Let's remember if we add string with other strings added can become a rope. You are all able to hear and see, yet you have chosen to be numb and complacent. I do not follow but lead; I say this is the way and cut the way through the jungle with a blade, knowing the way is not easy. I choose it because it is the direct path, the path of what is right, but no man can be alone and gain the path to open for all. We must all be able to see ahead to know we are in a direction that is good for our future; many of you have been lied to, and the media has placed you in their grip upon your life with lies directed to take control through those who desire to bring our nation down. Today, I say be as you can be, a better person, a smile, give someone help, and make that small effort to say- hello. These are bridges we make in our lives, and each bridge allows us to travel across divided ways of thinking. I can only say that I am like the man on the mountain; my view is not close or narrow; it is to see far away and wide and see what is coming. I feel I have been blessed to be one of discernment, as an inventor seeing the problems but also a visionary, not because I pretend but because it is my fabric to be able to look out from the mountain and see the storm clouds that are on the horizon or see the sunlight as it rises to meet a new day. If a man can warn his fellow man of what is horrid end at a "Y" in the road of life, and those who hear refuse as so many have done throughout history. Even Noah from the Bible warned those of the flood to come, I am not Noah , nor am I your savior, I am one given answers which I feel can benefit your life, and so when I warn you of anyone or anything, it is as a friend watching your back, yet even friends can be scorned, ridiculed and there are those who will attack my ideas and my person just as they do to our former President Trump, but truth will always surface, and they have made every attempt to call him a liar. His actions became well known, but it is you who need to open your heart, to attain the truth by seeking it, and not just allowing others to dictate their made up truth, you are listening to bought and paid for news media designed for socialists, all paid from China and Russia, all designed to give you their lies and distract you from real actions being done. Like the illegals entering. We are all very weak when it comes to being self- absorbed; we have our desires and goals, but I hope this for you and mean this with every fabric of my life. 1. Never join anyone who makes people into "THINGS," but be with those who cherish liberty, freedom of speech, and freedom of self-protection, and when you hear remove the guns, remember your right to protection is being removed, and those who would take your freedom know that once they have your protection removed, they now control "YOU." Do not follow those who say you have no rights or you to be treated as if you already did a crime, as they watch us as if we are criminals behind bars, and never take what you have not earned, for the deeper price will be paid when they control your life and your ability to live independently but strive to be what you see as what makes you happy. But remember, - no great people ever gain by not working hard. Be one who stands for what is right, and even give your life as others did for you to be able to be free; never forget their sacrifice was for you, so strive to be better above the average person, an individual with the care of others not because it is important to others, but because it is important to you, and when people mistreat you and gossip, laugh at them, I have felt that pain, it never is welcomed, but if we let those who try and hurt us know they are succeeding, they will only do more, so ignore their stupidity, as they are not worth your worry, and be ready each day as you meet a new day, say this as you head out the door. "I am seeking better today; I am going to do my best. I may fail to be better, but I will be better again. I am not alone; I have a friend, I can't see this friend who watches over me , but this friend is my guide, my help when I feel hurt, my help in to control my anger. I am better as a person, but not better than anyone else, we are all equal in my friend's eyes. But I always strive to be better every day. I thank the wisdom of my creator for I am filled at this moment in time with love that cannot be broken." One last wish I have for you, and I know this may seem slightly overkill, but- never judge a person by his or her color, or religion, or if they are from or not from your country, even if people have spoken badly about the person your meeting, but meet everyone as if they are your friend, give them a chance to show you they are worthy of your friendship, and if they turn out to be bad, and are takers who are only weak- minded who have no care of anything but self, then see it, and leave it behind without anger, without remorse, just leave it as life gave you a view of the wrong way to be. (Remember- you are going to be better today.)

143. SUBJECT: BEING TO BOLD or NOT BEING BOLD ENOUGH?

I am trying to be as direct as possible, and – many of you may say, this is a bad attempt to gain attention, well, okay, so when we meet in heaven, if you make it, you may be saying to me, Oh your Richard Dean Neff, - And I will reply, yes I am, and you will say, "Gee I guess you were right, look at this massive line of people! It just goes on forever!" Well I hope that conversation never happens. Instead, I wish it would be more like this: Hey aren't you Richard Dean Neff? Why yes I am – Thanks for the heads up about what you saw coming, it sure changed things down there. "Oh you're welcome". I am really sorry to paint such a bad picture of our world, but the storms are real, and we need to step up to the plate and swing a bat. That's a lot of metaphors, but, It's just very scary facts, these people who are wanting to go to war, have no idea that **nobody wins**. "THESE DANGERS WILL HAPPEN".

If "you" do nothing. I feel like the robot in the old T.V. show called "Lost in Space" the robot was always trying to warn Will Robinson, He would yell out: "Danger Will Robinson, Danger !" so he was warned , and I am warning you of many ills I see, and hope I have presented a clear image , but the people of the world are allowing these things to happen, and it's up to us to stop the roller coaster. So read on because the ride is almost over.

144. SUBJECT: NUCLEAR WAR

I am more than confident that the majority of people know about this subject, but what they don't know is that they themselves have allowed the mess to continue, the reasons are that people worldwide who have ill thoughts of hate or greed for what is not theirs continues even today, wars over land, food, and oil, and the desires of some nations to dominate other nations which do not live as they do , which to my observation is absurd to think that leaders just can't be happy with what they have. It is as they are childlike minded to have their toys and have what they use, yet want to take another child's toy just because it is not there toy, without a doubt they are not understanding this event I am talking about is a real potential danger, but it is you who have allowed this potential destruction without real concern. In fact, we can say you're willing to die this way and have all your children die and see the entire world of humans be gone as the dinosaur is now gone. But I am about to give you a view that even the greatest minds have never considered, and my reason? To hope that when you finish this book, you and millions like you finally come to the same conclusion- That: "These Devices of Extinction Need to go! ".

And here is why; as I stated, you are more than well aware of this threat, but are you really? No, and here are just a few reasons. 1st- The power shown in an attack made on Japan with the A-bomb shows the devastation of cities and even the size of the explosion in comparison to an H-Bomb is very small, yet these bombs were able to destroy the cities of Hiroshima and Nagasaki, leaving nothing but radiation and buildings in shambles. But we now have missiles with even more destructive power of the later developed H-Bomb which has the explosive ability of today to remove more than half of Japan with one single blast; the impact is around five times stronger and reaches hundreds of miles, and the air blast has enough power to level a city alone. Now, the really good news Russia has around which has been reported an estimated 6,000 of these, but the United States did them one better- The development of the MX-Missile which was deployed in 1986 is not a single warhead H-Bomb, no — it has three nuclear warheads in a single missile, gee what you estimate the distance, well never been tested, but- if it was, as close conclusion or probability, we are looking at a missile able to remove the State of New York, with one missile, and it has been recorded in the past that the United States has located in many areas- over 3,000 of these missiles. But today they say that there is 50, but would our government really tell us how many there really is? But these missiles are now out dated with newer faster response time as the top nations of the world now use even greater strength explosives that are able to blow up the entire world about three times over. This sounds reasonable that we as a world own the ability to dissolve (remove) all mankind forever. So here are a few facts for you to consider: The H-bomb a much larger explosive device and the distance it can reach, for a bomb that size, people up to 21 km (13 miles) as the direct kill area, and further away would experience flash blindness on a clear day, and people up to 85 km (52.8 miles) away would be temporarily blindness. But hey we only have 400 of these other much bigger toys to kill everyone and everything. The B83 nuclear bomb, currently the most powerful thermonuclear weapon in the inventory of the United States, has a yield of 1.2 megatons of TNT, making it **80 times more powerful than the H-Bomb.** There are an estimated 12,700 nuclear warheads on Earth.

So why are you sitting in your living room every day waiting for the end? And this end actually started when the United States began fielding the MX in 1986, a 32,600-square-kilometer missile complex there has served as home for all 50 deployed MX missiles and what is an MX? It has three nuclear war heads in one missile, gee how sweet. But today the United States has approximately 1,770 warheads that are deployed, 400 are on land-based intercontinental ballistic missiles, roughly 970 are on submarine-launched ballistic missiles. Now, this does not even include the many nations as hostel nations that own nuclear abilities, such as North Korea and China. In short, we live on a planet that is so fragile that at any moment on any given day, we as people could be wiped away as never been or with a history to never be read by anyone ever again. But hey who cares! It has no meaning right? The end of humanity, my children don't matter, heck we kill unborn babies today, so who cares about the future. Yes people, the world is some mess, and you, yes you, have allowed it! You who have accepted this horror, by being "NUMB" being unconcerned, be permissive, you are their sheep, and your ability to act is to ignore. Some foolish people believe that survivors will emerge after thousands of these missiles are released. Allow me to show you exactly the result, and show you how it starts. Well we have so many ways, but let's say, one of our enemies decide that since they got it, hey let's use it! The fools who do it who possibly believe that God is for this end, and will be placed in Hell when he meets them, no doubt, since Gods law states very clearly- "Thou Shalt Not Kill" But these with small brains and great bomb power won't care. Oh they will later, but they have just wiped out the entire planet and everyone they ever cared about. The devastation will be one city which is wiped out, and it takes 30 seconds from any city destroyed no matter where it is, why? Because that is called first response, and the entire world is now on **"Hair Trigger Alert"**. The explosion triggers for many military channels as the report is sent so is the order to those who are already authorized as "second response " as the keys are now turned and the skies are filled with instant release of thousands of missiles from land bases, and subs from all sides of the globe as each nation will track the other nations incoming doom, This will instantly be met with added launches, and now nations will be covered in explosives and fire and devastation as never seen, a living hell fire and destruction that rises from every point of land mass, nations who never even considered they would be involved are also destroyed as you will see here . Of course, the leaders and higher officials will be notified in minutes, and some will go underground, and the people who caused the attack will believe they are safe. "They are not safe either ", and so many will not make it to shelters and be wiped out, but those who make the shelters designed for the so-called elite will be in a safe haven for the time being, but not for long as they will survive the initial blasts because from where they are located they are miles from the explosives, but those outside, they are met with almost instant death as their children are in school, and they get the news that a city is about to be attacked by a nuclear missile traveling from the west. The news media breaks into every program to announce this event of catastrophic proportions, and a missile has been tracked less than 12 miles away, the military waits to see the outcome, is this real? Is it a dud? Military personnel from around the world are now at red alert, from around the world waiting to see if they need to launch, sirens are blasting from everywhere, the panic is off the chart, as people leave in their cars to go get there family, and the traffic in minutes is not moving but cars and trucks are driving even off the roads trying to flee to God knows where, the world is about to come to the end but the radio is still working inside the cars , the T.V. stations have interrupted all communication, and are telling people to take cover. To go to their basements or fallout shelter, it has only been 2 minutes from the sighting, but wait only lasts, about 1 more minute, yes 3 minutes have passed since the report, when this city is wiped from the face of the earth, but before this happens the sound millions who did live trying to hide, trying to run, and there is nowhere to run to, they are heard miles away. The sounds of the cries can be briefly heard as thousands looking up at the missile coming now directly descending downward, and then the deafening sound of the explosion can now be heard from areas not in the direct explosion radius. People miles away from the blast zone are instantly blinded by the blast, as a massive flash is followed by a deafening exposition. The missile is just of many to come, the blast is confirmed by the military, and the order goes out without delay, keys are turned and approved – "Launch all missiles", every sub, every land base missile is now launched everything they have it's now on their way to every nation which is a potential target. Thousands of missiles are crossing the globe. One after another which where the closest to their nations, some closer to targets will explode first in major cities which in the blink of an eye are -- Gone. Many people will be far from the actual blasts at first, a farmer who is working in the field will be first blinded by the explosion then hears the explosion as he sits on his tractor, he will be stunned to not be able to move. But he will soon be meeting death when the winds pick up his tractor which fly's about three football fields into trees that crush his life. People will be tossed from air blasts ten times stronger than a category 5 hurricane recorded as the highest winds, and this will be about a category 9 in comparison. Simple pieces of wood fragments from trees will shatter and become lethal killing bullets. The next group of missiles have now arrived, and they blanket the entire country side, not one inches of land has not been effected, and anyone who has been lucky to be away from the blasts or the winds , now are blind, and terrified, those only lucky to not a gain a direct look at the blast will regain their sight in about a few hours, to only be revealed that the world is getting darker and darker as if it is night time, they can hear even the furthest explosions some 100 miles away jolting the ground, and earth quakes are now rippling through the land, and buildings are falling with the massive ground being pounded. Finally after almost one hour the last missile has make land fall. The survivors who missed the blast and the winds from hiding in some area below the wind level of travel, are now feel stomach pains and start to see their skin is flaking, they are ill from radiation poisoning, the sky continues to get darker, the radiation fallout will kill millions who have been exposed. But the only good news is that death will soon become much faster because the temperature outside is dropping along with what looks like almost snowflakes, but these snowflakes are radioactive, but suddenly, the temperature is reaching 22 below zero. The survivor's worldwide are now freezing, and many are frozen standing upright in just a few hours. and this temperature continues to get colder and colder, as the dust from these massive bombs is now risen upward into the upper atmosphere and this had been increased with every explosion heaving dust and dirt into the sky as the missiles plumes of the explosions rise also, as the last few missiles have arrived the dust also continues to build even more dust higher and higher into the upper atmosphere, the people in fallout shelters , they have heaters, but the heat is not enough to save them. The temperature is now 150 below zero this is the main reason that not one person survives this event, because once the sky has been saturated with dust clear up to the point of space, this blocks the entire world from sun light, and we can potentially reach the same temperature as Neptune which is further than the sun but with the dust continues to rise, we will enjoy minus 346 degrees Fahrenheit (minus 210 degrees **Celsius**). The strangest thing that comes to mind which is that we have been warned and warned that these events could happen, yet we as people have stayed calm and ignore the soon to be everyone's future if people do nothing. But if we look at the movie called A.I. by Steven Speilburg, there is a scene which the little robot boy is now seeking to become a real boy and lands inside the ocean in front of a statue of the fairy who is in the story Pinocchio, and in that story the fairy made Pinocchio into a real boy, and we see this fairy under water that is surrounding the entire city of New York City high rises, which is most likely created from poles melted from global warming, but this is what they call a "Meca Child"(A.I. robot) who he believes the Fairy will make him a real boy is trapped inside his flying car under water. The next scene shows him being removed from being frozen, and is thawed out from ice which is everywhere, the entire world is filled with ice, in fact it is shown to cover almost every tallest city sky scrapper, only the tips of buildings are revealed, and the oceans are frozen solid ice. The world is covered with ice, and not one person is found alive, and in truth this is exactly what we would see if there is a nuclear war. Because it would take decades for the sun to be able to peek back down to the world, and the fact that ice expands water, the land mass would also be covered being our world is 75% water. So in the movie the end of the world was not shown of how the ice was formed, but it does correspond with my conclusion of the events to follow a nuclear attack, but we get to see life on earth without mankind, and the aliens who come here find our remains in ice. Gee how nice. and it is because not only will the radiation be worldwide as it travels on dust and in oceans depths to reach all life in the ocean and will kill every living thing. But the major event to follow is because of the continued temperature declining even people in submarines are frozen, and even at depths of 200 feet or more the ground with any water in it will become frozen ice, the planet is now void of all life. The oceans are now frozen completely. The people in bunkers, frozen, the world is void of all life. And-this "is" the end of mankind. So, I am asking all of you, all who have read my book, all who have a sense of right, a commitment to the survival of you and your loved ones, those who you cherish, and even to those you do not know or even like, the time has come to see your world, to see past your nose, to realize you live in seconds away from destruction and suffering beyond any time ever in history, your children, you, all you own, all your friends- gone in just a few minutes, I ask you, no I plead with you- I beg you, stand up, and be heard, we as humanity need these elements of destruction removed and all who have knowledge of this suffering devices also need to be removed from their ability to make these monsters of destruction and the creation of human extinction. Hear me world, hear my call to save our legacy, or history, or tomorrow. We as people need to awake, need to say no more, we need to refuse to be dragged into a period of never being again. You and I need to care, to speak, to tell, to unify, to stop their help for the desires of harming each other, for the change of trust is formed and replaces division , and thinking only of one's self, and this is done by all people who have the ability to see this horror is upon us, these who cherish their life, and they cherish their children, and they cherish the ability to know that when they leave this world, they have given those to come a chance to be here as we are. So - On June 6th, 2025, all people in your nation are to protest and gather by millions from your homes and your jobs, and let all your leaders know that you are tired of fear, tired of threat, tired of worry, tired of their illness that they desire to kill us all. And be aware that I have seen the call they have made to have welders build massive submarines the size of ocean liners, so is it possible for our leaders to survive, sorry folks, that is a joke, I explained the actual events, they will not be anything but what I have stated. So you actually believe that if you hide away in some sub under the water in the ocean while the land is being destroyed, they will survive a nuclear war? The subs are made of metal! The oceans will be filled with radiation that will only enter the hull of the subs. And you will also become human pop-cycles (frozen). So I am calling on you leaders also, it is time you see that you're not going to be leaders very much longer, that your end comes with our end, So who wins? NOBODY WINS. So you have a choice, to dismantle these elements and join as a world family and do it with the understanding that people all need the freedoms to be able to speak as freely as you do, Russia , China, North Korea, you're not the only people who want to have something good in this world, so read my proposal of a united world, join in , you could be elected if you care and do good for everyone, and follow the path of helping everyone, would that really be so hard? So June 6th 2025, the sounds of freedom will ring across every leaders home, every political persons home, every building will be rattled from the sounds of the peoples of the world massed together as family, no more caring about a man's faith as to divide and cause conflicts, peace is what God wants, he wants you to see that anyone who seeks him in their own way can be welcomed by him, you are not God and so stop pretending you know the mind of God, I only know he is love, so love of what he created was all of us or we would not be here. So faith is no longer to be pushed on others, it is told only when asked, with respect to everyone as people who want to think or care in the way they believe. So we all need to drop the color care, drop the difference of language, and focus that we are all needing good homes, good friends, good lives, and work together for the first time in history, --- WE CAN ALL BE FAMILY.

145. SUBJECT: NUCLEAR WAR ENDS \ D-Day was in 1944. Was on June 6th I am asking you to celebrate with me, the most positive action that may ever be recorded in history, and which actually may prevent the end of history.

June 6th, 2025 – N-DAY, the end of all nuclear designs. Because of you and others like you are reaching out each in their own way and language, and each spreading this message that "WE MATTER" WE ARE WORTH SAVING EACH OTHER! You are here in this world for a reason, and they who lead us, have failed to see the error of the design of war which only leads us all to a closure and the end of our lives, they refuse to see our hopes, and our dreams, they kill us without regard of what they do to achieve their demands from others who want their ability to live in peace. They send our youth off to die, for their greed, they obsess with the power they have as to control us, force us all to be afraid, force us all to be with mistrust, force us all to divide as these negatives grow into our lives, and into our children's minds. Hate is excepted and miss-trust is pushed down our throats as to choke us to death as the result. A death that does not need to happen or should happen. We have been told over and over how we are all doing our countries great service, they for men and women to offer their bodies to be in war is honor, and we are to respect this act as correctness, it is correct if your someone who leads and wants others to follow your desires of domination, or to think your people need protection form others, but WAIT! What if everyone in the entire world of youth and even older said – I refuse to die young, I refuse to follow others who command me to die for preservation of their power which says to me and to all who live under this person's power, "you must die for me to survive". To survive? How is death a means to survive? Death is a final action, and so to see death as a failure of living is correct is it not correct. So let's say every country refused to be in the military of their country, Russians facing Ukraine have actually done this, and many hundreds have just refused to fight and are alive. So if we all stop the want for military pay checks , I said all as every person worldwide walk away, then we can now meet face to face and bring all of us what we need , and what is that , we need each other to grow, to be, to live, to seek outward together into space, see the amazing universe that awaits our leaders and our businesses who will reach to new worlds and bring back gold, metals, food never enjoyed before, we will learn more and more, and life will be our mission, life of:

146. SUBJECT: PEACE

Many people don't believe that world peace can be attained. I must admit that just watching how people treat each other on a day-to-day basis tells me the quest may seem impossible, for the reasons we all seem to have our agendas, I need this, and I need that, and I want this, and Me, me, me, I am so much more important than you, a real issue that is a dead end. And making plans to be the top banana be the King of the Hill, my bullets are bigger than your bullets! I can kill you better! It seems to override the brain rather than what we could all be as to live together, (HOW BORING!! you mean we could be friends? Well I have friends and they like to kill, or build bombs, we make some serious money selling death, the attraction seems normal enough, you make billions of dollars and every person who gets into war and killing others is with big money to be made, okay I agree, it is serious green being made. But we also need to look at the end result, it all starts so simple, I invade you, you fight back, now the teams need more equipment to kill the other side, so big money is spent and everyone who supplies the stuff to kill better makes sorry for the pun: (A KILLING!) But then another war breaks out, and another country gets into the mix by mistake a drone miss's it's target and hits some leader's family by mistake. Oops! We now have another entry into the game, but this entry has friends who now supply his new enemy with paid troops, the game is growing folks, then it gets super messy when a few major players step in to the arena, the game is now gotten very serious, but guess what, the one of the big stake players just got a major black eye as their troops were wiped out almost to the last man in a massive battle. The loser well he isn't happy, and feels the other side is now entering into his living space. He is losing, and he knows it, his friends try and help, but they are now losing also. The next action is called failure to care because he knows he is about to die, so he gives the order to strike with his biggest card he's holding, the card of ACE of spades, the card of death. And he now launch's his missiles, and you who helped him and others fight in this war and I are now not anymore, and you who sold them the guns, the designs of killing, well your dead also. So to those who have made so much money making devices of war. You have a choice, you can continue, or as a strong business idea you can re-tool, re -design, and make something that could sell to the millions across the world, and close down the war junk, and now make something that won't put people in danger anymore, because when one of these who have nuclear weapons starts to lose, you're looking at just what I said, an end of us all. Now maybe you don't care about you, maybe you only care about money? Then how do you spend money if you're dead? It's time to rethink our lives. To improve, not waste it. So to everyone else who works at these places that build the destruction devices, maybe I would start to look for a new job, because I really think people may be tired of letting all of you build, an end point that can be very real. Very as in so real that the leader of Russia now is frustrated, and has made more than one threat to use nuclear missiles. So let me give you all an example if I may. Let's go to the local playground, and we are at the age of 12, there is this bully and he has been pushing kids around, and on this particular day he has now knocked down a kid named Johnny, and Johnny is crying with a black eye, all the children could have helped Johnny, but they didn't care. The reason Johnny is hurt is because the bully, well he's a pretty big kid for his age, and so he thinks he is so strong and so big that he can't be stopped by any other kids. So he pushes his way past other kids that are waiting to get on the slide, he gets up there, and suddenly a gun is fired and two shots are heard, all the kids run away, the bully is on the ground now and is bleeding to death, we see the gun and who is holding it, it isn't Johnny, it's his mother and father who both have just shot the bully. And the police arrive and take them off to jail to live there for the rest of their lives. You see the moral of this story is this, when people abuse others who cannot always protect themselves, we may have others who care and take action and the result to the bullies in life are never happy is what history shows us time and time again. Even the parents who used force found they could not be happy, instead they should have spoken with the bully and made an effort to show they cared, and the bully could have been reported to the school, and they could have given bully a detention, or reported the problem to his parents. But let's rewind, if the bully in this story of life had respect for others, and had been someone who helped other children get on the slide ahead of him, and treated kid's kind. Wouldn't he still be here alive? So if we want the bullies removed, we don't have to kill them, we just need to stop helping them, and be friends at life's playground, and nobody gets hurt or killed. So if we want peace in this world, my best advice is – play nice and find Love in life, a word scorned by many, to be considered weak and foolish, but in fact, the only true way to have a world safety net that if everyone adopted this attitude we could finally have peace. Peace, what a great word, if only people could live it everywhere. Okay call me dreamer, I don't care if you do, and I know I am right as rain. Peace.

147. SUBJECT: HOW TO START YOUR NEW LIFE

PEACE IS REAL IN ALL OF YOU; EVEN THE MOST ANGRY NEED PEACE.

Many are called to the table, but many will not attend, they refuse the feast so great, and the host has invited you to come and enjoy the fruits of his labor, the lamb was sacrificed, and the half cow of beef is basted over the hearth near the massive fireplace, the fruits of apples and peaches, and fresh dates and plum wine is almost as sweet as honey nectar mead, and the food is served on gold platters, and silk table cloth, we are served as Kings and Queens of old, the seats are labeled with your name on the plate. It says:

"welcome to those who see this worth"

Here is the teaching of true peace I hope you will live daily-

This should be read daily to end your day before bed time: This I call:

"THE WAY"

I am not alone, I am of the united human family.

I will respect you as your difference is unique even though it is not like mine.

I do not force my beliefs upon you- But if you ask me of my ways, then I will share them only.

I shall try my hardest to be kind always- Even though I know that I am not perfect, or that others are not perfect.

I will not be sarcastic or belittle you, but care of you, because that reflects like a mirror of life back to me.

I will not harm anyone in any way, because I respect my growth, and the growth of others.

I am responsible for taking care of my world and my home.

I may make mistakes, but I will work hard to improve to not repeat those mistakes,

I may hurt someone's feelings with or without knowing why, I will say I'm sorry to remove their pain, because I am strong to admit my error.

I will make my life an example of good to show joy for my mentors and the youth around me.

I will work daily to attain my life's goodness and respect all that you have as yours.

I will share whenever I have abundance, but do it in silence without boasting.

I may have more or less than another, and I am no better than they are and no less.

I am here as designed from birth; when I leave, my actions are accountable to enter the everlasting life made by our Creator.

The fact is the richest man goes to the grave the same way you do, and they can have a gravestone and the greatest send-off, but they rot in the ground, or are burned away just as easy. They are giving back to the world an equal amount of chemicals water, and other elements that make up the human body, so they are exactly no better than anyone. The only comfort I have and what I can give all of you is that if you consider what I call the 50/50, a 50% that Heaven is real, and a 50% that Hell is real. And then we have the last possibility that there is neither and you are only drained of all knowledge that you ever where here, and become worm food only, which really is the second choice to be in Hell, so you gain nothing good is my point but only from the first 50%.

So if you followed well, and did well for others, and did your best. You and I may meet one day in this place called Heaven and share a kind hello, a place boasted of good and paradise, and people treat you good everywhere, every day!, and I hope I make it there. I owe all of you an apology if I seemed overzealous with my hope for all of you to find peace, or seek the creator in your own way, many of you have no care for God, or Jesus, and I am well aware that the words can offend some people, but to acknowledge is my intent for you, is not for you to follow blindly in anything I have said here. But,- to see for "yourself", to test the waters, and for any subjects that I repeated, please consider that I made these thoughts over many years, which I admit that time can be an enemy of remembering. I do hope that through my gloom and doom that you found my true goal, which is to give you a better world or be a better person, as I strive for this every day, and it is my fabric of who I am even when others try and say this is not who I am, they have the lost hearts which I can only have pity on. If I have given you something better was my desire, and my hope for you is to see that there are people who are real, and actually care about "you" but let's face it, action is always better than words, so this is why I wrote this book. But let's also be with a firm understanding that being better takes work, if it's accomplished means that you and I may have a better world. As I have said, - I know I am not perfect, yet I strive to be so, I see my faults and I say, "Today I will try again". So thank you so much if you have seen my effort as one not to really preach, but to inspire us and including myself to make our world preserved, to be factual I consider my faith made from science not just from some religious stance. Again my goal is to have our children live in a place without street violence, or wars, or with hate for someone else. And I have stated items that seem to contradict these views, such as prisons, and military, and police, but to those institutions that are required at this juncture of time which is to protect us from others who have harmed people in many ways, all because they refused to care about another or allowed themselves to fail the ability to be kind, if world peace is achieved we can then say goodbye to those who place us in jail, say goodbye to wars, because people won't do wrong as their purpose, but help each other when they see someone is down and needs a friend. The world cannot toss away protection and controlling factors until the entire world has joined the movement of caring for each other, and then, if and when that happens, we won't need protection, but instead we live in freedom, peace, and love. For those of you who enjoyed this walk into the future, I do hope that you may recommend the book which will be a help to my 38 years of striving to build my newer inventions waiting in log books. With my warmest possible thanks I can give - Wishing you and yours the best of life. Am I the first to care about others ? No, even great people of the past tried to get your attention, a comedian named Charlie Chaplin, -the speech from his movie "The Great Dictator" go to : www.charliechaplin.com and listen: His speech is my speech, but imagine billions all over this planet making this happen. And the soldier is not one that fights with a gun, but with his heart reaching another heart. Listen to this and see it as your future with everyone. And if you see it, hold on to it, but share it, give it to others , and tell them to share it, until everyone has seen it. And live it as I do. Is it easy to love bad people?, no, it is hard, but we can ignore those who wish to be cruel, and gossip, or hate, we can stand with others and say – no more, I want to live, I want peace, I want to be safe, and happy. And I want it for everyone not just me.

Please know that I give all credit for my ideas as being a God given gift, and I could not have started this without Sandra L. Allen of Ripley New York. who gave 30 years as my secretary and worked two jobs one for me and the other outside our office, and never wanted a pay check form me, that is why with your support of telling others of this book, I am dedicating a portion of the funds for helping orphans, and homeless, and entrepreneurs seeking to build their inventions, I am well aware of the hardships that face a new idea, and so my efforts will be to be open to review ideas as to help support them, and I will sign a non-disclosure that protects not myself only, but makes it so I cannot take anyone's idea. Sandra was an orphan raised by her grandmother, and I was also an orphan raised by the Neff's as I stated before, and we both lived as homeless, and survived it together we also worked day and night non-stop for 30 years to achieve the dream of making new products to help people have jobs. Our start was on a picnic table in my parent's back yard, but what we went through together was no picnic, but we never lost sight of our hope to help others live better. Sandra died in my arms in our office, and I miss her deeply, when it came to being a friend, - she was the diamond. Sandra's last words to me, as I tried to help her up from her chair, "I can't move my foot". So project "STAND" will be supported in dedication to Sandy and my hope is we all see the need to give people a solid foundation to stand on.

You can look at a few of these products we worked on, and still not finished but trademarked and protected, but started with molds and packaging, and yes will be available for you or your family hopefully one day soon. Type this link to www.lovepricelow.com

God bless, & God Bless Freedom.

Richard Dean Neff

emailzoom@aol.com (Contact- Richard Dean Neff)

<u>officialrichardneff@gmail.com</u>   [For Investor relations.]

Quote by Richard Dean Neff: "I AM YOU- YOU ARE ME

<u>BOOK SIGNING:</u>

To have your book signed please contact email for information how you can have this added.

AUTHOR: ____.
SIGNATURE

Date:_______/_______/___________

<u>COMMENTS :</u>

SPECIAL THANKS TO : NYC PUBLISHING & ALSA INTERNATIONAL

<u>alsa@alsainternational.in</u>   Quality Printing since 1935

www.ingramcontent.com/pod-product-compliance
Lightning Source LLC
Chambersburg PA
CBHW051501150726
47997CB00001B/73